STRUCTURAL RELIABILITY
Analysis and Prediction

ELLIS HORWOOD SERIES IN CIVIL ENGINEERING

Series Editors: R. EVANS, Department of Civil Engineering, University College, Cardiff,
R. SELLIN, Department of Civil Engineering, University of Bristol, D. WOOD,
Department of Civil Engineering, University of Glasgow, and N. G. SHIRVE, Department
of Civil Engineering, University of Calgary, Canada

STRUCTURAL RELIABILITY
Analysis and Prediction

R. E. MELCHERS, B.E., M.Eng.Sc., Ph.D.
Professor of Engineering
Department of Civil Engineering and Surveying
University of Newcastle, N.S.W., Australia

ELLIS HORWOOD LIMITED
Publishers · Chichester

Halsted Press: a division of
JOHN WILEY & SONS
New York · Chichester · Brisbane · Toronto

First published in 1987 by
ELLIS HORWOOD LIMITED
Market Cross House, Cooper Street,
Chichester, West Sussex, PO19 1EB, England
The publisher's colophon is reproduced from James Gillison's drawing of the ancient Market Cross, Chichester.

Distributors:

Australia and New Zealand:
JACARANDA WILEY LIMITED
GPO Box 859, Brisbane, Queensland 4001, Australia

Canada:
JOHN WILEY & SONS CANADA LIMITED
22 Worcester Road, Rexdale, Ontario, Canada

Europe and Africa:
JOHN WILEY & SONS LIMITED
Baffins Lane, Chichester, West Sussex, England

North and South America and the rest of the world:
Halsted Press: a division of
JOHN WILEY & SONS
605 Third Avenue, New York, NY 10158, USA

© 1987 R. E. Melchers/Ellis Horwood Limited

British Library Cataloguing in Publication Data
Melchers, R. E.
Structural reliability: analysis and prediction. —
(Ellis Horwood series in civil engineering).
1. Structural Engineering   2. Reliability (Engineering)
1. Title
624.1'71   TA656.5
Library of Congress Card No. 87-9279
ISBN 0-85312-930-4 (Ellis Horwood Limited)
ISBN 0-470-20873-2 (Halsted Press)

Phototypeset in Times by Ellis Horwood Limited
Printed in Great Britain by The Camelot Press, Southampton

Table of contents

Preface

The aim of this book is to present a unified view of the techniques and theory for the analysis and prediction of the reliability of structures using probability theory. By reliability in this context will be understood not just reliability against extreme events such as structural collapse or facture, but against the violation of any structural engineering requirements which the structure is expected to satisfy.

In practice, two classes of problems may arise. In the first, the reliability of an existing structure at the 'present time' is required to be asssessed. In the second, and much more difficult class, the likely reliability of some future, or as yet uncompleted, structure must be predicted. One common example of such a requirement is in structual design codes, which are essentially instruments for the prediction of structural safety and serviceability supported by previous experience and expert opinion. Another example is the reliability assessment of major structures such as large towers, offshore platforms and industrial or nuclear plants for which structural design codes are either not available or not wholly acceptable. In this situation, the prediction of safety both in absolute terms and in terms of its interrelation to project economics is becoming increasingly important. This class of assessment relies on the (usually reasonable but potentially dangerous) assumption that past experience can be extrapolated into the future.

It might be evident from these remarks that the analysis (and prediction) of structural reliability is rather different from the types of analysis normally performed in structural engineering. Concern is less with details of stress calculations, or member behaviour, but rather with the uncertainties in such behaviour and how this interacts with uncertainties in loading and in material strength. Because such uncertainties cannot be directly observed for any one particular structure, there is a much greater level of abstraction and conceptualization in reliability analysis than is conventionally the case for structural analysis or design. Modelling is not only concerned with the proper and appropriate representation of the physics of any structural engineering problem, but also with the need to obtain realistic, sufficiently simple and workable models or representations of both the loads and the material strengths, and also their respective uncertainties. How such modelling might be done and how such models can be used to analyse or predict structural reliability is the central theme of this book.

In one important sense, however, the subject matter has a distinct parallel with conventional structural engineering analysis and its continual refinement; that is, that ultimately concern is with costs. Such costs include not only those of design, construction, supervision and maintenance but also the possible cost of failure (or loss of serviceability). This theme, although not explicitly pursued throughout the book, is nevertheless a central one, as will become clear in Chapter 2. The assessment or predictions obtained using the methods outlined in this book have direct application in decision-making techniques such as cost–benefit analysis or, more precisely when probability is included, risk–benefit analysis. As will be seen in Chapter 9, one important area of application for the methods presented here is in structural design codes, which, it will be recognized, are essentially particular (if perhaps rather crude and intuitive) forms of risk–benefit methodology.

A number of other recent books have been devoted to the structural reliability theme. This book is distinct from the others in that it has evolved from a short course of lectures for undergraduate students as well as a 30-h graduate course of lectures which the author has given periodically to (mainly) practising structural engineers during the last 8 years. It is also different in that it does not attempt to deal with related topics such as spectral analysis for which excellent introductory texts are already available.

Other features of the present book are its treatment of structural system reliability (Chapter 5) and the discussion of both simulation methods (Chapter 3) and modern second-moment and transformation methods (Chapter 4). Also considered is the important topic of human error and human intervention in the relationship between calculated (or "nominal") failure probabilities and those observed in populations of real structures (Chapter 2).

The book commences (Chapter 1) by reviewing traditional methods of defining structural safety such as the "factor of safety", the "load factor", "partial factor" formats (i.e. "limit state design" formats) and the "return period". Some consistency aspects of these methods are then presented and their limited use of available data noted, before a simple probabilistic safety measure, the "safety margin" and the associated failure probability are introduced. This simple one-load one-resistance model is sufficient to introduce the fundamental ideas of structural reliability assessment. Apart from Chapter 2, the rest of the book is concerned with elaborating and illustrating the reliability analysis and prediction theme.

While Chapters 3, 4 and 5 deal with particular calculation techniques for time-independent situations, Chapter 6 is concerned with extending the "return period" concept introduced in Chapter 1 to more general formulations for time-dependent problems. The three principal methods for introducing time, the time-integrated approach, the discrete time approach and the fully time-dependent approach, are each outlined and examples given. The last approach is considerably more demanding than the other two (classical) methods since it is necessary to introduce elements of stochastic process theory. First-time readers may well decide to skip rather quickly through much of this chapter. Applications to fatigue problems and structural vibrations are briefly discussed from the point of view of probability theory, but again the physics of these problems is outside the scope of the present book.

Modelling of wind and floor loadings is described in Chapter 7 whilst Chapter 8 reviews probability models generally accepted for steel properties. Both load and strength models are then used in Chapter 9. This deals with the theory of structural design codes and code calibration, an important area of application for probabilistic reliability prediction methodology.

It will be assumed throughout that the reader is familiar with modern methods of structural analysis and that he (or she) has a basic background in statistics and probability. Statistical data analysis is well described in existing texts; a summary of probability theory used is given in Appendix A for convenience.

Further, reasonable competence in applied mathematics is assumed since no meaningful discussion of structural reliability theory can be had without it. The level of presentation, however, should not be beyond the grasp of final-year undergraduate students in engineering. Nevertheless, particularly difficult theoretical sections which might be skipped on a first reading are marked with an asterisk (*).

For teaching purposes, Chapters 1 and 2 could form the basis for a short undergraduate course in structural safety. A graduate course could take up the topics covered in all chapters, with instructors having a bias for second-moment methods skipping over some of the sections in Chapter 3 while those who might wish to concentrate on simulation could spend less time on Chapter 4. For an emphasis on code writing, Chapters 3 and 5 could be deleted and Chapters 4 and 6 cut short.

In all cases it is essential, in the author's view, that the theoretical material be supplemented by examples from experience. One way of achieving this is to discuss particular cases of structural failure in quite some detail, so that students realize that the theory is only one (and perhaps the least important) aspect of structural reliability. Structural reliability assessment is not a substitute for other methods of thinking about safety, nor is it necessarily any better; properly used, however, it has the potential to clarify and expose the issues of importance.

ACKNOWLEDGEMENTS

This book has been a long time in the making. Throughout I have had the support and encouragement of Noel Murray, who first started me thinking seriously about structural safety, and also of Paul Grundy and Alan Holgate.

In more recent times, research students Michael Harrington, Tang Liing Kiong, Mark Stewart and Chan Hon Ying have played an important part.

The first (and now unrecognizable) draft of part of the present book was commenced shortly after I visited the Technical University, Munich, during 1980 as a von Humboldt Fellow. I am deeply indebted to Gerhart Schuëller, now of Universität Innsbruck, for arranging this visit, for his kind hospitality and his encouragement. During this time, and later, I was also able to have fruitful discussions with Rudiger Rackwitz.

Part of the last major revision of the book was written in the period November 1984–May 1985, when I visited the Imperial College of Science and Technology, London, with the support of the Science and Engineering Research Council. Working with Michael Baker was a most stimulating experience. His own book (with Thoft-Christensen) has been a valuable source of reference.

Throughout I have been extremely fortunate in having Mrs. Joy Helm and more recently Mrs. Anna Teneketzis turn my difficult manuscript into legible typescript. Their cheerful co-operation is very much appreciated, as is the efficient manner with which Rob Alexander produced the line drawings.

Finally the forbearance of my family was important, many a writing session being abruptly concluded with a cheerful "How's Chapter 6 going, Dad?"

Robert E. Melchers

Monash University
December 1985

1

Measures of structural reliability

1.1 INTRODUCTION

When an engineering structure is loaded in some way it will respond in a manner which depends on the type and magnitude of the load and the strength and stiffness of the structure. Whether the response is considered satisfactory depends on the requirements which must be satisfied. Such requirements might include safety of the structure against collapse, limitations on damage, or on deflections or any of a range of other criteria. Each of these requirements may be termed a "limit state". The "violation" of a limit state can then be defined as the attainment of an undesirable condition for the structure. Some typical limit states are given in Table 1.1.

Table 1.1 — Typical limit states for structures

Limit state type	Description	Examples
Ultimate (safety)	Collapse of all or part of structure	Tipping or sliding, rupture, progressive collapse, plastic mechanism, instability, corrosion, fatigue, deterioration, fire
Damage (often include in above)	Damage of structure	Excessive or premature cracking, deformation or permanent inelastic deformation
Serviceability	Disruption of normal use	Excessive deflections, vibrations, local damage, etc.

From observation it is known that very few structures collapse, or require major repairs, etc., so that the violation of the most serious limit states is a relatively rare occurrence. When the violation of a limit state does occur, however, the consequences may be extreme, as exemplified by the spectacular collapses of structures such as the Tay Bridge, Ronan Point Flats, Kielland Offshore Platform, etc.

The study of structural reliability is concerned with the *calculation* and *prediction* of the probability of limit state violation for engineered structures at any stage during their life. In particular, the study of structural safety is concerned with the violation of the ultimate or safety limit states for the structure.

The probability of occurrence of an event such as limit state violation is a numerical measure of the chance of its occurring. This measure either may be obtained from measurements of the long-term frequency of occurrence of the event for generally similar structures, or may be simply a subjective estimation of the numerical value. In practice it is not usually possible to observe for a sufficiently long period of time, and a combination of subjective estimation and frequency observations for structural components and properties is used to predict the probability of limit state violation for the structure as a whole.

In probabilistic assessments any uncertainty about a variable (expressed, as will be seen, in terms of its probability density function) is explicitly taken into account. This is not the case in traditional ways of measuring safety, such as the "factor of safety" or "load factor". These are "deterministic" measures in that the variables describing the structure and its strength are assumed to take on known (if conservative) values about which there is assumed to be no uncertainty. Precisely because of their traditional and really quite central position in structural engineering, it is appropriate to review the deterministic safety measures prior to developing probabilistic safety measures.

1.2   DETERMINISTIC MEASURES OF LIMIT STATE VIOLATION

1.2.1 Factor of safety

The traditional method to define safety is through a "factor of safety", usually associated with elastic stress analysis and which requires that:

$$\sigma_i(\varepsilon) \leq \sigma_{pi} \qquad (1.1)$$

where $\sigma_i(\varepsilon)$ is the ith applied stress component calculated to act at the generic point ε in the structure, and σ_{pi} is the permissible stress for the ith stress component.

The permissible stresses σ_p are usually defined in structural design codes. They are derived from material strengths (ultimate moment, yield point moment, squash load, etc.), expressed in stress terms σ_{ui} but lowered through a multiplier F:

$$\sigma_{pi} = \frac{\sigma_{ui}}{F} \tag{1.2}$$

where F is the "factor of safety". The factor F is usually selected on the basis of experimental observations, previous practical experience, economic and, perhaps, political considerations. The selection of F is one of the responsibilities of code committees.

According to (1.1), failure of the structure should occur when any stressed part of it reaches a local σ_{pi}. Whether failure does actually occur depends entirely on how well $\sigma_i(\varepsilon)$ represents the actual stress in the real structure at ε and how well σ_{pi} represents actual material failure. It is well known that observed stresses do not always correspond well to the stresses calculated by linear elastic structural analysis (as commonly used in design). Stress redistribution, stress concentration and changes due to boundary effects and the physical size effect of members all contribute to the discrepancies.

Similarly, the permissible stresses which are commonly associated with linear elastic stress analysis are not infrequently obtained by scaling down, from well beyond the linear region, the ultimate strengths from tests. From the point of view of structural safety this does not matter very much, provided that the designer recognizes that his calculations may well be quite fictitious and provided that (1.1) is a conservative safety measure.

By combining expressions (1.1) and (1.2) the condition of "limit state violation" can be written as

$$\frac{\sigma_{ui}(\varepsilon)}{F} \leqslant \sigma_i(\varepsilon) \quad \text{or} \quad \frac{\sigma_{ui}(\varepsilon)}{F}\bigg/\sigma_i(\varepsilon) \leqslant 1 \tag{1.3}$$

Expressions (1.3) are "limit state equations" when the inequality sign is replaced by an equality. Equations (1.3) can be given also in terms of stress resultants, obtained by appropriate integration:

$$\frac{R_i(\varepsilon)}{F} \leqslant S_i(\varepsilon) \quad \text{or} \quad \frac{R_i(\varepsilon)}{F}\bigg/S_i(\varepsilon) \leqslant 1, \text{ for all } i \tag{1.4}$$

where R_i is the ith resistance at location ε and S_i is the ith applied action (stress resultant). In general, the stress resultant S_i will be made up of the effects of one or more applied loads Q_j; typically

$$S_i = S_{iD} + S_{iL} + S_{iW}$$

where D is the dead load, L the live load and W the wind load.

The term "safety factor" has also been used in another sense, namely in relation to overturning, sliding, etc., of structures as a whole, or as in geomechanics (dams, embankments, etc.). In this application, expressions (1.3) are still valid provided that the stresses σ_{ui} and σ_i are interpreted simply as "resistance" and "applied force" respectively.

1.2.2 Load factor
The "load factor" λ is a special kind of safety factor developed for use originally in the plastic theory of structures. It is the factor by which a set of loads acting on the structure may by multiplied theoretically just to cause the structure to collapse. The loads chosen are commonly taken as those acting on the structure during service load conditions, while the strength of the structure is determined from the material strength properties for idealized plastic materials [Heyman, 1971].

For a given collapse mode, the structure is considered to have "failed" or collapsed when the plastic resistances R_{pi} are related to the factored loads λQ_j by

$$W_R(\mathbf{R}_p) \leq W_Q(\lambda \mathbf{Q}) \tag{1.5}$$

where $\mathbf{R}_p$ is the vector of all plastic resistances (e.g. plastic moments), $\mathbf{Q}$ the vector of all applied loads, $W_R()$ the internal work function and $W_Q()$ the external work function, with W_R and W_Q both determined by the collapse mode being considered.

If proportional loading is assumed, as is usual, the load factor can be taken out of the parentheses. Also the loads Q_j usually consist of dead, live, wind, etc., components, so that expression (1.5) may be rewritten in the form of a limit state equation as

$$\frac{W_R(R_{pi})}{\lambda W_Q(Q_D + Q_L + \ldots)} = 1$$

with "failure" denoted by the left-hand side being less than unity.

Clearly there is much similarity in formulation between the factor of safety and the load factor as measures of structural safety. What *is* different is the reference level at which the two measures operate: the first at the level of working loads and at the "member" level; the second at the level of. collapse loads and at the "structure level".

1.2.3 Partial factor ("limit state design")
A development of the above two measures of safety is the so-called "partial factor" approach. It is already in use in a number of structural design codes. Its general formulation, for structural inadequacy or failure, might be expressed at the level of stress resultants (i.e. member design level) for the limit state i as

$$\phi_i R_i \leqslant \gamma_{Di} S_{Di} + \gamma_{Li} S_{Li} + \dots \qquad (1.6)$$

where R is the member resistance, ϕ is the partial factor on R, S_D, S_L are the dead and live load effects respectively, and γ_D, γ_L are the partial factors on S_D, S_L respectively.

Expression (1.6) was originally developed during the 1960s for reinforced concrete codes. It enabled the live and wind loads to have greater "partial" factors than the dead load, in view of the former's greater uncertainty, and it allowed a measure of workmanship variability and uncertainty about resistance modelling to be associated with the resistance R [MacGregor, 1976]. This extension of earlier safety formats had considerable appeal since clear separation of uncertainties in loadings and resistances was achieved.

For a plastic collapse analysis at the structure level, formulation (1.6) becomes

$$W_R(\phi \mathbf{R}) \leqslant W_Q(\gamma_D \mathbf{Q}_D + \gamma_L \mathbf{Q}_L + \dots)$$

where $\mathbf{R}$ and $\mathbf{Q}$ are vectors of resistance and load components respectively. Naturally the set of partial factors (ϕ, γ) in this expression will be different from those of expression (1.6).

Example 1.1
The simple portal frame of Fig. 1.1(a) is subject to loads Q_1 and Q_2. If the relative moments of inertia of the members are known, the elastic bending moment diagram can be found as in Fig 1.1(b). The "limit states" for bending capacity are then

$$\text{section 2:} \qquad \phi M_{C2} = \gamma_1 \frac{3l}{16} Q_1 + \gamma_2 \frac{3l}{16} Q_2$$

$$\text{sections 1 and 3:} \quad \phi M_{C1,3} = \gamma_1 \frac{l}{16} Q_1 + \gamma_2 \frac{l}{16} Q_2$$

where ϕ, γ_1 and γ_2 are partial factors described by a structural design code. The M_{Ci} are the ultimate moment capacities required at sections i for the structure to be considered "just safe".

If the frame is to be designed or analysed assuming rigid–plastic theory, the relative distribution of the plastic moments M_i around the frame must be known or assumed. If all M_{pi} are equal, the plastic bending moment diagram of Fig. 1.1(c) is obtained and only one limit state equation is needed for sections 1–3:

$$\phi_p M_{pi} = \gamma_{p1} \frac{l}{8} Q_1 + \gamma_{p2} \frac{l}{8} Q_2$$

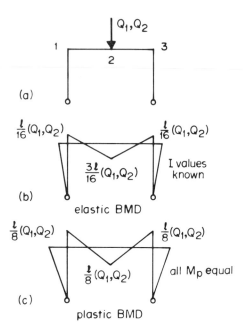

Fig. 1.1 — Example 1.1: bending moment diagrams.

where now M_{pi} is the required plastic moment capacity at sections 1, 2 and 3 and where ϕ_p, γ_{p1} and γ_{p2} are again code-prescribed partial factors, different in values from ϕ, γ_1 and γ_2.

1.2.4 A possible deficiency in deterministic safety measures: lack of invariance

From the example it will be evident that the partial factors ϕ, and γ_i ($i=1,\ldots$) in equation (1.6) depend on the limit state being considered, and hence on the definitions of R, S_D and S_L. However, even for a given limit state, the definitions of R, S_D, S_L, etc., are not necessarily unique, and therefore the partial factors may not be unique either. This phenomenon is termed the "lack of invariance" of the safety measure. It arises because there are different ways in which the relationships between resistances and loads may be defined. Some examples of this are given below. Ideally, the safety measure should not depend on the definition of loads and resistances.

Example 1.2
The structure shown in Fig. 1.2 is supported on two columns. The capacity of column B is $R = 24$ in compression. The safety of the structure can be measured in three different ways using the traditional "factor of safety":

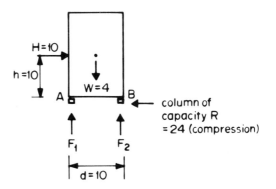

Fig. 1.2 — Example 1.2: structure subject to overturning.

(a) *Overturning resistance about A*

$$F_1 = \frac{\text{resisting moment about A}}{\text{overturning moment about A}} = \frac{dR}{Hh + Wd/2}$$

$$= \frac{10 \times 24}{10 \times 10 + 4 + 5} = \frac{240}{120} = 2.0$$

(b) *Capacity of column B*

$$F_2 = \frac{\text{compression resistance of column B}}{\text{compressive load on column B}}$$

$$= \frac{R}{Hh/d + W/2} = 2.0$$

(c) *Net capacity of column B* (resistance minus load effect of W)

$$F_3 = \frac{\text{net compressive resistance of column B}}{\text{compressive load on column B}}$$

$$= \frac{R - W/2}{Hh/d} = 2.2$$

All three of these factors of safety for column B apply to the same structure and the same loading, so that the difference in the values is due entirely to the definitions of resistance and applied load (or stress). However, it is easily seen that the calculations give the identical result $F_1 = F_2 = F_3 = 1.0$ if a partial safety factor $\phi = \frac{1}{2}$ is applied to the resistance R; thus

$$F_1 = \frac{d\phi R}{Hh + Wd/2}, \quad F_2 = \frac{\phi R}{Hh/d + W/2}, \quad F_3 = \frac{\phi R - W/2}{Hh/d}$$

Of course, the result $F_1 = F_2 = F_3 = 1$ would also be achieved if the loads H and W were factored by $\gamma = 2$. In fact any combination of ϕ and γ could be chosen provided that $F_i = 1$:

$$F_1 = \frac{d\phi R}{\gamma(Hh + Wd/2)} = F_2 = \frac{\phi R}{\gamma(Hh/d + W/2)} = F_3 =$$
$$= \frac{\phi R - \gamma W/2}{\gamma Hh/d} = 1$$

An alternative and useful measure of safety is the "safety margin", which measures the excess resistance compared with the stress resultant (or loading); thus

$$Z = R - S \tag{1.7}$$

For the present example, the safety margins are

$$Z_1 = dR - \left(Hh + \frac{Wd}{2}\right)$$

$$Z_2 = R - \left(\frac{Hh}{d} + \frac{W}{2}\right)$$

and

$$Z_3 = R - \frac{W}{2} - \frac{Hh}{d}$$

It is readily verified that when $Z = 0$, i.e. at the point of failure, these three expressions are equivalent.

Example 1.3 (adapted from Ditlevsen (1973))
The reinforced concrete beam shown in Fig. 1.3(a) has a moment capacity R when it is subject to an axial force N and a moment M applied at the beam centroid $\xi = 0$. N and M are each composed of the effects of a dead load and a live load: $N = N_D + N_L$ and $M = M_D + M_L$. The moment capacity calculated about $\xi = a$ is $R_1 = R + aN$, from simple statics. (Note that the moment capacity of the beam is not changed!) Also, at $\varepsilon = a$, the applied moment is given by $M_1 = M + aN$. The state of "just safe" can now be

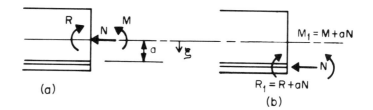

Fig. 1.3 — Example 1.3: reinforced concrete beam.

defined for given moment capacity R, and given axial force N, by the factor of safety as

$$F_0 = \frac{R}{M} \qquad\qquad \text{at } \xi = 0 \qquad (1.8a)$$

$$F_1 = \frac{R_1}{M_1} = \frac{R + aN}{M + aN} \qquad\qquad \text{at } \xi = a \qquad (1.8b)$$

In this format, $F_1 = F_0$ only when $\xi = a = 0$, so that the factor of safety depends on the convention chosen for the origin of the applied actions and the resistance. If, as in Example 1.2, R is replaced by the factored term ϕR, such that $F_0 = 1$, then it follows readily that F_1 is also unity. Hence, provided that "partial factor" ϕ is chosen in such a way that the "factor of safety" F is unity, the origin chosen to define R, N and M is immaterial. A similar result holds if N and M are replaced by γN and γM, where γ is an appropriately chosen partial factor on the loading.

The state of "just safe" can also be written in the partial factor format of expression (1.6). Indeed, noting that $M = M_D + M_L$ and $N = N_D + N_L$, at $\xi = 0$

$$\phi R = \gamma_D M_D + \gamma_L M_L \qquad (1.9a)$$

and at $\xi = a$, treating, as before, $R_1 = R + aN$ as the resistance to bending,

$$\phi(R + aN_D + aN_L) = \gamma_D(M_D + aN_D) + \gamma_L(M_L + aN_L) \qquad (1.9b)$$

Subtracting (1.9a) from (1.9b) and dividing out by a leaves

$$(\phi - \gamma_D)N_D + (\phi - \gamma_L)N_L = 0 \qquad (1.10)$$

Since in general N_D, $N_L > 0$, it follows that (1.10) will only be satisfied if either $\gamma_D \leqslant \phi \leqslant \gamma_L$ or $\gamma_L \leqslant \phi \leqslant \gamma_D$. Except for $\phi = \gamma_D = \gamma_L = 1$, these expressions are both inconsistent with the conventional interpretation that $\phi \leqslant 1$ (to reduce the calculated resistance) and γ_D, $\gamma_L \geqslant 1$ (to increase the loads or applied stresses).

The reason for this result should be clear. In expression (1.9b) the term $(aN_D + aN_L)$ on the left-hand side was treated as a resistance, *per se*, whereas it is strictly a resistance effect caused directly by the applied loading (note that it is not affected by workmanship, material strength, etc., as is R). The key to an invariant safety measure is thus at hand. Partial factors such as ϕ should be directly applied to resistances only, and partial factors such as γ to loads only, and the direct application of expression (1.6) to a mixed variable $R_1 = R + aN$ is not correct.

It is important to note that the safety margin Z (equation (1.7)) is invariant for both definitions of resistance in this example. In the first case $Z_0 = R - M$ while in the second case $Z_1 = (R + aN) - (M + aN) = R - M$.

1.2.5 Invariant safety measures

As can be seen from the above examples, one form of invariant safety measure is obtained if the resistances R_i and the loads Q_j acting on the structure are so factored that the ratio between any relevant pair $\phi_i R_i$ and $\gamma_j Q_j$ is unity at the point of limit state violation. In simple terms, this requires that all variables be reduced to a common base before being compared. This is the case for the permissible stress measure of structural safety expressed by equation (1.3). Another and important form of invariant safety measure is the safety margin $Z = R - S$ defined in equation (1.7). It will be used extensively in the sections to follow because of its invariant properties.

Some readers may recognize a parallel between the above discussion and the decision criteria in cost–benefit analysis. The safety margin corresponds to the "net present value of benefits" criterion and the problem of safety factors to the "numerator–denominator" problem [e.g. Prest and Turvey, 1965].

1.3   A PARTIAL PROBABILISTIC SAFETY MEASURE OF LIMIT STATE VIOLATION — THE RETURN PERIOD

In the historical development of engineering design, loads due to natural phenomena such as winds, waves, storms, floods, earthquakes, etc., were recognized quite early as having a randomness in time as well as in space. Frequently these loads have only a "remote" chance of occurring during the lifetime of a structure or engineering project. For these relatively rare loads, the probabilistic notion of the "return period" has been developed. It describes the probability of occurrence of a given loading event on a structure. The design of the structure itself is then usually considered deterministically, i.e. using conventional design procedures.

The return period is defined as the average (or expected) time between two successive statistically independent events. The actual time T between events is a random variable. In most practical applications an event constitutes the exceedance of a certain threshold associated with loading (e.g. wind velocity > 100 m/s).

For independent samples from a population (i.e. for a Bernoulli trial sequence), the trial T on which the first occurrence of an event takes place is given by the Geometric distribution (equation (A.23)), which states that the probability that the first occurrence occurs on the tth trial is:

$$P(T=t)=p(1-p)^{t-1}, \quad t=1,2,\ldots \tag{1.11}$$

where p is the probability of occurrence of the event (e.g. $X>x$) in any one trial and $1-p$ is the probability that the event does not occur. If trials are now interpreted as time intervals, during each of which only the occurrence of events $X>x$ is recorded, the first occurrence of an event becomes the "first occurrence time", given by expression (1.11). The "mean occurrence time" or the "return period" is then the expected value of T (see equation (A.10)):

$$E(T) = \overline{T} = \sum_{t=1}^{\infty} tp(1-p)^{t-1} =$$
$$= p[1 + 2(1-p) + 3(1-p)^2 + \ldots]$$
$$= \frac{p}{1-(1-p)^2} \quad \text{for} \quad (1-p)<1.0$$
$$= \frac{1}{p} \quad \text{or} \quad [1-F_X(x)]^{-1} \tag{1.12}$$

where $F_X(x) = P(X \leqslant x)$ is the cumulative distribution function of X.

Thus the return period $\overline{T}$ is equal to the reciprocal of the probability of the occurrence of the event in any one time interval. For most engineering problems, the chosen time interval is 1 year, so that p is the probability of occurrence of the event $X>x$ in any one year (e.g. the probability that a load $>x$ will occur (at least once) during the year). Then $\overline{T}$ is the number of years, on average, between events.

Because the exceedance events that occur during a time period (e.g. during a year) are associated with the end of that period, $\overline{T}$ is dependent on the time period chosen [Borgman, 1963]. This is illustrated in Fig. 1.4, where four exceedance events, A, B, C and D are shown occurring after an arbitrary initial event O. The mean recurrence time $\overline{T}_1$ for the actual observations is as shown in Fig. 1.4(a) and is given by the average of the distance (i.e. time) between the events, i.e. by $\overline{T}_1 \approx 1.5$ years. In Fig. 1.4(b) with the time period taken as 1 year, and the events counted at the end of each time period, it follows easily that $\overline{T}_2 = 7/4 = 1.75$ years. Similarly, $\overline{T}_3 = 2$ years. However, when a 4-year time period is used (Fig. 1.4(d)) two of the events in each period are counted as one at the end of the period, and $\overline{T}_4$ in this case becomes 4 years.

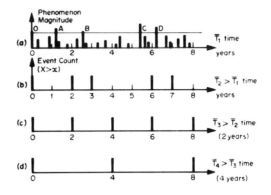

Fig. 1.4 — Idealizations of actual load phenomenon for the "return period" concept.

This rather artificial example shows three things. Firstly, that the return period depends, as noted, on the definition of the time scale, and secondly that the possible occurrence of more than one event within a time period is ignored. This means that, where event occurrence is relatively frequent compared with the time period employed, the return period measure is not accurate.

The third and a most important point is that the probability distribution of the magnitude of X (i.e. the phenomenon being considered) is not considered. Only magnitudes $X > x$ are counted. This means that the return period is a probabilistic measure in terms of time only, but not in terms of the magnitude of the loading and its interaction with the resistance.

It should be clear that in practice the events may not be independent, as postulated, particularly if the events occur rather frequently. Fortunately, the return period concept is used mainly for quite rare events (i.e. the level X is quite high) and it is then reasonable to assume event independence. Time scale dependence is then not a significant issue.

Example 1.4
For a structure subject to a "50-year wind" of 60 km/h velocity:

(a) the return period for a 60-km/h wind = $\overline{T}$ = 50 *years*;
(b) the probability of exceeding 60 km/h in any one year is $p = 1/\overline{T} = 1/50 = 2\%$;
(c) the probability of exceeding the design wind velocity (i.e. $V > 60$) for the first time during the fourth year, is (geometric distribution, equation (A.23)) $P_T(T = 4) = (0.02)(0.98)^3 = 0.01882$;
(d) the probability of exceeding the design wind velocity in any one year in a 4-year period is given by the binomial distribution (equation (A.17))

$$P_X(x=1) = \binom{4}{1}(0.02)(0.98)^3 = \frac{4 \times 3 \times 2 \times 1}{1(3 \times 2 \times 1)}(0.02)(0.98)^3 =$$
$$= 0.0753$$

(e) the probability of exceeding the design wind velocity (i.e. $V > 60$) during one year within 4 years is given by the geometric distribution (equation (A.23))

$$P_T(T \leqslant 4) = \sum_{t=1}^{4} P_T(T=t)$$

$$= \sum_{t=1}^{4} (0.02)(0.98)_{t-1}$$

$$= 0.02 + 0.0196 + 0.01921 + 0.01883$$

$$= 0.0776$$

or alternatively,

$$P_T(T \leqslant 4) = 1 - [P(V < 60)]^4$$
$$= 1 - (1 - 0.02)^4 = 0.0776$$

(note that the period 4 years can be generalized to "design life" t_L and the question rephrased to "the probability of exceeding the design velocity within the design life":

$$P_T(T \leqslant t_L) = \sum_{t=1}^{t_L} P_T(T=t)$$

$$= 1 - (1 - p)^{t_L} \qquad (1.13)$$

and some typical values for the relationship between the exceedance probability $P_T(T \leqslant t_L)$, the return period $\overline{T} = 1/p$ and the design life t_L are given in Table 1.2 [Borgman, 1963]);

(f) the probability of exceeding the design wind velocity within the return period is

$$P_T(T \leqslant \overline{T}) = 1 - [P(V < 60)]^{\overline{T}}$$

but

$$P(V < 60) = 1 - P(V \geqslant 60) = 1 - \frac{1}{p}$$

Table 1.2 — Return period $\overline{T}$ as function of design life t_L and exceedance
probability $P_T(T \leqslant t_L)$

Design life t_L	Return period $\overline{T}$ for the following exceedance probabilities (see equation (1.13))								
	0.02	0.05	0.10	0.15	0.20	0.30	0.40	0.50	0.70
1	50	20	10	7	5	3	3	2	1
2	99	39	19	13	9	6	4	3	2
3	149	59	29	29	14	9	6	5	3
4	198	78	38	25	18	12	8	6	4
5	248	98	48	31	23	15	10	8	5
6	297	117	57	37	27	17	12	9	6
7	347	137	67	44	32	20	14	11	6
8	396	156	76	50	36	23	16	12	7
9	446	176	86	56	41	26	18	13	8
10	495	195	95	62	45	29	20	15	9
12	594	234	114	74	54	34	24	18	10
14	693	273	133	87	63	40	28	21	12
16	792	312	152	99	72	45	32	24	14
18	892	351	171	111	81	51	36	26	15
20	990	390	190	124	90	57	40	29	17
25	1238	488	238	154	113	71	49	37	21
30	1485	585	285	185	135	85	59	44	25
35	1733	683	333	216	157	99	69	51	30
40	1981	780	380	247	180	113	79	58	34
45	2228	878	428	277	202	127	89	65	38
50	2475	975	475	308	225	141	98	73	42

where $p = 1/\overline{T}$.
Hence

$$P_T(T \leqslant \overline{T}) = 1 - (1-p)^{\overline{T}}$$

$$= 1 - (1 - \overline{T}p + \frac{\overline{T}(\overline{T}-1)}{2!}p^2 - \ldots)$$

$$\approx 1 - e^{-(\overline{T}p)} \text{ for large } \overline{T} \text{ (i.e. small } p\text{)}$$

$$\approx 1 - e^{-1} = 1 - 0.3679 = 0.6321$$

Note that even for smaller $\overline{T}$, this result is a good approximation; thus,
for $\overline{T} = 5$,

$$P_T(T \leqslant 5) = 1 - \left(1 - \frac{1}{5}\right)^5 = 0.6723$$

This shows that there is a chance of about 2 in 3 that the exceedance event will occur within a design life equal to the return period.

1.4   A PROBABILISTIC MEASURE OF LIMIT STATE VIOLATION

1.4.1   Introduction

The return period concept considers only the probability that a loading exceeds a set limit and assumes such exceedings (or "level crossings") to be randomly distributed in time. This is a useful improvement over deterministic descriptions of loading but ignores the fact that, even at a given point in time, the actual loading is uncertain. This is illustrated in Fig. 1.5 for floor

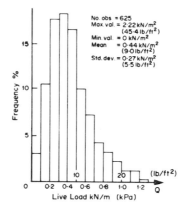

Fig. 1.5 — Histogram of private office Live loads [Culver, 1976].

loading. The histogram shows, for example, that the probability that the floor loading lies between 0.6 and 0.7 kPa is about 7%. Such information is obtained from actual surveys of floor loads (see Chapter 7), and can be represented by the probability density function $f_Q(q)$. (Recall that $f_Q()$ denotes the probability that the load Q will take on a value between q and $q+\Delta q$ as $\Delta q \to 0$ (see also section A.3). The load Q can be converted to a load effect S by structural analysis so that the probability density function $f_S()$ can, in principle, also be obtained, using, if necessary, methods such as outlined in section A.10. However, details of this need not be of concern for the present.

Resistance, geometric and workmanship variables and many others may similarly be described in probabilistic terms. For example, a typical resistance histogram and the inferred probability distribution for the yield strength of steel are shown in Fig. 1.6. Naturally, material strengths such as

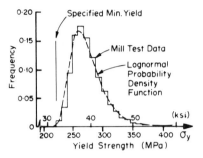

Fig. 1.6 — Histogram and inferred distribution for structural steel yield strength [Alpsten, 1972].

steel yield strength can be converted to member resistance R by multiplying by section properties (such as A, the cross-sectional area). Then it is possible to determine a probability density function $f_R()$.

Both load effect S and resistance R are generally functions of time. Usually the uncertainty of prediction of both S and R increases with time so that the probability density functions $f_S()$ and $f_R()$ become wider and flatter. In addition, the mean values of S and R may also change with time. Loads have a tendency to increase, and resistance to decrease, with time. The general reliability problem can therefore be represented as in Fig. 1.7.

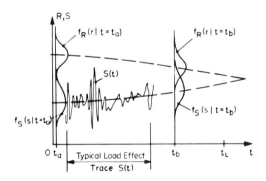

Fig. 1.7 — Time-dependent reliability problem.

The safety limit state will be violated whenever, at any time t,

$$R(t) - S(t) < 0 \quad \text{or} \quad \frac{R(t)}{S(t)} < 1 \tag{1.14}$$

The probability that this occurs for any one load application is the probability of limit state violation, or simply the probability p_f of failure. It may be roughly represented by, but is not actually equal to, the amount of overlap of the probability density function f_R and f_S in Fig. 1.7. Since this overlap may vary with time, p_f may also be a function of time.

In many situations, however, it is convenient to assume that neither Q nor R is a function of time. This will be the case if the load Q is applied once only to the structure and the probability of limit state violation is sought for that load application only.

However, if the load is applied many times (e.g. a single time-varying load might be considered this way) then it is the maximum value of that load (within a given time interval $[0, T]$) which is of interest. The load is then more properly represented by an extreme value distribution such as the Gumbel (*EV-I*) or Frechet (*EV-II*) distributions (see Appendix A). If this is done, the effect of time may be ignored in the reliability calculations. This approach is not satisfactory when more than one load is involved or when the resistance changes with time. Discussion of these matters will be deferred until Chapter 6.

1.4.2 The basic reliability problem

The basic structural reliability problem considers only one load effect S resisted by one resistance R. Each of S and R are described by a known probability density function, $f_S()$ and $f_R()$ respectively. As noted, S may be obtained from the applied loading Q through a structural analysis (either deterministic or with random components). It is important that R and S are expressed in the same units.

For convenience, but without loss of generality, only the safety of a structural element will be considered here and as usual, that structural element will be considered to have failed if its resistance R is less than the stress resultant S acting on it. The probability p_f of failure of the structural element can be stated in any of the following ways:

$$p_f = P(R \leqslant S) \tag{1.15a}$$

$$= P(R - S \leqslant 0) \tag{1.15b}$$

$$= P\left(\frac{R}{S} \leqslant 1\right) \tag{1.15c}$$

$$= P(\ln R - \ln S \leqslant 1) \tag{1.15d}$$

or in general

$$= P[G(R, S) \leqslant 0] \tag{1.15e}$$

where $G()$ is termed the "limit state function" and the probability of failure is identical with the probability of limit state violation. Equations (1.15) could, of course, also have been written in terms of R and Q for the structure as a whole.

Quite general (marginal) density functions f_R and f_S for R and S respectively are shown in Fig. 1.8 together with the joint (bivariate) density

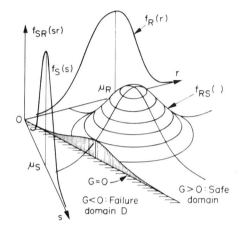

Fig. 1.8 — Region D of integration for failure probability determination.

function $f_{RS}(r, s)$ (see also section A.6). For any infinitesimal element $(\Delta r \, \Delta s)$ the latter represents the probability that R takes on a value between r and $r + \Delta r$ and S a value between s and $s + \Delta s$ as Δr and Δs approach zero. In Fig. 1.8, equations (1.15) are represented by the hatched failure domain D, so that the failure probability becomes:

$$p_f = P(R - S \leqslant 0) = \int_D \int f_{RS}(r, s) \, dr \, ds \qquad (1.16)$$

When R and S are independent, $f_{RS}(r, s) = f_R(r) f_S(s)$ (see section A.6.3) so that (1.16) becomes:

$$p_f = P(R - S \leqslant 0) = \int_{-\infty}^{\infty} \int_{-\infty}^{s \geqslant r} f_R(r) f_S(s) \, dr \, ds \qquad (1.17)$$

For any random variable X, the cumulative distribution function $F_X(x)$ is given by (A.8) or

$$F_X(x) = P(X) \leqslant x) = \int_{-\infty}^{x} f_X(y) \, dy$$

provided that $x \geqslant y$. It follows that for the common, but special case when R and S are independent, expression (1.17) can be written in the form

$$p_f = P(R - S \leqslant 0) = \int_{-\infty}^{\infty} F_R(x) f_S(x) \, \mathrm{d}x \tag{1.18}$$

This integral is also known as a "convolution integral" with meaning easily explained by reference to Fig. 1.9. $F_R(x)$ is the probability that $R \leqslant x$, or the

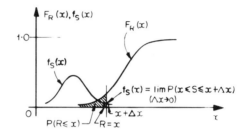

Fig. 1.9 — Basic $R-S$ problem: $F_R(\ ) f_S(\ )$ representation.

probability that the actual resistance R of the member is less than some value x. Let this represent failure. The term $f_S(x)$ represents the probability that the load effect S acting in the member has a value between x and $x + \Delta x$ in the limit as $\Delta x \to 0$. By considering all possible values of x, i.e. by taking the integral over all x, the total failure probability is obtained. This is also seen in Fig. 1.10 where the (marginal) density functions f_R and f_S have been drawn

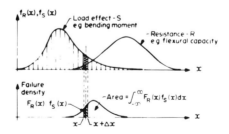

Fig. 1.10 — Basic $R-S$ problem: $f_R(\ ) f_S(\ )$ representation.

along the same axis.

Through integration of $f_R(\ )$ in expression (1.17), the order of integration was reduced by one. This is convenient and useful, but not general. It was only possible because R was assumed independent of S. In general dependence between variables may be possible.

An alternative to expression (1.18) is:

$$p_f = \int_{-\infty}^{\infty} [1 - F_S(x)]f_R(x)\,dx \qquad (1.19)$$

which is simply the sum over all the cases of resistance for which the load exceeds the resistance.

The lower limit of integration shown in expressions (1.17)–(1.19) may not be totally satisfactory, since a "negative" resistance is not usually possible. The lower integration limit should therefore strictly be zero, although this may be inconvenient and slightly inaccurate if R or S or both are modelled by distributions unlimited in the lower tail (such as the normal or Gaussian distribution). The inaccuracy arises strictly from the modelling of R and/or S, and not from the theory involved with expressions (1.17)–(1.19).

1.4.3   Special case: normal random variables

For a few distributions of R and S it is possible to integrate the convolution integral (1.18) analytically. The most notable example is when S and R are normal random variables with means μ_R and μ_S and variances σ_R^2 and σ_S^2 respectively. The safety margin $Z = R - S$ then has a mean and variance given by:

$$\mu_Z = \mu_R - \mu_S \qquad (1.20a)$$
$$\sigma_Z^2 = \sigma_R^2 + \sigma_S^2 \qquad (1.20b)$$

using well-known rules for addition (subtraction) of normal random variables. Equation (1.15b) then becomes

$$p_f = P(R - S \leqslant 0) = P(Z \leqslant 0) = \Phi\left(\frac{0 - \mu_Z}{\sigma_Z}\right) \qquad (1.21)$$

where is the standard normal distribution function (zero mean and unit variance) extensively tabulated in statistics texts (see also Appendix C). The random variable $Z = R - S$ is shown in Fig. 1.11, in which the failure region $Z \leqslant 0$ is shown shaded. Using equations (1.20) and (1.21) it follows that [Cornell, 1969]

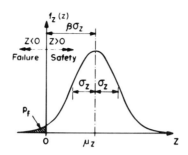

Fig. 1.11 — Distribution of safety margin.

$$p_f = \Phi\left[\frac{-(\mu_R - \mu_S)}{(\sigma_S^2 + \sigma_R^2)^{1/2}}\right] = \Phi(-\beta) \tag{1.22}$$

where β is defined as the "safety index"

$$\beta = \mu_Z/\sigma_Z \quad \text{(see expression (1.21))}$$

If either σ_S or σ_R or both are increased the term in square brackets in (1.22), will become smaller and hence p_f will increase, as might be expected. Similarly if the difference between the mean of the load effect and the resistance is reduced, p_f increases. These observations may also be deduced from Fig. 1.7, taking the amount of overlap of $f_R()$ and $f_S()$ as a rough indicator of p_f at any point in time.

Example 1.5
A simply supported timber beam of length 5 m is loaded with a central load Q having mean $\mu_Q = 3$ kN and variance $\sigma_Q^2 = 1$ kN². The bending strength of similar beams has been found to have a mean strength $\mu_R = 10$ kN m with a coefficient of variation (COV) of 0.15. It is desired to evaluate the probability of failure.
    Assume that the beam self-weight and any variation in the length of the beam can be ignored, and that the applied moment at the centre of the beam is given by

$$S = \frac{QL}{4} = \frac{5}{4}Q$$

Therefore

$$\mu_S = \frac{5}{4}\mu_Q = \frac{5}{4} \times 3 \qquad\qquad = 3.75 \text{ kN m} \qquad \text{(see (A.160))}$$

$$\sigma_S^2 = \left(\frac{5}{4}\right)^2 \sigma_Q^2 = \frac{25}{16} \times 1 \qquad = 1.56 \ (kN \ m)^2 \quad (see \ (A.162))$$

$$\mu_R = 10 \ kN \ m$$

and

$$\sigma_R^2 = \{(COV)\mu_R\}^2 = (0.15 \times 10)^2 = 2.25 \ (kN \ m)^2$$

Hence

$$\mu_Z = \mu_R - \mu_S = 10 - 3.75 = 6.25$$
$$\sigma_Z^2 = \sigma_R^2 + \sigma_S^2 = 2.25 + 1.56 = 3.81$$

Therefore

$$\frac{\mu_Z}{\sigma_Z} = \frac{6.25}{1.95} = 3.20$$

From (1.21)

$$p_f = \Phi(-3.20) = 7 \times 10^{-4} \ from \ Appendix \ C$$

1.4.4 Safety factors and characteristic values

The traditional deterministic measures of limit state violation, namely the factor of safety and the load factor, can be directly related to the probability p_f of limit state violation. This is most easily demonstrated analytically for the basic one-resistance one-load-effect case, when R and S (or Q) are each normally distributed.

Consider a convenient simple safety measure sometimes referred to as the central safety factor λ_0 and defined as

$$\lambda_0 = \frac{\mu_R}{\mu_S} \ or \ \frac{\mu_R}{\mu_Q} \tag{1.23}$$

This definition does not accord with conventional usage, since generally some upper range value of applied load or stress is compared with some lower range value of strength of material. Such values might be termed "characteristic" values, reflecting that in conventional usage (e.g. design) the load or strength is described only by this value. Thus the characteristic yield strength of steel bars is the strength that most (say 95%) of bars will exceed. There is a finite (but small) probability that some bars will have a lower strength.

For resistances, the design or "characteristic" values are defined on the low side of the mean resistance (see Fig. 1.12):

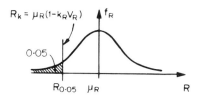

Fig. 1.12 — Definition of characteristic resistance.

$$R_k = \mu_R(1-k_R V_R) \tag{1.24}$$

where R_k is the characteristic resistance, μ_R the mean resistance, V_R the coefficient of variation for R and k_R a constant. This description is based on the normal distribution. R_k is the value of resistance below which only, say, 5% of samples will fail. Then, from the standardized normal distribution function (see section A.5.7),

$$0.05 = \Phi\left(-\frac{R_k - \mu_R}{\sigma_R}\right)$$

or

$$k_{0.05} = 1.645 = \frac{\mu_R - R_k}{\sigma_R}$$

from tables of Φ (see Appendix C). Expression (1.24) follows directly, noting that the standard deviation σ_R can be expressed as $\sigma_R = \mu_R V_R$.

Similarly for load effect the characteristic value is estimated on the high side of the mean:

$$S_k = \mu_S(1 + k_S V_S) \tag{1.25}$$

where S_k is the characteristic load effect (a design value), μ_S the mean load effect, V_S the coefficient of variation for S and k_S a constant.

If design values are defined, for example, as not being exceeded 95% of the time a load effect is applied, then $k_S = 1.64$ if S is normally distributed (see Fig. 1.13). Where loads (actions) are used, Q replaces S in expression (1.25).

In codified design, the percentiles used (such as 5% and 95% above) either are explicitly specified or may be deduced from the characteristic value specified. Other percentile characteristic values can be obtained in the manner indicated above for normal distributions, and also for non-normal distributions. Example 1.6 below shows a typical calculation, while Table

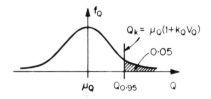

Fig. 1.13 — Definition of characteristic load.

1.3 summarizes 5 and 95 percentile values for some common distributions. Similar results may be derived for other percentile values.†

Using the characteristic values for the basic variables, it is now possible to define the so-called "characteristic safety factor" λ_k:

$$\lambda_k = \frac{R_k}{S_k} \text{ or } \frac{R_k}{Q_k} \tag{1.26}$$

which corresponds closely to the conventional understanding of the factor of safety if the characteristic values are taken to correspond to the usual design values.

A relationship can be established between the characteristic safety factor λ_k (or the central λ_0) and the probability p_f of limit state violation. This will obviously depend on the distributions of R and S, so that no general result can be given. Again, a particularly simple case is when both R and S are described by normal distributions.

From equation (1.22), the probability of failure is

$$p_f = \Phi\left[\frac{-(\mu_R - \mu_S)}{(V_R^2\mu_R^2 + V_S^2\mu_S^2)^{1/2}}\right] \tag{1.27}$$

and dividing through by μ_S

† This definition of the characteristic load is adequate where only one load application of a random load value occurs. As already noted, for time-varying loads an extreme value description may be more appropriate (see section 1.4.1 and Chapter 6). In this case an appropriate fractile to use would correspond to the mean of the extreme value distribution.

Table 1.3 — 5% and 95% values of X_k/μ_X

Distribution type	q (%)	X_k/μ_X for the following coefficients V_X of variation				
		0.1	0.2	0.3	0.4	0.5
Normal	5	0.8355	0.6710	0.5065	0.3421	0.1176
	95	1.164	1.329	1.493	1.658	1.822
Lognormal	5	0.8445	0.7080	0.5910	0.4927	0.4112
	95	1.172	1.358	1.552	1.750	1.945
Gumbel	5	0.8694	0.7389	0.6083	0.4778	0.3472
	95	1.187	1.373	1.560	1.746	1.933
Frechet	5	0.8802	0.7809	0.6999	0.6344	0.5818
	95	1.187	1.367	1.534	1.681	1.809
Weibull	5	0.8169	0.6470	0.4979	0.3736	0.2747
	95	1.142	1.305	1.489	1.689	1.903
Gamma	5	0.8414	0.6953	0.5608[a]	0.4355[a]	0.3416
	95	1.170	1.350	1.541[a]	1.752[a]	1.938

[a] Note that values are for $V_X = 0.302$ and 0.408 respectively, since the gamma distribution normally only allows discrete values of $V_X^2 (= k)$ (see section A.5.6).

$$= \Phi\left[\frac{-(\lambda_0 - 1)}{(V_R^2\lambda_0^2 + V_S^2)^{1/2}}\right] = \Phi(-\beta), \text{ say} \tag{1.28}$$

where λ_0 is given by (1.23) (β is the "safety index"; see (1.22)). It is easily shown that

$$\lambda_0 = \frac{1 + \beta(V_R^2 + V_S^2 - \beta^2 V_R^2 V_S^2)^{1/2}}{1 - \beta^2 V_R^2} \tag{1.29}$$

Also equations (1.24)–(1.26) give

$$\lambda_k = \frac{1 - k_R V_R}{1 + k_S V_S}\lambda_0 \tag{1.30}$$

so that a relationship between p_f, λ_0 and λ_k for given V_R, V_S, k_R and k_S follows immediately. Some typical relationships obtained by numerical integration are given in Fig. 1.14.

Expressions (1.29) and (1.30) indicate that the factors λ_0 and λ_k are dependent on the variability or uncertainty associated with R and S; with greater V_R and V_S requiring greater factors if the failure probability p_f is to

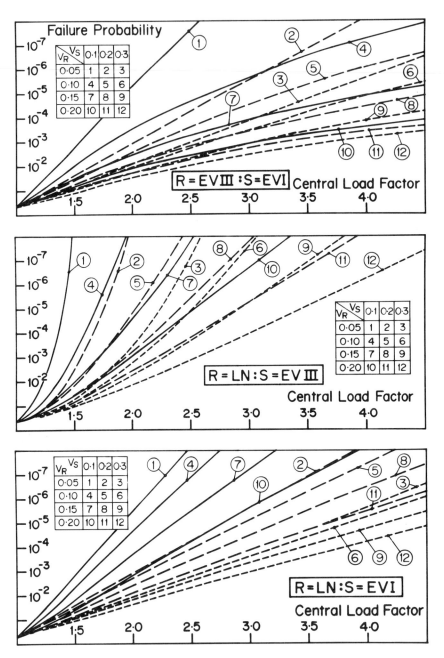

Fig. 1.14 — Failure probability versus central safety factor.

be kept constant (Fig. 1.14). This demonstrates again the deficiencies of the deterministic measures of limit state violation. They ignore much information which may be available about uncertainties in structural strength or applied loads.

Example 1.6

For a random variable S, with $\mu_S = 60$, $V_S = 0.2$, the 95 percentile for the Gumbel distribution, for example, may be determined as follows. From equation (A.77)

$$0.95 = F_Y(y) = \exp[-e^{-\alpha(y-u)}]$$

where, from equation (A.79), $\alpha^2 = \pi^2/6\sigma_Y^2$, and, from equation (A.78), $u = \mu_Y - \gamma/\alpha$ with $\gamma = 0.57722$.

Now $\sigma_Y = \sigma_S = 0.2 \times 60 = 12$, $\mu_Y = \mu_S = 60$, so that $\alpha = (\pi/\sqrt{6})/12 = 0.1069$ and $u = 60 - 0.57722/0.1069 = 54.60$. Hence

$$0.95 = \exp[-e^{-0.1069(S-54.60)}]$$

or

$$S_{0.95} = 82.38$$

From Table 1.3, it is found that, for the Gumbel distribution, $S_{0.95}/\mu_S = 1.373$, so that the 95 percentile value of S is $S_{0.95} = 1.373\mu_S = 82.38$.

1.4.5 Integration of the convolution integral

As noted above, closed-form integration of expressions (1.16) or (1.18) is only possible for some special cases. One of these, when both R and S are normally distributed, has already been considered. When both R and S are lognormal in distribution, and failure is defined as $Z=R/S<1$, an exactly parallel result can be obtained (see Example 1.7).

In general, recourse must be made to numerical integration. The simplest approach, the trapezoidal rule, is often quite effective [Dahlquist and Björck, 1974; Davis and Rabinowitz, 1975]. Step sizes around $x=0.2\sigma_R$ have proved sufficiently accurate together with an integration range of about $\pm5\sigma_Z$ instead of $\pm\infty$ [Ferry-Borges and Castenheta, 1971].

Some typical results obtained by numerical integration are given in Fig. 1.14. Other and similar results have been given by Freudenthal (1964) and Ferry-Borges and Castenheta (1971).

Example 1.7

Use the probability density function (A.61) for the lognormal variable $Z = R/S$, where R and S are each lognormal, to show that

$$p_f = \Phi(-\beta_1)$$

where

$$\beta_1 = \frac{\ln\{(\mu_R/\mu_S)[(1+V_S^2)/(1+V_R^2)]^{1/2}\}}{\{\ln[(1+V_R^2)(1+V_S^2)]\}^{1/2}}$$

which simplifies to

$$\beta_1 \approx \frac{\ln(\mu_R/\mu_S)}{(V_R^2 + V_S^2)^{1/2}} \quad \text{for} \quad V_R < 0.3,\ V_S < 0.3$$

Also derive an expression for the central safety factor $\lambda_0 = \mu_R/\mu_S$ and show that this simplifies to

$$\lambda_0 \approx \exp[\beta(V_R^2 + V_S^2)^{1/2}]$$

1.4.6 Basic variables

For many problems the simple formulations (1.15a)–(1.15e) are not entirely adequate, since it may not be possible to reduce the structural reliability problem to a simple R versus S formulation with R and S independent random variables.

In general, R is a function of material properties and element or structure dimensions, while S is a function of applied loads Q, material densities and perhaps dimensions of the structure, each of which may be a random variable. Also, R and S may not be independent, such as when some loads act to oppose failure (e.g. overturning) or when the same dimensions affect both R and S. In this case it is not valid to use the convolution integral (1.18). It is also not valid when there is more than one applied stress resultant acting at a section, or more than one factor contributing to the resistance of the structure. A more general formulation is required.

The fundamental variables which define and characterize the behaviour and safety of a structure may be termed the "basic" variables. They are usually the variables employed in conventional structural analysis and design. Typical examples are dimensions, densities or unit weights, materials, loads, material strengths. The compressive strength of concrete would be considered a basic variable even though it can be related to more fundamental variables such as cement content, water-to-cement ratio, aggregate size, grading and strength, etc. However, structural engineers do not normally use these latter variables in strength or safety calculations.

It is often possible to choose the basic variables such that they are independent. However, this is not always the case. Thus the compressive and tensile strengths and the elastic modulus of concrete are related; yet in a particular analysis they might each be treated as a basic variable. Dependence between basic variables usually adds complication to a reliability analysis. It is important that the dependence structure between dependent variables be known and expressible in some form. This will usually be in the form of a correlation matrix; however, as noted in Appendix A, this can at best provide only limited information.

The probability distributions for the basic variables can be obtained by direct inference from observed data, by subjective assessment or by some combination of these techniques. In practice some subjective influence is nearly always present since there are seldom sufficient data available to identify unambiguously only one distribution as appropriate. All this assumes furthermore that past observations and experience for similar structures can be validly used for the structure under assessment.

Sometimes physical reasoning may be used to suggest an appropriate probability distribution. Thus, where the basic variable consists of the sum of many other variables (which are not explictly considered), the central limit theorem (see section A.5.8) can be invoked to suppose that a normal distribution (see section A.5.7) is appropriate. This reasoning would be appropriate for the compressive strength of concrete (many component strengths) and for the dead load of a beam or slab (again many components of weight and several dimensions). In another example, the maximum wind velocity per year might be represented by the Gumbel (*EV-I*) distribution (see section A.5.11), as this is based on an underlying wind phenomenon which is described as essentially normal in probability distribution (see Chapter 7).

The parameters of the distribution may be estimated from the data using one of the usual methods, e.g. methods of moments, maximum likelihood, order statistics. These are well described in standard statistics texts and will not be considered here [e.g. Ang and Tang, 1975]. However, it must be emphasized that such techniques should not be used blindly. Critical examination of the data for trends and outliers is always necessary and the reasons for these phenomena established. It is quite possible for such behaviour to be the result of data recording and storage procedures rather than the behaviour of the variable itself.

Finally, when model parameters have been selected, the model should be compared with the data if at all possible. A graphical plot on appropriate probability paper is often very revealing, but analytical "goodness of fit" tests (e.g. Kolmogorov–Smirnov test) can also be used.

It will not always be possible to describe each basic variable by an appropriate probability distribution. The required information may not be available. In such circumstances a "point estimate" of the value of the basic variable might be used, i.e. the best estimate given the known information. If some uncertainty information about the variable is also available, it might be appropriate to represent it by an estimate of its mean and its variance only. This is then known as a "second moment" representation. One way in which such a representation might be interpreted is that in the absence of more precise data, the variable might be assumed to have a normal distribution (as this is completely described by the mean and variance, i.e. the first two moments (see section A.5.7)). However, other probability distributions might be more appropriate, even if only the first two momemts are known.

When more than second-moment, but less than full distribution information is available, (e.g. some correlation data) the incorporation of this in the reliability analysis becomes rather more difficult. Some inital progress in this direction has been made in connection with the methods discussed in Chapter 4 [der Kiureghian and Liu, 1985]. Much of the difficulty can be avoided, however, by choosing independent basic variables.

1.4.7 Generalized limit state equations
With the basic variables and their probability distributions established, the simple $R-S$ form of limit state function may be replaced with a generalized version expressed directly in terms of basic variables.

If the vector $\mathbf{X}$ represents the basic variables of the problem then the resistance R could be expressed as $R = G_R(\mathbf{X})$ and the loading or load effect as $S = G_S(\mathbf{X})$. The functions G_R and G_S may be non-linear, and in general the cumulative distribution function F_R, for example, must be obtained by multiple integration over the relevant basic variables (see (A.155)):

$$F_R(r) = \int^r \ldots \int f_{\mathbf{X}}(\mathbf{x}) \, d\mathbf{x}$$

A similar expression applies for S and F_S. Fortunately, it is seldom necessary to derive F_R (or F_S) specifically, since the limit state function $G(R, S)$ in (1.15e) can also be generalized. When the functions $G_R(\mathbf{X})$ and $G_S(\mathbf{X})$ are used in $G(R, S)$, the resulting limit state function can be written as $G(\mathbf{X})$, where $\mathbf{X}$ is the vector of all relevant basic variables and $G(\ )$ is some function expressing the relationship between the limit state and the basic variables. The limit state equation $G(\mathbf{x}) = 0$ now defines the boundary between the satisfactory "safe" $G > 0$ and unsatisfactory "unsafe" $G \leqslant 0$ domains in n-dimensional basic variable space. It is usually easily derived from the physics of the problem. (Note that $\mathbf{X} = \mathbf{x}$ defines a particular "point" $\mathbf{x}$.)

Where some loads may influence resistance (e.g. in overturning situations (see Fig. 1.2)) care should be taken that $G(\mathbf{X})$ is properly defined. Again by analogy with the simple case of Fig. 1.2 a useful rule is that any basic variable adding resistance to the limit state should have a positive gradient: $\partial G / \partial X_i > 0$.

Example 1.8
Consider a simple pin-ended strut supporting one end of a simply supported beam of length L_1, loaded at midpoint by a load Q. The actual load on the strut is thus $QL_1/2$. The strength of the strut is governed by its length L_2, its radius r of gyration, its cross-sectional area A and either the yield strength σ_Y of the steel or some combination of axial load capacity and bending capacity, usually expressed by an interaction rule in structural design codes.

Such rules are based, usually, on experimental observations and are then modified for code users by adding conservative assumptions and factors of safety. It is apparent, therefore, that code rules must be used with great caution in reliability analyses. A better approach is to use the original data and/or original relationships for ultimate strength.

For the squash load limit state, it follows easily that

$$G_1(\mathbf{X}) = \sigma_Y A - \frac{QL_1}{2}$$

is the limit state equation; all variables may be random, or A, say, might be considered closely deterministic.

For the interaction case, the limit state equation would be

$$G_2(\mathbf{X}) = FN\left(\sigma_Y A, \frac{L_2}{r}\right) - \frac{QL_1}{2}$$

where $FN()$ is an appropriate interaction equation for ultimate strength of pin-ended struts.

1.4.8   Generalized limit state violation problems
It follows that, with the limit state function expressed as $G(\mathbf{X})$, the generalization of (1.16) is:

$$p_f = P[G(\mathbf{X}) \leq 0] = \int \ldots \int_{G(\mathbf{X}) \leq 0} f_{\mathbf{X}}(\mathbf{x}) \, d\mathbf{x} \tag{1.31}$$

Here $f_{\mathbf{X}}(\mathbf{x})$ is the joint probability density function for the n vector $\mathbf{X}$ of basic variables. Note that the resistance R and load effect S need no longer be independent, as each is now described in terms of the basic variables $\mathbf{X}$. In general it is preferable for the basic variables themselves to be independent; if so, then (see (A.117))

$$f_{\mathbf{X}}(\mathbf{x}) = \prod_{i=1}^{n} f_{X_i}(x_i)$$

$$= f_{X_1}(x_1) f_{X_2}(x_2) f_{X_3}(x_3) \ldots \tag{1.32}$$

where $f_{X_i}(x_i)$ is the "marginal" probability density function for the basic variable X_i.

The region of integration $G(\mathbf{X}) \leq 0$ in expression (1.31) denotes the space of limit state violation and is directly analogous to that for the simple $R - S \leq 0$ limit state shown in Fig. 1.8. Except for some special cases, the integration over the failure domain $G(\mathbf{X}) \leq 0$ cannot be performed analytically. Methods to deal with the integration are considered in Chapters 3 and 4. They are essentially of two types:

(a) multidimensional integration of the original problem using numerical approximations such as simulation;
(b) transformation of the original problem such that the probability density function for each variable is approximated by a normal distribution.
  The (multi)normal distribution has some remarkable characteristics which may then be used to determine, approximately, the probability of failure.

  The limit state violation problem (1.31) can be further extended if the limit state function is generalized. For convenience, introduce the indicator function I defined such that (Fig. 1.15(a))

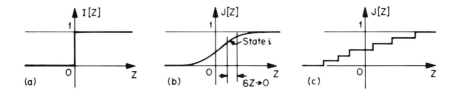

Fig. 1.15 — Limit state violation indicators.

$$I(x) = 0 \ if \ x \leq 0$$

$$= 1 \ if \ x > 0$$

(1.33)

It follows that $I[G(\mathbf{X})]$ may then be interpreted as a "utility function" with the failure state $G(\mathbf{X}) \leq 0$ denoting a "utility" of zero, and the safe state $G(x) \leq 0$ having unit "utility".
  In practice, and in particular in questions of lack of serviceability, the distinction between full utility and zero utility may not always be clear cut, and values between zero and unity may be appropriate (see Fig. 1.15). Thus it may be that utility depends inversely on concrete crack size, with no cracks having a utility of 1, cracks $< 10^{-3}$ mm a utility of 0.5 and greater cracks zero utility. Clearly many other possibilities exist. If now $J(x)$ denotes this more general interpretation of the indicator function (see Fig. 1.15(b)) and $J^c()$ $= 1 - J()$ defines the complement of J, the generalization of (1.31) becomes:

$$p_f = P\{J^c[G(\mathbf{X})]\} = \int \ldots \int_{\mathbf{x}} J^c[G(\mathbf{x})]f_{\mathbf{X}}(\mathbf{x})\,d\mathbf{x} \qquad (1.34)$$

As might be imagined, evaluation of (1.34) is not a simple matter. In addition, of course, the function $J[G(\mathbf{x})]$ must be known. Some initial work to establish $J[G(\mathbf{x})]$ has been reported by Reid and Turkstra (1980).

1.5 CONCLUSION

Various ways in which structural reliability may be defined have been reviewed in this Chapter. To do so it is necessary to introduce the concept of "limit states", which were seen to be a formalization of the criteria under which the structure can be considered to have "failed" or have reached an unsatisfactory state.

Traditional measures of limit state violation were reviewed, including the factor of safety, the load factor and possible "limit state design" concepts. It was shown that care is required in their definition; otherwise the measure would not be "invariant".

Another common measure, the return period, was reviewed prior to the introduction of a fully probabilistic measure of limit state violation. Several aspects of this were then outlined and generalizations given.

2

Structural reliability assessment

2.1 INTRODUCTION

Before proceeding to elaborate the concepts introduced in Chapter 1, it is necessary to address fundamental questions about the meaning of the calculated probability of limit state violation (whether for ultimate limit states or otherwise). Specifically, what does the calculated probability p_f mean? Can it be related to observed rates of failure for real structures? How can a knowledge of p_f help in achieving better (safer?) structures or more economical structures? These are important questions and ones about which a degree of controversy and disagreement still exists.

It will be helpful to examine the meaning of a few terms. "Probability" has already been used in Chapter 1. It denotes the chance that a particular, predefined event occurs. Classically, the probability of event occurrence was considered to be obtainable only from many repeated observations of the process which led to the event, the so-called "frequentist" (or objective) definition. The fact that the events must be observed means that there is immediately an element of subjectivity even in the frequentist meaning, in much the same way that observations in, say, physics are always partly subjective [de Finetti, 1974; Popper, 1959]. This aspect is sometimes (erroneously) ignored, and relative frequency data assumed to be purely "objective" information.

An alternative interpretation is that probability expresses a "degree of belief" about the occurrence of an event, rather than the actual (but unknown) frequency. It is therefore a "subjective" or "personal" probability. This interpretation is much wider than the relative frequency definition, and in its extreme form could be based on no previous data or experience of any sort to express degree of belief. Such a possibility is rather rare, however. In most situations some relevant or related frequentist information is available to form a basis for subjective probability estimates.

Generally, as more frequentist data become available the subjective probability estimates will tend to be adjusted so as to be in broad agreement with such data. The manipulation of subjective estimates and use of compatible data to amend and improve the estimates may be achieved through Bayes' theorem (A.7). Consequently, subjective probability assessments are sometimes known as "Bayesian" probabilities [Benjamin and Cornell, 1970; Lindley, 1972]. It is sometimes noted that a subjective probability estimate reflects the degree of ignorance about the phenomenon under consideration. Reconciliation of the two interpretations of the meaning of probability is controversial [e.g. Fishburn, 1964, Chapter 5; Hasofer, 1984].

The term "reliability" is commonly defined as the complement of the probability of failure ($= 1 - p_f$) but more properly it is the probability of safety (or proper performance) of the structure over a given period of time.

"Risk" has two meanings. In the context of structural engineering, the first meaning is equivalent to the probability of "structural failure" from all possible causes, both from violation of predefined limit states and from other causes.

The second interpretation of "risk" refers to the magnitude of the "failure" condition, usually expressed in money terms, and is commonly used in connection with insurance. This meaning will not be used herein.

"Structural failure" might be considered to be the occurrence of one or more types of undesirable structural response including the violation of predefined limit states. Thus collapse of all or part of a structure, major cracking and excessive deflection are some possible forms of failure (see Table 1.1).

Fortunately, real structures only rarely fail in a serious manner, but when they do it is often due to causes not directly related to the predicted nominal loading or strength probability distributions considered in Chapter 1. Other factors such as human error, negligence, poor workmanship or neglected loadings are most often involved [Melchers *et al.*, 1983]. To a large extent these factors are foreseeable and predictable; their occurrence might be considered as the occurrence of "imaginable" events. They must obviously be accounted for in any analysis of structural reliability which attempts to replicate or predict reality with some degree of confidence. However, not all possible reasons for structural failure are always imaginable [Ditlevsen, 1982a]. Events that must once have been "unimaginable" have led to structural failure; examples include the collapse of the Tay Bridge (1879) due (mainly) to underestimation of wind loading in storm conditions, and the Tacoma Narrows Bridge (1940) due to wind excitation of the deck. Even in these cases there is some evidence to suggest that the phenomena involved were not totally "unimaginable" before the accidents occurred. They were, however, apparently unimaginable to those involved with the projects [Sibly and Walker, 1977].

It is clearly not possible to make estimates of failure probabilities for truly unimaginable events, but care is obviously required that designers in particular are at least aware of the state of the art of their specialization.

Further, it is suggested that the public is more likely to accept the consequences of truly unimaginable events than it is likely to accept those of imaginable (and therefore foreseeable) events.

The theory needed to carry out a reliability analysis for imaginable events is the topic of this book. To do so it will be necessary to consider the various types of uncertainty which must be taken into account. These will be considered in section 2.2. Some uncertainties will be describable in terms of probability density functions. Others may only be describable in less precise ways, such as by "point" estimates of probabilities (see also Example 2.3). For the most part, the latter type of uncertainty will not be considered herein, although how it might be incorporated into a reliability assessment is described in section 2.3.

Reliability analysis considering only a subset of uncertainties will result in a probability estimate which herein will be termed a "nominal" or "formal" measure. The meaning of such a measure will be considered in section 2.5, together with the implications of using such a measure for deriving rules such as those given in structural design codes.

Finally, some criteria which may be used to decide the acceptability of calculated reliabilities will be discussed in section 2.4.

2.2   UNCERTAINTIES IN RELIABILITY ASSESSMENT

2.2.1   Identification of uncertainties

The uncertainties considered in Chapter 1 were the load acting on a structural element and its resistance. More generally a range of uncertainties may need to be considered. These might include various environmental conditions, workmanship and human error, and prediction of future events.

Identification of uncertainties for complex systems may be difficult. It may be advantageous to use a systematic scheme to help to enumerate all operational and environmental loading states and for each consider possible combinations of error or malfunction. This is essentially "event-tree analysis" [Henley and Kumamoto, 1981]. Rather similarly, the systematic development of all possible forms of hazard to which a structure might be subjected has been termed "hazard scenario analysis" [Schneider, 1981]. More generally, techniques such as "brain storming" [Osborn, 1957] may be of use.

Various other techniques have also been suggested [Henley and Kumamoto, 1981]. All amount essentially to a critical analysis of the problem to be analysed, consideration of all imaginable consequences and all imaginable possibilities and retaining only those with some finite probability of occurrence. Further, all techniques rely on having available expert opinion for the various assessments to be made and up-to-date information on which to base assessments.

The types of uncertainty which might be identified are shown schematically in Fig. 2.1. Each of these will now be discussed. Extra attention will be

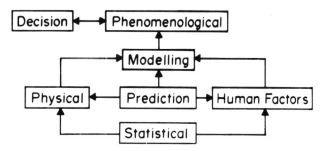

Fig. 2.1 — Uncertainties in reliability assessment.

given to human factors as these have been found to be particularly important.

2.2.2 Phenomenological uncertainty

As already noted, sometimes an apparently "unimaginable" phenomenon occurs to cause structural failure. The Tacoma Narrows Bridge noted earlier was apparently one such case. It was also of a design which departed considerably from earlier suspension bridge designs.

Phenomenological uncertainty may be considered to arise whenever the form of construction or the design technique generates uncertainty about any aspect of the possible behaviour of the structure under construction, service and extreme conditions. It is therefore of particular importance for novel projects, or those which attempt to extend the "state of the art" [Pugsley, 1962]. It should be quite clear that for such projects only subjective estimates of the effect of this type of uncertainty can be given.

2.2.3 Decision uncertainty

Decision uncertainty arises in connection with the decision as to whether a particular phenomena has occurred. In terms of limit states it is concerned purely with the decision as to whether a limit state violation has occurred.

A typical example concerns crack widths or deflections. It is unlikely, in general, that a slight increase in either will suddenly render the structure unsafe or unserviceable. At most, it is a question of relative loss of structural usefulness. One way in which decision uncertainty might be formulated has already been suggested in section 1.4.8 using the indicator function $J( )$. This might also be taken as a measure of utility (see Fig. 1.15).

2.2.4 Modelling uncertainty

Modelling uncertainty is caused by the use of a simplified relationship between the basic variables to represent the "real" relationship or pheno-menon of interest. In its simplest form, modelling uncertainty concerns the uncertainty of physical models, such as limit state equations.

Modelling uncertainty can be incorporated into a reliability analysis by introducing a modelling variable X_i, say, to represent the ratio between actual and predicted model response or output (see Fig. 2.2). This is a

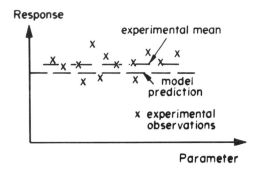

Fig. 2.2 — Modelling error (schematic).

convenient approach where both the model and the actual phenomenon are measured by the same type of variables (e.g. deflection, or crack width). However, if structural behaviour is modelled by member behaviour, say (as is implicit in permissible stress design philosophy), there usually will be a lot of variability, depending on the type and the degree of static indeterminacy of the structure. It would be preferable in this case to include the structural analysis more formally in the reliability analysis (see Chapter 5).

A special case of model uncertainty relates to the treatment of human errors and human intervention effects. If sufficient information about these effects is known, allowance could, in principle, be made for them in the modelling. Because of their special nature, a separate and detailed discussion of human factors will be given in section 2.2.8 below.

2.2.5 Prediction uncertainty

Many problems in structural reliability assessment involve the prediction of some future state of affairs; in this case the prediction of the reliability of some structure at some time $t > 0$ in the future.

An estimate of structural reliability depends on the state of knowledge available to the analysist(s). As new knowledge related to the structure becomes available, the estimate will become more refined, with, usually but not necessarily, a concomitant reduction in uncertainty. This applies particularly during the construction phase of a project, when information about actual strengths of materials, workmanship, etc., becomes available to replace estimates based on the past performances of, and the experiences with, similar structures. When the structure is placed in service, its response to initial loading (or perhaps to "proof loading") will constitute further information from which the reliability estimate may be revised.

It will be clear, therefore, that a probability estimate is not only a function of the properties of the structure, but also a reflection of the analysist's knowledge of the structure and the forces and influences likely to act on it. Similarly, if a reliability estimate for a particular structural lifetime is required, the analyst's uncertainty in the prediction of the lifetime (and the loadings which might be expected during that time) enter into the uncertainty of the reliability estimate (see Chapter 6).

2.2.6 Physical uncertainty

Physical uncertainty is that identified with the inherent random nature of a basic variable. Examples include;

(1) the variation in steel yield strength;
(2) the variability of wind loading;
(3) the variability of actual floor loading;
(4) the physical dimensions of a structural component.

Physical uncertainty can be reduced but not eliminated with greater availability of data, or greater effort in quality control. It is a "fundamental" property of the basic variable.

The physical uncertainty for any basic variable is generally not known *a priori* and must be estimated from observations of the variable or be subjectively assessed (see section 1.4.6).

2.2.7 Statistical uncertainty

Statistical estimators such as the sample mean and higher moments can be determined from available data and then used to suggest an appropriate probability density function and associated parameters. Generally the observations of the variable do not perfectly represent it and as a result there may be bias in the data as recorded. In addition, different sample data sets will usually produce different statistical estimators. This causes statistical uncertainty. The process of using sample statistic estimators to infer (subjectively) a probability distribution for a variable is adequately described in standard texts [e.g. Benjamin and Cornell, 1970] and will not be considered here.

Statistical uncertainty can be incorporated in a reliability analysis by letting the parameters such as the mean and variance (and other parameters which describe the probability distribution) themselves be random variables. Alternatively, the reliability analysis might be repeated using different values of the parameters to indicate sensitivity.

Example 2.1

Consider concrete cylinder test results for a concrete of specified nominal strength supplied by a number of concrete suppliers. To achieve the required nominal strength, some suppliers, with lesser control or quality, will aim for a higher mean strength to counter the greater variability of their product. There will also be variability due to between-batch variability for any one supplier, and variability due to the casting and testing of the test cylinders.

It is clear that the probability distribution which can be developed for concrete cylinder strength will depend on the manner in which each of these uncertainties contributes to the strength being measured. Evidently, if one supplier is changed, or deleted, the recorded histogram and hence the inferred probability distribution function for concrete strengths would be expected to change. Ideally, sampling should be on homogenous samples (i.e. those for which there is no *evidence* of non-homogeneity), with other factors separately sampled and perhaps included in a combined distribution model if required. Thus, restricting attention to only two different suppliers, the probability density function f_c for concrete strength, irrespective of supplier, is given by

$$f_c(\ ) = q_1 f_1(\ ) + q_2 f_2(\ )$$

with

$$q_1 + q_2 = 1$$

where q_1, q_2 represent the respective contributions of suppliers 1 and 2, having concrete strength probability distribution f_1 and f_2 respectively. Clearly a change in q_1 and q_2 changes f_c.

2.2.8   Uncertainties due to human factors

The uncertainties resulting from human involvement in the design, construction, use, etc., of structures may, for convenience, be considered in two categories:

(1) human errors
(2) human intervention.

In practice there is interaction between these, as will be evident.

2.2.8.1   *Human error*

Human errors can be divided, roughly, into errors V due to natural variation in task performance and gross errors E, G (see Table 2.1). As shown, gross error might be considered again in two categories: those errors which occur in the normal processes of design, documentation, construction and use of the structure within accepted procedures, and those which are a direct result of ignorance or oversight of fundamental structural or service requirements.

The relative importance of these types of errors can be gauged from Tables 2.2 and 2.3 which summarize the findings of Matousek and Schneider (1976) and Walker (1981). Other surveys suggest similar findings [Melchers *et al.*, 1983]. Note that unimaginable events, as interpreted from the information in Tables 2.2 and 2.3, constitute a rather low percentage of error factors.

Table 2.1 — Classification of human errors

Error type	Human variability V	Human error E	Gross human error G
Failure process	In a mode of behaviour against which the structure was designed		In a mode of behaviour against which the structure was *not* designed
Mechanism of error	One or more errors during design, documentation construction and/or use of the structure		Engineer's ignorance or oversight of fundamental behaviour. Profession's ignorance of fundamental behaviour
Possibility of analytic representation	High	Medium	Low to negligible

Adapted from Baker and Wyatt (1979).

Table 2.2 — Error factors in observed failure cases

Factor	%
Ignorance, carelessness, negligence	35
Forgetfulness, errors, mistakes	9
Reliance upon others without sufficient control	6
Underestimation of influences	13
Insufficient knowledge	25
Objectively unknown situations (unimaginable?)	4
Remaining	8

Adapted from Matousek and Schneider (1976).

The overriding conclusion from these surveys is that human error is involved in the majority of cases of recorded failure. Human error must, it seems, be considered if a reliability assessment is to relate to reality.

Unfortunately understanding of human error is limited, and much of that understanding is qualitative. It is known that humans perform best at an appropriate level of arousal, as indicated schematically in Fig. 2.3 [e.g. Warr, 1971]. If the level of arousal is too high or too low, performance deteriorates, although at the extremes there are rather vague barriers due to legal sanctions (see section 2.2.8.2 below) [Melchers, 1980]. Further,

Table 2.3 — Prime "causes" of failure

Cause	%
Inadequate appreciation of loading conditions or structural behaviour	43
Mistakes in drawings or calculations	7
Inadequate information in contract documents or instructions	4
Contravention of requirements in contract documents or instructions	9
Inadequate execution of erection procedure	13
Unforeseeable misuse, abuse and/or sabotage, catastrophe, deterioration (partly "unimaginable"?)	7
Random variations in loading, structure, materials, workmanship, etc.	10
Others	7

Adapted from Walker (1981).

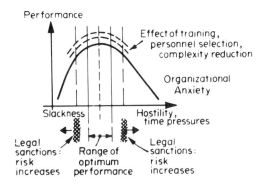

Fig. 2.3 — Human performance function [Melchers, 1980].

different persons operate at different arousal levels and, even at peak arousal, their performance will vary (cf. IQ tests).

According to Pugsley (1973), the main factors which may affect "proneness to structural accidents" are:

(1) new or unusual materials;
(2) new or unusual methods of construction;
(3) new or unusual types of structure;
(4) experience and organization of design and construction teams;
(5) research and development background;
(6) financial climate;

(7) industrial climate;
(8) political climate.

These factors will clearly have an influence on individual arousal and hence performance; they will also affect human interaction. Similar observations are given in the management, psychology and sociology literature [e.g. Luthans, 1981]. Because of its complexity, human behaviour cannot yet be related to all the various factors which influence it.

However, some specific empirical results for operator error in the nuclear, aircraft and chemical process industries, for example, are available [Joos *et al.*, 1979; Harris and Chaney, 1969; Drury and Fox, 1975]. Typical rates for human error in psycho-motor tasks (such as monitoring or active control of a process) are of the order of 10^{-2} per stimulus, but with quite wide variations [e.g. Meister, 1966].

Preliminary investigation of typical (micro)tasks used in detail structural design has found that errors occur in numerical desk-top calculator computation at a rate of about 0.02 per mathematical step. As the mean length of computation involves about two mathematical steps, the average error per calculation is about 0.04 [Melchers and Harrington, 1984]. Somewhat similar error rates were found to occur in "table look-up" tasks and for "table interpolation" tasks.

Such errors, while indicative of those which may occur in detail design may not, by themselves, be very significant. What needs to be considered is the effect that an error may have on the structure as built and this depends largely on error magnitude [Nowak, 1979]. It has also been suggested that usually more than one gross error needs to be committed before a structure is likely to fail [Lind, 1983].

For these reasons, and because errors may be detected, the integrated effect of several micro-tasks, of error occurrence and of error detection is of interest. The output from performance of a "macro-task" such as design loading calculation has been studied and tends to show rather less variation than that observed for micro-tasks. For example, Fig. 2.4 shows the histo-

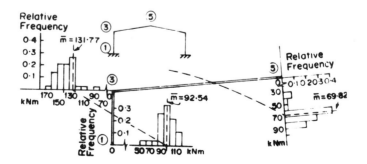

Fig. 2.4 — Typical histogram of bending moments from analysis of loading.

grams obtained for the calculated required moment capacity at three locations of a steel portal frame subject to wind, live and dead load. Designers participating in the study were required to decide on the loads acting on the various surfaces, and these were then converted to actions [Melchers and Harrington, 1984]. It is seen that most designers tend to err on the conservative side, but that some low results were recorded. Examination of individual responses showed that much of the variation was due to differences in assumed unit wind loads as a result of using "short cuts" in following structural design code requirements.

The studies at present available are not sufficient to give definitive information about the direct effect of human error on structural resistance. It is evident, however, that human error will usually increase the uncertainty in the resistance of structures or structural members above that conveyed by uncertainty in material strength and geometric properties alone.

2.2.8.2 *Human intervention*

There is little doubt that many existing adequate structures remain in service despite (many minor) errors committed during their design and construction. The main reason for this is that, apart from committing errors, humans also intervene in the processes of design, documentation and construction and, to some extent, also in the use of a structure [Bosshard, 1979; Knoll, 1985]. Some forms of intervention are institutionalized, e.g. design checking to obtain building approval, and sanctions for violations of the law (contract, criminal or tort). Intervention may also be informal, such as may result from the observation that "something is wrong".

Safety factors or other nominal measures of safety can cater for some degree of human variability. However, it is unlikely that merely strengthening a structure against well-understood hazards, as would be achieved by increasing factors of safety, will be effective in obviating the effects of gross human error (see Table 2.2). Some form of positive action is required; a number of strategies are shown in Table 2.4. Brief comments about each of

Table 2.4 — Human intervention strategies

Facilitative measures	Control measures
Education	Self-checking
Work environment	External checking and inspection
Complexity reduction	Legal (or other) sanctions
Personnel selection	

these follow but, as will be obvious, a detailed discussion is well beyond the scope of this book.

(*i*) *Education*   Education is widely recognized as important, particularly continuing professional education. Some of this occurs quite informally "on the job", and through the technical press. Of particular interest in this regard would be balanced accounts of failure or poor performance of structures; yet it is precisely this information which has been difficult to obtain. Proposals for data banks are being implemented [Melchers, *et al.*, 1983]

(*ii*) *Work Environment*   Work environment is recognized as an important factor in the effectiveness of the people working within an organization [e.g. Luthans, 1981]. Thus an open-minded goal-oriented environment is probably more likely to aid identification of all appropriate uncertainties. There are many examples in the literature where organizational problems created work environments which contributed to structural failure [Melchers, 1977].

(*iii*) *Complexity reduction*   The simplification of complex tasks is a recognized strategy for error reduction. However, oversimplification can lead to boredom with concomitant increase in error rate. Some design (and other) processes lend themselves to extensive computerization, and this, together with the use of check lists and standardization, may well reduce certain types of error. However, different types of error are likely to arise in their place. A related development is that of "expert systems", computer-based storage of available expert information.

A possible problem associated with standardization is that an undetected error may become "institutionalized", with possible widespread effects, such as has occurred with so-called "system built" housing and some high-rise flats projects.

(*iv*) *Personnel selection*   The effectiveness of a design or construction team will depend on the skills and abilities of the team members. In practice various constraints (such as seniority, lack of experience, existing staff commitments) may exist to prevent completely appropriate personnel selection. Conventional management theory lays considerable stress on having appropriate personnel [e.g. Luthans, 1981].

(*v*) *Self-checking*   Self-checking of a task appears to occur to some degree in all human actions [e.g. Warr, 1971]. Together with checking by subsequent individuals or organizations in the design and construction processes, self-checking has been identified as an important factor in structural engineering. In design, for example, a designer is reasonably likely to spot significant errors that he has committed. This happens mainly when the member being designed is not of the "right" proportions, or the reinforcement is somehow not as expected. Obviously not all errors will be detected since some experience is necessary to make the necessary judgements, but it seems plausible that the large errors will be more likely to be detected than small errors. No data appear to be available, however.

(*vi*) *External checking and inspection*   From the point of view of ensuring societal safety levels, external controls such as checking and inspection, and legal sanctions (see below) are often held to be far more powerful strategies to control human error than the ones described so far [CIRIA, 1977; Melchers, 1980]. Such procedures are well recognized in structural engineering.

According to the studies by Matousek and Schneider (1976), only about 15% of errors would not be detected either by existing control arrangements working with greater care or through additional control measures. This last would, according to their estimate, have been required in about half the cases of failure which they studied.

A typical model for checking effectiveness is a "filter" which removes part of the error [Rackwitz, 1977]. Thus, if x_i represents the error rate in design calculations before checking, the error rate after checking might be given by

$$x_{i+1} = (1 - \gamma_i)x_i \qquad (2.1)$$

where γ_i is the error detection possibility. Values of γ_i for inspection of electrical and other small components in factory production are in the range 0.3–0.9 with 0.75 for simple visual tasks under good conditions and with trained inspectors [Drury and Fox, 1975]. Preliminary data for checking of structural designs suggest an efficiency of about 0.7–0.8.

Search theory [Kupfer and Rackwitz, 1980; Nessim and Jordaan, 1983] suggests that the detection probability increases with the time t spent on checking according to an exponential relationship, but Stewart and Melchers (1985) have found that a "learning curve" of the form

$$\gamma = 1 - \exp[-\alpha(t - t_0)], \quad t > t_0 \text{ (familiarization time)} \qquad (2.2)$$

or alternatively

$$\gamma = \frac{1}{1 + A \exp(-Bt^{1/2})} \qquad (2.3)$$

better fits empirical data. Here α is a parameter which depends on the degree of detailed examination and on the size of the task. A and B are constants with a similar function. All constants are dependent on the person performing the checking task. The checking time t could be replaced by the cost of checking, with appropriate new constants.

In each case the checking function is preceded by a learning function required for the checker to familiarize himself with the material to be checked. This is distinctly different to the search theory models originally proposed.

Some data are available to suggest that checking efficiency increases with error size but tends to level off to about 85% for large errors [Stewart and Melchers, 1985].

(*vii*) *Legal sanctions*   The doctrine that legal sanctions deter is firmly entrenched, it usually being taken for granted that fear of sanctions acts as a motivator and inhibitor of human conduct [Hagen, 1983]. There is evidence to suggest that sanctions may well be effective for "premeditated" crime but that in general the effect is likely to be most pronounced on those least likely to be involved. It is reasonable to suggest that few engineers premeditate to perpetrate errors, so that the most likely result of excessive threat of legal sanction is inefficiency, overcaution and conservatism in the execution of work.

The available research on deterrence relates mainly to cases for which the boundary between lawful and unlawful behaviour or action is relatively clearly defined. Except for negligence and deliberate malpractice, this does not appear to be the case for human errors.

2.2.8.3   *Modelling of human error and intervention*

Human error may be incorporated in a reliability analysis as follows. If the human error phenomenon can be represented by a (subjective) random variable for which a probability density function can be postulated, it can be incorporated directly in the analysis for p_f, i.e. it becomes another basic variable (see section 1.4.6). This is often possible for the type V and E errors of Table 2.1. Otherwise, a point estimate of the value of the phenomenon may be used to represent it.

If R_m represents the structural resistance modified for human error, the probability density function for R is then likely to be further modified, in the lower tail region, to R_I to allow for human intervention effects, as shown in Fig. 2.5. The precise form of the modification cannot be given at the present time owing to lack of data, although the discussion above suggest that an appropriate form might be

$$f_{R_I}(r) = K(r)f_{R_M}(r) \qquad\qquad\qquad r \leqslant R_d$$
$$= f_r(r) + f_{R(2R_d} - r)[1 - K(R_d - r)] \qquad r > R_d \quad (2.4)$$

with

$$K(r) = \exp[A(r - R_d)] \qquad\qquad\qquad r \leqslant R_d$$

and where R_d is a so-called "discrimination" level, a mean value of the resistance at which errors are first noted and A is a constant. The second

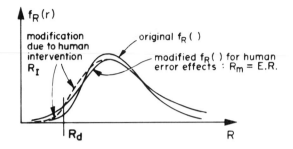

Fig. 2.5 — Modification of resistance probability density function for human error
and human intervention effects.

equality in (2.4) is necessary to ensure that the area under the probability
density function for R remains unity.

Values of A can be estimated from data on checking and inspection
effectiveness. A reasonable first choice might be to assume that, with R_d
selected at one standard deviation, the reduction in the distribution at two
standard deviations below the mean is, say, 90%. This would correspond in
general terms to the initial estimates of checking effectiveness.

Example 2.2
The resistance R when modified for human error E, a random variable,
becomes ER. The error term E may involve bias ($\mu_E \neq 1.0$) and will have a
probability density function $f_E(\ )$ which, in the absence of contrary information, might be taken as normal. This is in agreement with the theory of
errors and the central limit theorem (see section A.5.8) since E is likely to be
the result of a number of error processes.

An alternative formulation for modified resistance is $R+E$, with $\mu_E = 0$
for zero bias. If R is $N(100, 20)$ and E is $N(0, 10)$, the distribution of the
modified R in this case is (see (A.56) and (A.57)) given by $N(100, \sigma)$, where
$\sigma = (20^2 + 10^2)^{1/2}$.

2.2.8.4   *Quality assurance*
The various approaches to the reduction of gross human error described in
the previous section should not be seen in isolation. In any engineering
project they should be viewed as complementary techniques to achieve a
desired goal, usually that of a safe and satisfactory project of sufficiently low
cost and sufficiently high utility over the required design life. Integration of
the above techniques together with the establishment of appropriate mana-
gerial and organization structures, and systems for material compliance
testing, etc., together, of course, with the selection of appropriate nominal
safety measures may be given the umbrella title of quality assurance (QA).
In the broadest sense this is concerned with the management, coordination
and monitoring of all stages of a project so as to achieve a desired set of

objectives. In the particular context of a building or construction project these functions are applied to the conceptualization, design and analysis, documentation, construction, use and maintenance aspects of a structure.

Formalized methods of QA must take into account the need to meet the requirements of structural safety, serviceability and durability. One approach to achieve this is the institution of a "safety" plan. This is based on a detailed "hazard scenario" analysis. The safety plan is then used to lay down the requirements to be met by the QA procedures. These would include [Schneider, 1981]:

(1) proper definition of functions;
(2) definition of tasks, responsibilities, duties;
(3) adequate information flow;
(4) structural design brief;
(5) control plans and check lists;
(6) documentation of accepted risks and supervision plan;
(7) inspection and maintenance plan;
(8) user instructions.

In addition it would be necessary to have adequate and systematic feedback of information to management at all stages of the project.

There is a real danger that excessive formalization of QA procedures will lead to unacceptable and self-defeating generation of paperwork. An appropriate level of QA is needed, tailored to each project and preferably designed in consultation with all concerned. This will probably mean that, for minor projects, little or no change from current procedures is necessary. For major projects there is the further threat of complacency and adhering unthinkingly to instituted QA procedures.

In principle it is possible to arrive at an optimal set of QA measures using cost–benefit (or risk–benefit) analysis. The principles for this will be considered in section 2.4.2 below. In practice such an approach may be difficult, partly because of the lack of appropriate models for human error effects, partly because of the lack of models for the effect of QA measures and partly because of the difficulty of assigning costs associated with QA measures [Schneider, 1983].

2.3 ANALYSIS AND PREDICTION OF STRUCTURAL RELIABILITY

2.3.1 Calculation of the probability of failure

The probability estimates considered this far, and in particular in Chapter 1, rest on the idea that the uncertainty associated with a basic variable can be represented by a probability density function. In section 1.4.8 it was pointed out that this is not always the case, and that point estimates may need to be used for likely values of basic variables. Similarly, it may not always be possible to calculate probabilities in the manner suggested in Chapter 1, owing to lack of information. Again point estimates may need to be used instead. These will be termed "non-analytic" in contrast with the previous "analytic" values, indicating that they are usually indicative, perhaps subjective, values not obtained rigorously from probability density information. For example, the observed failure rate per metre of high-pressure piping of given diameter is of this type, since no information is given about the type or extent of damage. Failure rates of electrical components (which either fail or do not) is another example, as are failure probability estimates due to human error.

A comprehensive estimate of failure probability must be able to incorporate both types of probability information. Assuming that the events described by point estimates are independent of those described by probability density functions, the combined (subjective) estimate of the structure failure probability is given by [CIRIA, 1977; Melchers, 1978] (see (A.2))

$$p_f = p_{fv} + p_{fu} \tag{2.5}$$

where p_{fv} represents the failure probability as a function of "analytic" variables and p_{fu} the failure probability as a function of the "non-analytic" events. Where the independence assumption is not valid, the more general form with (+) replaced by the union should be used. However there are then seldom sufficient data to make evaluation possible.

There are considerable differences of opinion as to the validity of (2.5). Some hold that p_{fv} represents essentially "objective" information and as such should not be combined with the largely "subjective" probability p_{fu} on the grounds that subjective probabilities express a "degree of belief" which is not compatible with objective or frequentist probabilities. There is a school of thought which believes that other methods of combining objective and subjective information should be used. One such procedure is based on fuzzy set theory [e.g. Blockley, 1980, 1985]. The view taken here, however, is that all probabilities are "subjective" to some degree (see section 2.1), combining both objective and subjective knowledge, and are therefore additive. This accords essentially with interpretations made in decision theory [e.g. Fishburn, 1964; Lindley, 1972].

The term p_{fv} can be evaluated with the aid of the total probability theorem (A.6):

$$p_{fv} = \sum_{i=1}^{n} p_{fi} p_i = \sum_{i=1}^{n} P(F|N_i) P(N_i) \tag{2.6}$$

where $p_{fi} = P(F|N_i)$ is the probability of failure given that the ith "state of nature" N_i occurs and $p_i = P(N_i)$ is the probability of occurrence of the ith state of nature, with

$$\sum_{i=1}^{n} p_i = 1$$

The state of nature refers to the set of conditions, qualifications, assumptions and state of knowledge which is implicit in the evaluation of $P(F)$; this may include assumptions (and hence predictions) about human error, workmanship, etc., as well as, for example, an estimate of how well the calculation p_{fi} represents the actual failure probability [e.g. Tribus, 1969]. Expression (2.6) assumes independence of N_i; appropriate selection of the set $\{N_i\}$ may be required to achieve this.

The term p_{fu} is simply the sum of all point estimate probabilities of failure not otherwise accounted for in (2.6). Generally the events whose probability of occurrence are included in this term are not well understood, so that questions of dependence or correlation are of little relevance. Also, as understanding grows about these events, they will be more easily expressed in analytical terms. Hence p_{fu} would be expected to be reduced, and p_{fv} generally increased. Conversely, poorly understood problems would be represented mainly by p_{fu}.

Example 2.3.
It is estimated that the probability of accidental traffic overload (event N_1) for a small bridge is 0.01. When that occurs, the probability $P(F|N_1)$ of failure of the bridge, estimated using the method of Chapter 1, is 0.1. Under normal loading conditions the probability of failure is 0.002. In addition, it is possible for flooding to occur independently of the traffic loads applied to the bridge; the probability that this occurs at a magnitude sufficient to cause failure is estimated at 0.005. All probabilities are for a 50-year life. The predicted probability of failure is then estimated as follows:

$$P(F|N_1) = 0.1, \qquad P(N_1) = 0.01$$

$$P(F|N_2) = 0.002, \qquad P(N_2) = 1 - P(N_1) = 0.99$$

$$p_{fu} \qquad = 0.005$$

Substituting into (2.5) and (2.6),

$$p_f = (0.1)(0.01) + (0.002)(0.99) + 0.005 = \underline{0.008}$$

2.3.2 Analysis and prediction
The determination of the probability of failure can be carried out from two viewpoints: (i) analysis of a given state of affairs, and (ii) prediction of failure probability for some time period in the future.

For analysis, the probability density functions of all analytic variables are presumed known for the time at which the analysis is required; these may, of course, have been obtained from direct observation or from subjective estimates. Similarly, non-analytic random variables or events are assumed to be representable for the time at which the analysis is required. Using such information in expression (2.5) together with the analytic techniques described in this book will allow an estimate of the current probability of failure to be obtained. If it is assumed that there are no changes to be expected in any of the parameters with time, then the calculated probability estimate may also be used to describe the probability of failure for future times.

More generally, however, it would be expected that probability density functions for some of or all the random variables change with time (see Fig. 1.7). The problem of prediction of probability of failure is thus one of predicting the future probabilistic descriptions of the relevant random variables. Not only will means and variances change with time, but also the type of probability density function may change.

For a project in the planning stage, the predicted failure probability for the completed structure will be quite uncertain, since details of actual loadings, material strengths, etc., will not be well defined. Hence coefficients of variation are typically rather large. This is relevant, for example, in structural design code calibration work (see Chapter 9). When construction is under way, details of actual material suppliers become available and this should, generally, allow a reduction in coefficients of variation for material strengths, and better estimates of workmanship variability. Similarly, when the structure is put into service, better definition of loading uncertainties should be possible.

It follows directly that, as better estimates of the various properties and assumptions become available, the (subjective) probability prediction can be refined. It would be expected that in the limit the probability of failure so estimated would approach the probability of failure for similar structures; this assumes that subjective estimates tend to approximate objective frequentist data as the information available improves (which may not always be correct [Fishburn, 1964]).

Further, in principle, it would be expected that with the inclusion of all relevant variables and with perfect probabilistic modelling of all the variables and the appropriate limit states, the estimated failure probability would approach the failure rate which might be observed for sufficient large samples of nominally identical structures.

In practice this is unlikely to be closely achieved. Apart from severe problems with historical records and observed rates of failure (see section 2.4), the failure probability estimate for a particular structure is based on past experience with, at best, similar structures and with probability distributions for materials, sizes and loadings (etc.) subjectively selected. Although frequentist data for these may be available from historical records for other structures, the validity of their application to the structure being considered still requires (subjective) judgement. In addition the accuracy of design, the standard of workmanship and the effectiveness of inspection and control must be estimated.

It follows that the quality of a structural reliability estimate is very much a function of the data and modelling used to derive it but that it is unlikely to agree with rates of failure for real structures (assuming that such rates can actually be observed).

2.4 CRITERIA FOR RISK ACCEPTABILITY

When a risk assessment has been performed it must be decided whether the risk assessed by (2.5) is acceptable. Two criteria for making this decision will be outlined below. More complete discussions are available, particularly in the literature related to nuclear accidents [e.g. Rowe, 1977].

2.4.1 Risk criterion

One criterion is to compare the calculated probability of structural failure with societal risks. Estimates of selected risks in society are reproduced in Table 2.5. It is evident that there is a difference of about one order of

Table 2.5 — Selected risks in society

Activity	Approximate death rate[a] ($\times 10^{-9}$ deaths/h exposure)	Typical exposure[b] (h/year)	Typical risk of death ($\times 10^{-6}$/year) (rounded)
Alpine climbing	30 000–40 000	50	1500–2000
Boating	1500	80	120
Swimming	3500	50	170
Cigarette smoking	2500	400	1000
Air travel	1200	20	24
Car travel	700	300	200
Train travel	80	200	15
Coal mining (UK)	210	1500	300
Construction work	70–200	2200	150–440
Manufacturing	20	2000	40
Building fires[c]	1–3	8000	8–24
Structural failures[c]	0.02	6000	0.1

[a] Adapted from Allen (1968) and CIRIA (1977).
[b] For those involved in each activity (estimated values).
[c] Exposure for average person (estimated).

magnitude between the so-called "voluntary" risks and the "involuntary" (or background) risks, and that risk depends on the degree of exposure to a hazard. Since structures are used by people in the expectation that they will not fail (compare by contrast with aircraft use), the probability of structural failure may be related to involuntary risk. However, opinion about this varies.

Some broad indicators of tolerable risk levels have also been suggested [Otway *et al.*, 1970]:

10^{-3} per person per year uncommon accidents; immediate action is taken to reduce the hazard;

10^{-4} per person per year people spend money, especially public money to control the hazard (e.g. traffic signs, police, laws);

10^{-5} per person per year mothers warn their children of the hazard (e.g. fire, drowning, firearms, poisons), also air travel avoidance;

10^{-6} per person per year not of great concern to average person; aware of hazard, but not of personal nature; act of God.

Typical failure rates for building structures and for bridges are given in Tables 2.6 and 2.7 respectively. For comparison purposes, the rates have

Table 2.6 — Typical "collapse" failure rates for building structures

Structure type	Data cover	Number of structures	Average life (year)	Estimated lifetime p_f
Apartment floors	Denmark	5×10^6	30	3×10^{-7}
Mixed (housing	The Netherlands (1967–1968)	2.5×10^6	—	5×10^{-4}
Controlled domestic housing	Australia (New South Wales)	145 500	—	10^{-5}
Mixed (housing)	Canada	5×10^6	50	10^{-3}
Engineered structures	Canada			10^{-4}

Sources: Allen, 1981a; Ingles, 1979; Melchers, 1979.

Table 2.7 — Typical "collapse" failure rates for bridges

Bridge type	Data cover	Number of structures	Average life (year)	Estimated lifetime p_f
Steel railway	USA (<1900)		40	10^{-3}
Large suspension	World (1900–1960)	55	40	3×10^{-3}
Cantilever and suspended span	USA	—	—	1.5×10^{-3}
Bridges	USA	—	—	10^{-3}
Bridges	Australia	—	—	10^{-2}

Sources: Pugsley, 1962; Ingles, 1979.

been adjusted, where necessary, to attempt to extract the limit state of "collapse".

The rates must be viewed with caution for a number of reasons. There is some evidence and much qualitative conceptual support that "collapse" failures account for only about 10–20% of all cases of failure. Many failures of lesser magnitude involve serviceability limit state violation [Melchers *et al.*, 1983]. Further, many of the "collapse" failure cases occurred during construction. It is not always clear whether the structure itself failed (in a sense a case of not surviving a proof load) or whether temporary works failed. Also, the probability of failure during construction may not always be considered in a reliability assessment and it would therefore be quite erroneous to compare failure statistics including construction with predictions excluding construction.

A further matter to be taken into account is subjectivity in deciding whether a structure has "failed". This may very well depend on the consequences, with large consequence failure gaining much publicity and being subject to major inquiries, etc. However, a failure with little consequence may never be recorded. It is likely that the rather higher rates of failure for bridges in Table 2.7 are due to this effect. It may also be due to phenomenological uncertainty (see section 2.2.2) associated with novel forms of construction.

2.4.2 Socio-economic criterion

If failure consequences are to be taken into account, a more general criterion for assessing acceptability of structural failure probability is cost–benefit analysis (also termed "risk–benefit analysis", using "risk" in the insurance sense):

$$\max(B - C_T) = \max(B - C_I - C_{QA} - C_c - C_{INS} - C_M - p_f C_F) \quad (2.7)$$

where B is the total benefit of project, C_T the total cost of project, $C_I(\lambda)$ the initial cost of project, $C_{QA}(\mathbf{e})$ the cost of n QA measures which is given by

$$C_{QA}(\mathbf{e}) = \sum_{i=1}^{n} C_{QAi}(\mathbf{e})$$

$C_c(\mathbf{e})$ the cost of corrective actions in response to QA measures, $C_{INS}(\lambda, \mathbf{e})$ the cost of insurance, $C_M(\mathbf{e})$ the cost of maintenance, $p_f(\lambda, \mathbf{e})$ the probability of failure of the project, C_F the costs associated with failure, λ the nominal factor of safety and $\mathbf{e}$ the vector of QA efforts. All costs and the benefit B must be discounted to allow for the effect of time as in standard cost–benefit analysis [e.g. de Neufville and Stafford, 1971]. Formulation (2.7) can also be rewritten as a minimization problem in costs, since the benefit B is generally neither a function of the degree e of QA nor of the nominal safety factor λ.

Both p_f and C_F depend on the mode(s) of failure being considered. The value of C_F is uncertain. It depends on how much damage is likely to be caused, how many lives lost, etc. Hence C_F might be modelled as a random variable, although appropriate data are scarce. In addition, the long-term effects of project failure should be included in C_F. The evaluation of C_F may well vary for different parties involved in the ownership, use and responsibility for the structural project, at various stages during its life, so that (2.7) is strictly a multiobjective optimization problem with C_F varying for different interest groups.

Before the maximization of (2.7) can be carried out, relationships between $p_f(\lambda, \mathbf{e})$ and $\mathbf{e}$ and between the costs $C_{QA}(\mathbf{e})$, $C_c(\mathbf{e})$, $C_{INS}(\mathbf{e})$ and $C_M(\mathbf{e})$ and $\mathbf{e}$ must be established. Models such as these outlined in section 2.2.8 may be of some help here, particularly for $C_{QA}(\ )$. The cost of insurance should, in theory, vary inversely with the amount of QA, although this does not appear to be the case in practice [CIRIA, 1977] except in a special sense in the French decennial form of insurance and building control [Cibula, 1971]. Maintenance costs would be expected to decrease with $\mathbf{e}$.

In principle it is possible to use expression (2.7) to derive the optimal probability of failure for a structure, given a certain level of quality assurance. Assuming a constant benefit B and minimizing the total cost, the variation of the various costs is shown in Fig. 2.6. Evidently, a minimum

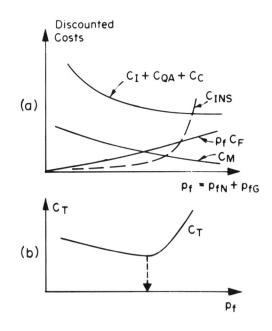

Fig. 2.6 — (a) Component costs and (b) total costs as functions of p_f.

$dC_T/dp_f=0$ exists, and it is clear that this minimum is highly sensitive to changes in the slopes of the cost curves. Nevertheless, it has been estimated that the optimal probability of failure for buildings is about 10^{-4} per year, and 10^{-3} per year for bridges, apparently assuming that the cost C_F of failure is negligible and that human error effects may be ignored [Rüsch and Rackwitz, 1972]. This means that these values are probably better regarded as nominal failure probabilities rather than more realistic ones (see section 2.5 below).

Appropriate levels of QA efforts can be similarly determined. The result depends strongly on the effectiveness and unit costs of each QA measure, as well as the ratio C_F/C_I. Typical values for the latter are 30–75 for beam elements and 350–700 for columns supporting one floor to 900–1800 for columns supporting ten floors [CIRIA, 1977]. Unless C_F/C_I is very high (which is not normally the case), optimal total cost is only achieved with high QA effectiveness; otherwise $C_{QA}(e) > p_f C_F$. The full implications of this for practical quality control schemes does not appear to have been fully realized.

Finally, it must be noted that increased investment in structural safety whether in terms of C_{QA} or C_I may not be as effective in terms of benefit gained, as additional expenditure in, say, road safety or medical research. The allocation of resources between competing sectors of the economy is essentially a socio-political matter. An appropriate framework for structural engineering decisions may well be to assume that C_T is fixed and that any optimization must be a matter of allocating available resources within this constraint.

A more general decision framework than risk–benefit analysis is utility theory. However, formulation of appropriate utility functions is often difficult. Some initial studies in this area have been reported [Augusti *et al.*, 1984].

2.5 NOMINAL PROBABILITY OF FAILURE

2.5.1 General

The discussion in section 2.2 indicated the importance of accurate modelling, and in particular the inclusion of human error and human intervention effects in determining a good estimate of structural failure probability. When these matters are ignored, and also sometimes when approximations have been made in the calculations, the corresponding failure probability becomes a nominal one, P_{fN} (cf. section 2.1). It is pertinent to question whether such a probability measure has any useful meaning. This question is relevant since p_{fN} has been used in a comparative sense, with a lower value preferred to a higher one. A particular application is in code calibration work (see Chapter 9). There, uniform nominal failure probabilities expressed through the safety index β (see section 1.4.3) are sought for different structural elements [Ellingwood *et al.*, 1980]. More generally, the possibility of the use of p_{fN} as a surrogate for p_f in decision making is normally discussed with the proviso that it should not be interpreted as a relative frequency, but rather as a "formal" failure probability measure, interpreted as a "degree of belief" [e.g. Ditlevsen, 1983a].

As noted in section 2.1, the interpretation of probability as a degree of belief is an acceptable one. However, whether p_{fN} can be used to make valid comparisons needs to be examined. For p_{fN} to be used as a surrogate for p_f and hence to be used validly to compare two alternatives, 1 and 2, it is required that

$$\frac{p_{f1}}{p_{f2}} \approx \frac{p_{fN1}}{p_{fN2}} \tag{2.8}$$

The validity of this statement can be examined on at least two (somewhat interrelated) grounds:

(1) axiomatic definition;
(2) influence of gross and other errors on design.

2.5.2 Axiomatic definition
In expression (2.8) the p_{fi} ($i = 1, 2$) may be replaced by the sum p_{fNi} of the nominal failure probabilities and a term p_{fGi} expressing the contribution of uncertainties not absorbed by p_{fNi}. The latter are mainly due to (gross) human error, while p_{fNi} includes mainly physical and statistical uncertainties. Then,

$$\frac{p_{f1}}{p_{f2}} \approx \frac{p_{fN1} + p_{fG1}}{p_{fN2} + p_{fG2}} \tag{2.9}$$

which reduces to (2.8) only if (a) $p_{fGi} = kp_{fNi}, i = 1, 2, k =$ constant, or (b) $p_{fG1} = p_{fG2}$ and $p_{fN1}/p_{fN1} = p_{f1}/p_{f2} = 1$ (impartiality).

In view of the discussion in section 2.2.8, condition (a) is unlikely to be true in general. Condition (b) indicates that equal human and other error effects on probability must be assumed if the (nominal) failure probabilities are to be identical. This restriction on p_{fG} is unlikely to be satisfied except, possibly, if the components being compared are essentially similar and therefore subject to similar human and other errors. This may well be true for certain components in building structures. More generally condition (b) will not be met and its acceptance can only be seen as axiomatic.

2.5.3 Influence of gross and other errors
A favoured argument to suggest that p_{fN} is an adequate measure of structural safety is to show that the choice of design (e.g. its safety factors, or p_{fN}) is unaffected by the knowledge that (gross) human and other errors might occur (i.e. by the knowledge of p_{fG}). This question has been addressed by Baker and Wyatt (1979) and Ditlevsen (1983a) among others,

using a rather more numerical and mathematical approach than that given here.

Consider the total cost C_T (in present value) as in (2.7) and let the failure probability p_f be represented, as before, by $p_{fN}+p_{fG}$, with $p_{fG} \gg p_{fN}$ (see section 2.3.8.1). Then (2.7) can be written in simplified form as

$$C_T = C_I(p_{fN})+(p_{fN}+p_{fG})C_F \qquad (2.10)$$

where the initial cost C_I (in present value) depends, reasonably, only on the nominal probability p_{fN} of failure.

For a very low likelihood of failure, the initial cost would be expected to be very high, reducing progressively less as p_{fN} increases, as shown in Fig. 2.7. The cost of C_F failure would usually contain a term representing

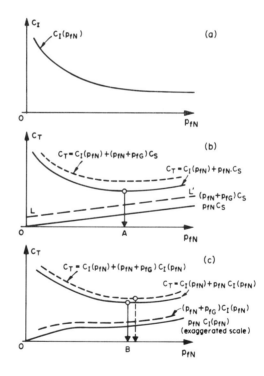

Fig. 2.7 — (a) Initial cost curve; (b) optimal p_{fN} for $C_F=C_S=$constant; (c) optimal p_{fN} for $C_F=C_I$.

reconstruction costs, assumed here for convenience equal to the initial cost $C_I(p_{fN})$ plus a term C_S representing other or sequential costs. Two extreme cases may be considered: $C_F=C_S(\ )$ and $C_F=C_I$. These are shown in Figs. 2.7(b) and 2.7(c) respectively.

In the first case it is seen that changing the value of p_{fG} has no effect on the optimal value p_{fN} (at A) since this merely moves the line LL' parallel to itself, provided that p_{fG} is independent of p_{fN}, a reasonable first approximation. In the second case, changing the value of p_{fG} has only a small effect on the optimal value of p_{fN} (at B).

The true cost of failure will lie somewhere between the two extremes shown in Figs. 2.7(b) and 2.7(c); note also that the actual values of $p_{fN}C_S$ and $p_{fN}C_I$ respectively do not change the conclusion.

In reality p_{fG} could be expected to be positively correlated to p_{fN}. One example is in the design against earthquakes, where an increase in ductility capacity would reduce p_{fN} for the structure and also its sensitivity to (gross) human error [Rosenblueth, 1985a]. It is easily verified that this makes little difference to the conclusion.

Although p_{fN} is insensitive to p_{fG}, clearly C_T and $p_f=p_{fN}+p_{fG}$ are not. This is important in comparing alternatives. Consider two members which have the same $C_I(p_{fN})$ curve and the same C_F. It follows that p_{fN} will be the same for each (see Fig. 2.7) but the costs C_T and the failure probabilities p_f will depend on the relative p_{fG} values. Clearly the members with the lowest cost C_T (and hence also lowest p_f) will be preferred. The value of p_{fN} is non-informative for this choice, unless the values of p_{fG} are identical for each member. This corresponds to the conclusion reached in the previous section.

2.5.4   Practical implications

It is clear that the use of p_{fN} as a surrogate for p_f is strictly only acceptable in a comparative sense if the failure probability component due to the influence of human error and other effects not included in p_{fN} affect each of the alternatives in a manner roughly proportional to p_{fN}.

For practical purposes, however, p_{fN} can be accepted as a measure of a more accurately determined probability of failure if it is interpreted in the same sense that factors of safety and load factors have traditionally been used, as a purely nominal measure. This is then an axiomatic definition.

When p_{fN} is not used in a comparative manner between components but as a measure of the (nominal) safety of a structure (however defined), the argument of section 2.5.3 shows that knowledge of gross and other error effects may well have a significant influence on QA measures since the total cost is affected. However, such knowledge does not change the appropriate value for the nominal measure p_{fN} very much, if at all. For this reason it is rational to use nominal safety measures (such as p_{fN}) for design and in design codes to obtain structural sizes.

2.5.5   Target values for p_{fN}

Just as conventionally a factor of safety of about 1.7–2.0 appears to be acceptable, depending on the structure, materials, consequences, etc., the nominal probability of failure might also be expected to have a "target" value. As will be seen in Chapter 9, for structural design code writing purposes it is often convenient and appropriate to back-calculate p_{fN} from existing practice, and to use a similar value for a modified or new code. However, proposals for determining target values of p_{fN} have been given on purely empirical grounds and these will now be briefly reviewed.

If p_{fN}^{*} denotes the target value, CIRIA (1977) proposed

$$p_{fN}^{*} = 10^{-4} \mu T n^{-1} \tag{2.11}$$

for a structural design life of T (years). Here n is the average number of people within or near the structure during the period of use and μ is a social criteria factor (see Table 2.8).

Table 2.8 — Social criteria factor

Nature of structure	μ
Places of public assembly, dams	0.005
Domestic, office, trade, industry	0.05
Bridges	0.5
Towers, masts, offshore structures	5

Source: CIRIA, 1977.

A somewhat different proposal has been made by Allen (1981a):

$$p_{fN}^{*} = 10^{-5} A W^{-1} T n^{1/2} \tag{2.12}$$

where T and n have the same meaning as in (2.11) and where A and W are "activity" and "warning" factors respectively (see Table 2.9). The use of

Table 2.9 — Activity and warning factors

Activity factor	A	Warning factor	W
Post-disaster activity	0.3	Fail-safe condition	0.01
Normal activities:		Gradual failure with some warning	0.1
buildings	1.0	likely	
bridges	3.0	Gradual failure hidden from view	0.3
High exposure structures	10.0	Sudden failure without previous	1.0
(construction, offshore)		warning	

Source: Allen, 1981a.

$n^{-1/2}$ rather than n^{-1} clearly suggests the influence of utility theory notions in which the rate of risk aversion decreases with the number of fatalities [de Neufville and Stafford, 1971; Rowe, 1977].

Comparison of the two formulae without specific information is not possible. Both lack accounting for injuries and other economic costs of failure. These and n are extremely difficult to predict in any case so that both formulae must be considered to be indicative only, to be used with expert guidance.

2.5.6 The "tail sensitivity" problem

The fact that the probability distribution assigned to any one basic variable can have a marked influence on the calculated failure probability can be illustrated by comparison of curves with similar coefficients of variation in Fig. 1.14. The reason for the difference lies, of course, in the shape and extent of the overlap of the probability density functions shown in Fig. 1.10. This dependency of p_f on the assumptions made for the probability density functions is known as the "tail sensitivity problem", since it was considered to be a major obstacle in providing meaningful estimates of p_f. However, in view of the discussion in the preceding sections, it is evident that the tail sensitivity problem is merely a reflection of the various uncertainties which arise in quantifying the variables to be considered in a reliability analysis. Further, if p_f is seen as merely a "nominal" or "formal" measure of structural failure probability, in the sense described in the present section, the tail sensitivity problem essentially becomes inconsequential, as no frequentist meaning is now attached to such a measure.

2.6 HIERARCHY OF STRUCTURAL RELIABILITY MEASURES

The discussion in Chapter 1 and this chapter may be conveniently summarized in terms of various levels at which safety (or more generally limit state violation) can be defined. This is set out in Table 2.10. At the lowest, and simplest level, level 1, there is the partial factor approach of section 1.2.3. It is a non-probabilistic generalized version of the traditional safety factor and load factor formats. It is the format most commonly advocated for limit state design codes at the present time. How this format relates to level 2 procedures will be discussed in Chapter 9.

Level 2 procedures deal with nominal probabilities based on the use of the normal distribution and simple forms for the limit state function. Chapter 4 gives a further discussion.

Level 3 procedures attempt to obtain the best estimate of the probability of failure, using accurate probability models as well as the use of human error and intervention data if available. Structural system effects, and the influence of time may be of importance. Chapters 3, 4, 5 and 6 deal with these aspects.

Table 2.10 — Hierarchy of structural reliability measures

Level	Calculation methods	Probability distributions	Limit state functions	Uncertainty data	Result
1: code level methods	(Calibration to existing code rules using level 2 or 3)	Not used	Linear functions (usually)	Arbitrary factors	Partial factors (ϕ, γ factors)
2: "Second-moment methods"	Second-moment algebra	Normal distributions only	Linear, or approximated as linear	May be included as second data	"Nominal" failure probability p_{fN}
3: "exact methods"	Trans-formation	Related to equivalent normal	Linear, or approximated as linear	May be included as random variables	Failure probability p_f
	Numerical integration and simulation	Fully used	Any form		
4: decision methods	Any of the above, plus economic data				Minimum cost, or maximum benefit

2.7 CONCLUSION

The combined and interrelated effects of human error (gross and otherwise) and human intervention have been shown to be a major consideration in the estimation of the probability of structural failure. Better understanding of these factors, coupled with appropriate data from similar structures and design and construction practices, should allow better predictions of failure probability to be made; as noted repeatedly, all such estimates must be seen as subjective or nominal.

For any particular structure the applicability of probability models for basic variables and the conditions under which the evaluation is valid can only be judged subjectively. Measures of failure probability do not, there-fore, have an absolute objective relative frequency interpretation. Nominal measures of failure probability can be used as surrogates for more accurately determined measures if the effects of human error in particular are assumed to be similar for similar situations or structural components. Such nominal measures are particularly useful for design-code-writing purposes.

Finally QA measures must be used to control both the total probability of failure (including gross error estimates) and the total discounted cost of a structural engineering project.

3

Integration and simulation methods

3.1 INTRODUCTION

In this chapter some ways of integrating the probability expressions (1.18) and (1.31) will be discussed. In particular, it will be assumed that the probability density function of each basic variable is known and that it will not be approximated. Any approximation will be in performing the actual integration. In Chapter 4 the converse problem will be addressed, namely using approximations to the actual distributions to enable the integration to be performed.

A numerical solution of the convolution integral (1.18) is easily obtained through the use of the trapezoidal rule (see section 1.4.5). This approach has been found to give satisfactory results, mainly because any underestimation of the exact integrand around the mean is compensated by slight but extensive overestimation elsewhere [Dahlquist and Björck, 1974]. However, more refined methods, such as Simpson's rule, or the methods based on polynomials, such as Laguerre–Gauss or Gauss–Hermite quadrature formulae may be more appropriate [e.g. Davis and Rabinowitz, 1975]. Standard routines for such numerical integration are available on most computer systems.

When R and S in the convolution integral (1.18) are not independent, or there are more than two variables, the probability of failure must be obtained from the general formulation (1.31). The computation of this integral in all but very special cases cannot be achieved in closed form and numerical integration is not always feasible owing to the growth of round-off errors and excessive computation times. Even then, the integration region is confined to one of the following: hypercube, n-dimensional solid sphere, or the surface thereof, n-dimensional simplex (generalization of triangle and tetrahedron) and the semi-infinite half-space [Davis and Rabinowitz, 1975;

Stroud, 1971; Johnson and Kotz, 1972]. Apart from some brief comments in section 3.3.2 numerical integration of (1.31) will not be discussed herein.

In the special case when the limit state function is a linear function

$$G(\mathbf{X}) = \sum_{i=1}^{n} c_i X_i$$

where c_i are known constants, it is possible to reduce the multiple integral (1.31) to a recursive series of single integrals [Stevenson and Moses, 1970]. However, the evaluation still requires extensive numerical work or rather drastic simplifications. In a similar vein, it has been suggested that it may be possible to invoke the divergence theorems of Stokes and Gauss to convert two- and three-dimensional integrals to one- and two-dimensional contour and surface integrals, respectively [Shinozuka, 1983].

It has already been seen in section 1.4.3 that, when both load effect S and resistance R are described by normal distributions, the (two-dimensional) integration of the probability integral (1.31) is effectively bypassed through the special properties of the normal distribution.

More generally, the special properties of the normal distribution can be applied to the general case of n-dimensional integration of (1.31), provided that the integration region is described by a linear limit state function

$$G(\mathbf{x}) = Z = 0 = a_1 x_1 + a_2 x_2 + \ldots + a_n x_n$$

In this case, the mean μ_Z of the safety margin is given by (A.160) and the variance by (A.162). Also, dependence between the random variables X_i presents no difficulties.

However, limit state functions more general than linear functions also occur and for this reason the integration of the normal distribution in two-dimensional and multidimensional regions will be briefly considered in sections 3.2 and 3.3. Particular attention will be given to some of the special properties of correlated multinormal distributions and their integration over hypercubes.

The main part of this chapter, however, is devoted to Monte Carlo methods. These form a class of approximate numerical solutions to the probability integral (1.31) applicable to problems for which the limit state functions $G(\mathbf{X})$ may have any form, and for which the probabilistic description of the random variables X_i is unrestricted.

3.2 SPECIAL CASE I: BIVARIATE NORMAL INTEGRAL

3.2.1 Format

Consider the special case of eqn. (1.31) for which the vector $\mathbf{X}$ consists of just two components, X_1 and X_2, each normally distributed, and dependent through the correlation coefficient ρ (cf. (A.124)). Let the probability of interest and hence the failure region (and limit state equations) be defined by the region $X_1 > x_1$, $X_2 > x_2$, so that (1.31) becomes

$$P_f = \int_{X_2 > x_2}^{\infty} \int_{X_2 > x_1}^{\infty} f_{\mathbf{X}}(\mathbf{x}, \rho) \, d\mathbf{x} \tag{3.1}$$

The joint probability density function for $\mathbf{X}$ is given by (A.125):

$$f_{\mathbf{X}}(\mathbf{x}, \rho)) = \frac{1}{2\pi\sigma_{X_1}\sigma_{X_2}(1-\rho^2)^{1/2}} \exp\left[-\frac{\frac{1}{2}(h^2 + k^2 - 2\rho hk)}{1-\rho^2}\right] \tag{3.2}$$

for $-\infty \leqslant x_i \leqslant \infty$, $i = 1, 2$ and with $h = (x_1 - \mu_{X_1})/\sigma_{X_1}$ and $k = (x_2 - \mu_{X_2})/\sigma_{X_2}$.

The joint cumulative distribution function, $F_{\mathbf{X}}(\ )$ is obtained directly by integration of (3.2) (cf. (A.127)):

$$F_{\mathbf{X}}(\mathbf{x}, \rho) \equiv P\left[\bigcap_{i=1}^{2}(X_i \leqslant x_i)\right] \equiv \int_{-\infty}^{x_2} \int_{-\infty}^{x_1} f_{\mathbf{X}}(u, v, \rho) \, du \, dv \tag{3.3}$$

As in the univariate case, it is convenient to work with standardized normal variables, $Y_i = (X_i - \mu_{X_i})/\sigma_{X_i}$ ($\mu_{Y_i} = 0$, $\sigma_{Y_i} = 1$). Making these substitutions in (3.2) and (3.3) produces

$$f_{\mathbf{X}}(\mathbf{x}, \rho) = \frac{1}{\sigma_{X_1}\sigma_{X_2}} \phi_2(\mathbf{y}, \rho) \text{ and } F_{\mathbf{X}}(\mathbf{x}, \rho) = \Phi_2(\mathbf{y}, \rho) \tag{3.4}$$

where $\phi_2(\ )$ and $\Phi_2(\ )$, respectively, are the probability density function and cumulative distribution function for the standardized variables. The coordinates (h, k) in the reduced, but correlated, Y space are shown in Fig. 3.1; the

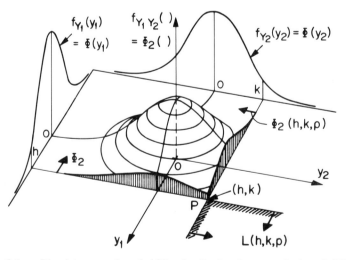

Fig. 3.1 — Bivariate normal probability density function, marginal probability density functions and regions of integration Φ and L.

probability content described by (3.3) is that to the upper left of the lines $y_1 = h, y_2 = k$, (i.e. h, P, k). The actual shape of the distribution, and hence the probability content described by $\Phi_2(\mathbf{y}, \rho)$, depends on the correlation coefficient ρ, as indicated schematically in Fig 3.2.

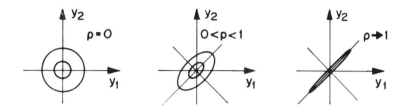

Fig. 3.2 — Effect of correlation ρ between y_1 and y_2 on the form of the bivariate normal probability density function.

The region of interest for the calculation of p_f is that given by $y_1 > h$, $y_2 > k$ i.e. that lying to the lower right in Fig 3.1. The probability content in that region is usually denoted by (A.132):

$$L(h, k, \rho) \equiv \frac{1}{2\pi} (1 - \rho^2)^{-\frac{1}{2}} \int_h^\infty \int_k^\infty \exp\left[-\frac{\frac{1}{2}(u^2 + v^2 - 2\rho uv)}{1 - \rho^2} \right] du\, dv \quad (3.5)$$

It is not difficult to verify from Fig. 3.1 that $L(h, k, \rho) = \Phi_2(-h, -k, \rho)$, so that, if $L(\ )$ can be evaluated, so can Φ_2 and vice versa. Tables for $L(\ )$ and various other functions exist (see Appendix A). A particularly simple result is that $\Phi_2(y_1, +\infty, 0) = \Phi(y_1)$, the marginal distribution of y_1, as is easily verified. This is also shown in Fig 3.1.

3.2.2 Reductions of form

Although expression (3.5) can be directly integrated numerically, it can also be reduced to a single integral prior to numerical integration (Sheppard, 1900) (cf. (A.140a) with $\rho = \cos \theta$):

$$L(\mathrm{h, k,} \rho) = \frac{1}{2\pi} \int_{\arccos \rho}^{\pi}$$

$$\exp\left[-\frac{\frac{1}{2}(h^2 + k^2 - 2hk \cos \theta)}{\sin^2 \theta} \right] d\theta, \qquad h, k \geqslant 0 \qquad (3.6)$$

Expression (3.6) is obtained by substituting (3.2) into (3.3) and standardizing it to $\Phi_2(\ )$, then successively:

(a)   differentiating with respect to ρ;
(b)   integrating the result with respect to y_1 and y_2;
(c)   integrating that result with respect to [Owen, 1956].

Similarly, $\Phi_2(\ )$ can also be reduced to a single integral, a result given by Owen (1956) (A.140b):

$$\Phi_2(h,\ k,\ \rho) = \frac{1}{2\pi}\int_0^\rho (1-\zeta^2)^{-\frac{1}{2}} \exp\left[-\frac{\frac{1}{2}(h^2+k^2-2hkz)}{1-z^2} \right] dz +$$

$$\Phi(h)\Phi(k), \qquad h,\ k \geqslant 0 \qquad\qquad\qquad (3.7)$$

Other equivalent formulations exist [Johnson and Kotz, 1972], but (3.6) and (3.7) have particular simplicity, although they are unlikely to be accurate if $\rho \to 1$.

3.2.3   Bounds
Rather than carry out numeric integration of (3.6) or (3.7), in some applications it may be sufficient to bound the probability content expressed by Φ_2 or L. This can be done quite simply. Consider first the relationship between the random variables Y_1 and Y_2 shown in Fig. 3.3(a) for a typical

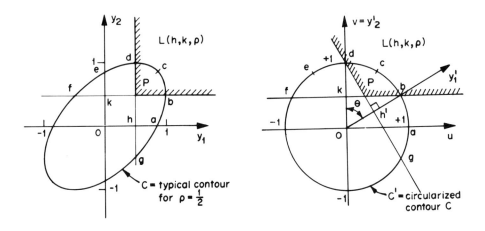

Fig. 3.3(a) — Integration region in original (correlated) standardized bivariate normal space; (b) transformed integration region in transformed (independent) standardized bivariate normal space.

probability density contour ($\rho = \frac{1}{2}$). It is more convenient to work in the circularized standard normal space, U, V, so that either of the transformations (A.142) or (A.143) must be applied. If the latter is used, the transformed contour C' (a circle) is obtained and the location of the y_1 axis is as shown in Fig. 3.3(b). Typical points (a)–(f) and the shaded zone $L(h, k, \rho)$ transform as shown. In the circularized normal space the axes y_1, y_2 are at an angle $\theta < \pi/2$ to each other. The probability content $L(\ )$ enclosed by the lines $y_1 = h$ and $y_2 = k$ (lines dg, bf) in Fig. 3.3(a) is transformed to the shaded region shown in Fig. 3.3(b). It follows directly from (3.6) that the correlation coefficient and the angle shown in Fig. 3.3(b) are related by $\rho = \cos\theta$. Thus, if X_1 and X_2 are uncorrelated and independent, $\rho = 0$ (see Fig. 3.2) and $\theta = \pi/2$, which accords with Fig. 3.3.

Essentially the same information has been shown in Fig. 3.4. Let now CP

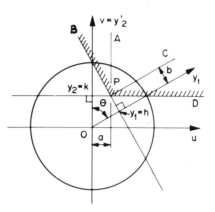

Fig. 3.4 — Bounds BPC and APD for $\Phi_2(\ )$ over region BPD.

be perpendicular to PB and DP perpendicular to AP. In the standardized normalized space u, v the probability in any right-angled region such as APD is given by the product of the probabilities given by $\Phi(-a)$ and $\Phi(-k)$ (see (A.4)). It then follows immediately that the probability content in the hatched region (i.e. $L(h, k, \rho) = \Phi_2(-h, -k, \rho)$) is more than either that above BPC (i.e. $\Phi(-b)\,\Phi(-h)$) or that to the upper right of APD (i.e. $\Phi(-a)\,\Phi(-k)$) but less than that contained in these two components summed [Ditlevsen, 1979b]:

$$\max[\Phi(-b)\Phi(-h),\ \Phi(-a)\Phi(-k)] \leqslant \Phi_2(-h,\ -k,\ \rho) \leqslant$$
$$\Phi(-b)\Phi(-h) + \Phi(-a)\Phi(-k) \tag{3.8}$$

It is not difficult to show that with $\rho = \cos\theta$

$$a = \frac{h - \rho k}{(1 - \rho^2)^{\frac{1}{2}}}$$
$$b = \frac{k - \rho h}{(1 - \rho^2)^{\frac{1}{2}}} \tag{3.9}$$

The above holds for $\rho \geqslant 0$, i.e. θ as shown. If $\rho < 0$, the y_1 axis is rotated clockwise and the bounds (3.9) become:

$$0 \leqslant \Phi_2(h, k, \rho) \leqslant \min[\Phi(b)\Phi(h), \Phi(a)\Phi(k)] \tag{3.9a}$$

3.3 SPECIAL CASE II: THE MULTIVARIATE NORMAL INTEGRAL

3.3.1 Format

Extending directly from the bivariate normal distribution, it follows readily that the joint probability density function for the n-vector $\mathbf{X}$ is given by

$$f_{\mathbf{X}}(\mathbf{x}, \mathbf{C}_X) = (2\pi)^{-n/2} |\mathbf{C}_X|^{-\frac{1}{2}} \exp[-\tfrac{1}{2}(\mathbf{x} - \boldsymbol{\mu}_X)^T \mathbf{C}_X^{-1}(\mathbf{x} - \boldsymbol{\mu}_X)] \tag{3.10}$$

where $\mathbf{C}_X = \{\sigma_{ij}\}$ is the covariance matrix of $\mathbf{X}$, with $\sigma_{ii} \equiv \sigma_{Xi}^2 = \mathrm{var}(X_i)$, and, $i \neq j$, $\sigma_{ij} = \mathrm{cov}(X_i, X_j)$ (cf. (A.124)). Also $|\mathbf{C}_X|$ is the determinant of $\mathbf{C}_X$, and $\mathbf{C}_X^{-1}$ is its inverse. It is not difficult to verify that this expression reduces to (3.2) for $n = 2$.

The probability content defined by $X_i \leqslant x_i, i = 1, \ldots, n$ is then given by the joint cumulative distribution function

$$F_{\mathbf{X}}(\mathbf{x}, \mathbf{C}_X) = P\left[\bigcap_{i=1}^{n} (X_i \leqslant x_i)\right] = \int_{-\infty}^{x_n} \cdots \int_{-\infty}^{x_1} f_{\mathbf{X}}(\mathbf{v}, \mathbf{C}_x) \, d\mathbf{v}$$

$$\tag{3.11}$$

For use in actual applications, the equivalent standardized expressions are particularly useful since, if these can be evaluated, so can the original expressions. When the variables are standardized (as usual according to $Y_i = (X_i - \mu_{X_i})/\sigma_{Xi})$], $\mathbf{C}_X$ becomes $\mathbf{R}$, the correlation matrix, and $F_{\mathbf{X}}(\mathbf{x}, \mathbf{C}_X) = \Phi_n(\mathbf{y}, \mathbf{R})$, where $\mathbf{Y}$ is the standardized normal variate vector, with zero mean, unit variance and correlation matrix $\mathbf{R}$.

3.3.2 Numerical integration of multi-normal integrals

Direct numerical integration of the multinormal integral (3.11) is not usually practical if $n > 5$. Iterative reduction formulae are reviewed by Johnson and Kotz (1972) who describe them as somewhat laborious in practice when $\rho > \frac{1}{2}$, even with the aid of computers. A more recent procedure due to Milton (1972) uses the fact that the distribution of $X_1, X_2, \ldots, X_{r-1}$ conditional on X_r is itself multivariate normal, so that computation from 2 to r dimensions can be applied iteratively. The method was slightly modified for bi and trivariate integrals by Daley (1974).

A direct integration algorithm for the bivariate normal integral due to Drezner (1978) is noted as giving high accuracy for polygonal regions of integration but is claimed to be slower in computation than another direct integration algorithm which integrates over angular regions after coordinate transformation [Didonato et al., 1980]. Really useful comparisons between all the various approaches do not appear to have been made. Some early comparisons are reviewed in Johnson and Kotz (1972), as are some series approximations of lower accuracy.

3.3.3   Reduction to a single integral*

In the special case where the correlation matrix $\mathbf{R} = \{\rho_{ij}\}$ has the special form $\rho_{ij} = b_i b_j$ $(i \neq j)$ with $-1 \leqslant b_i \leqslant +1$, the n-dimensional integral (3.11) can be reduced to a single integral.

Consider the transformation between the correlated standardized normal n-vector $\mathbf{Y}$ and the $n+1$ vector $(U_0, \mathbf{U})$ of independent standardized normal random variables, given by

$$Y_i = b_i U_0 + (1 - b_i^2)^{\frac{1}{2}} U_i, \qquad i = 1, \ldots, n \tag{3.12}$$

It is a simple matter (using (A.163)) to show that the correlation between Y_i, Y_j is then given by $\rho_{ij} = b_i b_j$ $(i \neq j)$ as desired.

The standardized multivariable joint cumulative distribution function can now be written as

$$\Phi_n(\mathbf{y}, \mathbf{R}) \equiv P\left[\bigcap_{i=1}^{n}(Y_i \leqslant y_i)\right] = P\left[\bigcap_{i=1}^{n}\left(U_i \leqslant \frac{y_i - b_i U_0}{(1 - b_i^2)^{1/2}}\right)\right] \tag{3.13}$$

which must hold for all U_0 values. Hence, integrating over U_0 (noting that U_0 is distributed as $\phi(u)$, and remembering that the U_i, $i = 1, \ldots, n$ are statistically independent), it follows that [e.g. Dunnett and Sobel, 1955; Curnow and Dunnett, 1962]

$$\Phi_n(\mathbf{y}, \mathbf{R}) = \int_{-\infty}^{\infty} \left\{\prod_{i=1}^{n} \Phi\left[\frac{y_i - b_i u}{(1 - b_i^2)^{1/2}}\right]\right\} \Phi(u) \, \mathrm{d}u \tag{3.14}$$

Expression (3.14) represents the probability content contained in the hypercube $Y_i \leqslant y_i$. This is easily evaluated using one-dimensional numerical integration and is of particular interest for $y_i < 0$ since the "tails" of the distribution are then evaluated. Evidently, numerical problems may arise if $\rho_{ij} \to 1$. Expression (3.14) can also be applied for negative correlations, although the integrand will then be complex [Johnson and Kotz, 1972].

3.3.4   Bounds on the multivariate normal integral*

For the standard normal n-vector $\mathbf{Y}$ with correlation matrix $\mathbf{R}$, Slepian (1962) has shown that for the probability given by (3.13) the derivative with respect to the correlations ρ_{ij} in $\mathbf{R}$, is a non-decreasing function. This means that for another standard normal n-vector, $\mathbf{V}$, say, with correlation matrix $\mathbf{K} = \{k_{ij}\}$ such that $k_{ij} \leqslant \rho_{ij}$ for all (i, j).

$$\Phi_n(\ ) = P_Y\left[\bigcap_{i=1}^{n}(Y_i \leqslant y_i)\right] \geqslant P_V^l\left[\bigcap_{i=1}^{n}(V_i \leqslant v_i)\right] \tag{3.15}$$

If $\mathbf{V}$ is selected to be identical with $\mathbf{Y}$, and k_{ij} selected such that $k_{ij} \leqslant \rho_{ij}$ for all (i, j), this inequality represents a lower bound on $\Phi_n(\ )$; by direct analogy an upper bound can be given [Gupta, 1963]. When these expressions are used in conjunction with (3.14), the bounds can actually be evaluated.

If the b_i values in (3.14) are chosen such that $b_i b_j \lessgtr \rho_{ij}$ then the lower and upper bounds respectively are obtained. Choosing $b_i^2 = \min_j(\rho_{ij})$, $b_i^2 = \max_j(\rho_{ij})$ results in wide bounds [Gupta, 1963]; an improvement is obtained by selecting, respectively, $b_i = \min_j(\rho_{ij}/b_j, 1)$ and $b_i = \max_j(\rho_{ij}/b_j,)$ for $i \neq j$ and $b_j \neq 0$ [Curnow and Dunnett, 1962]. Other choices of b_i are also possible; however, an appropriate selection is not straightforward. It follows that, in the case of negative correlation coefficients, replacing them with zero will produce a valid upper bound, without having to resort to complex integration (cf. section 3.2.2.1).

In all cases care must be taken to ensure that the resulting correlation coefficient matrix is nonnegative definite (see Appendix B).

3.4 MONTE CARLO SIMULATION

3.4.1 Introduction

As the name implies, Monte Carlo simulation techniques involve "sampling" at "random" to simulate artificially a large number of experiments and to observe the result. In the case of analysis for structural reliability, this means, in the simplest approach, sampling each random variable X_i randomly to give a sample value $\hat{x}_i$. The limit state function $G(\hat{x}) \leq 0$ is then checked. If the limit state function is violated, the structure or structural element has "failed". The experiment is repeated many times, each time with a randomly chosen vector $\hat{x}$ of $\hat{x}_i$ values. If N trials are conducted, the probability of failure is given approximately by

$$p_f \approx \frac{n(G \leq 0)}{N} \qquad (3.16)$$

where $n(G \leq 0)$ is the number of trials for which $G \leq 0$. Obviously the number N of trials required is related to the desired accuracy for p_f.

It is clear that in the Monte Carlo method a game of chance is constructed from known probabilistic properties in order to solve the problem many times over, and from that to deduce the required result (i.e. the failure probability).

In principle, Monte Carlo methods are only worth exploiting when the number of trials or simulations is less than the number of integration points required in numerical integration. This is achieved for higher dimensions by replacing the systematic selection of points by "random" selection, under the assumption that the points so selected will be in some way unbiased in their representation of the function being integrated.

To apply Monte Carlo techniques to practical problems it is necessary:

(a) to develop systematic methods for numerical "sampling" of the basic
     variables **X**;
(b) to select an appropriate economical and reliable simulation technique
     or "sampling strategy";
(c) to consider the effect of the complexity of calculating $G(\mathbf{X})$ and the
     number of basic variables on the simulation technique used;
(d) for a given simulation technique to be able to determine the amount of
     "sampling" required to obtain a reasonable estimate of p_f.

It may also be necessary to be able to allow for dependence between all
or some of the basic variables. All these matters will be considered in the
sections to follow.

3.4.2 Generation of uniformly distributed random numbers

In a physical experiment it might be possible to select a sample value of each
basic variable by means of some arbitrary random selection process, such as
putting a sequence of numbers in a lot and selecting one. Provided that the
lot size is large and the interval between numbers small, the probability
distribution for these numbers would be the "uniform" or "rectangular"
distribution ((A.73) or (A.74)), given by

$$
\begin{aligned}
F_R(r) = P(R \leqslant r) = r, & \quad \text{for } 0 \leqslant r \leqslant 1 \\
f_R(r) = 1, & \quad \text{for } 0 \leqslant r \leqslant 1 \\
= 0, & \quad \text{elsewhere}
\end{aligned}
\tag{3.17}
$$

It is possible to generate uniformly distributed random numbers through
automated roulettes or the noise properties of electronic circuits. These
generators tend to be slow and non-reproducible, so that an "experiment"
can never be checked. Tables of random numbers [e.g. Rand Corporation,
1955] (see Appendix D) can be stored in computer systems, but their
recovery for use is also very slow.

The most common practical approach is to employ a suitable "pseudo"
random number generator (PRNG), available on virtually all computer
systems. They are "pseudo" since they use a formula to generate a sequence
of numbers. This sequence is reproducible and repeats after (normally) a
long cycle interval. For most practical purposes, a sequence of numbers
generated by a suitable modern PRNG is indistinguishable from a sequence
of strictly true random numbers [Rubinstein, 1981]. A PRNG allows
production of a reproducible sequence, which in certain problems can be an
advantage. However, this reproducibility can be destroyed simply by
(randomly) changing the "seed number" required for most PRNGs. A
simple device is to use the local time as a seed value.

There are some mathematical reservations about the terminology
"random sampling". As soon as a table of "random numbers" or a PRNG is
used, the sequence of numbers is determined, and so no longer random. It
follows that "tests of randomness" are therefore strictly misnomers; these

tests are usually applied only for a one-dimensional sequence anyway, and may not "guarantee randomness" in more dimensions [e.g. Deak, 1980].

To avoid the philosophical difficulties, it has been suggested that Monte Carlo methods using PRNGs be dubbed "quasi Monte Carlo" methods and that the only justification for the use of sequences of pseudo-random numbers be simply the equi-distribution (or uniform distribution) property implied by their use. In essence, the numbers so selected give a reasonable guarantee of accuracy in the computations [Zaremba, 1968].

Example 3.1
A simple random number generator can be made as follows. Take a die and let the "six" denote an invalid sample Also take a coin and let "heads" be 5 and "tails" be 0. Then tossing the die and the coin repeatedly will produce (apart from invalid samples) a series of digits between 1 and 10, which should be random (if the die and coin are unbiased).

"Lotto" results can also be used as random number generators.

3.4.3   Generation of random variates
Basic variables only seldom have a uniform distribution. A sample value for a basic variable with a given (nonuniform) distribution is called a "random variate" and can be obtained by a number of mathematical techniques.

The most general of these is the "inverse transform" method. Consider the basic variable X_i for which the cumulative distribution function $F_X(x_i)$ must, by definition, lie in the range $(0, 1)$. The inverse transform technique is to generate a uniformly distributed random number r_i $(0 \leqslant r_i \leqslant 1)$ and equate this to $F_X(x_i)$ as shown in Fig 3.5:

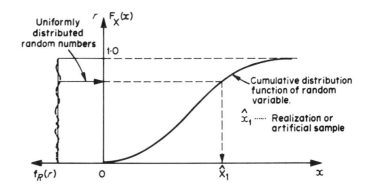

Fig. 3.5 — Inverse transform method for generation of random variates.

$$F_X(x_i) = r_i \quad \text{or} \quad x_i = F_X^{-1}(r_i) \tag{3.18}$$

This uniquely fixes the sample value $x_i = \hat{x}_i$, provided that an analytic expression for the inverse $F_{X(ri)}^{-1})$ exists (as it does for the Weibull, exponential, Gumbell and rectangular distributions among others). In these cases the inverse transform method is likely to be the most efficient technique. The technique can also be applied to basic variables whose cumulative distribution function has been obtained from direct observation.

Specialized techniques for generating random variates from specific distributions are often computationally more efficient that the inverse transform method. Most computer systems have standard subroutines available. One such procedure is due to Box and Muller (1958). It produces a pair of "exact" independent standardized normal variates, u_1 and u_2, given by

$$u_1 = (-2 \ln r_1)^{1/2} \sin 2\pi r_2 \qquad (3.19a)$$

$$u_2 = (-2 \ln r_1)^{1/2} \cos 2\pi r_2 \qquad (3.19b)$$

where r_1, r_2 are uniformly distributed independent random variables in the interval $(0,1)$. Lognormally distributed random variables V_i may be obtained directly from expressions (3.19) since, for V lognormally distributed, $v_i = \ln u_i$.

3.4.4 Direct sampling ("crude" Monte Carlo)

The technique sketched in section 3.4.1 is the simplest Monte Carlo approach for reliability problems; it is probably also the most widely used but not the most efficient. The basis for its application is as follows. The probability of limit state violation (1.31) may be expressed as

$$p_f = J = \int \cdots \int I[G(\mathbf{x}) \leqslant 0] f_{\mathbf{X}}(\mathbf{x}) \, d\mathbf{x} \qquad (3.20)$$

where $I[\ ]$ is an "indicator function" which equals 1 if [ ] is "true" and 0 if [ ] is "false". Here the indicator function has taken on the role of identifying the integration domain, but it can also take on the more general role discussed in Section 1.4.8.

From comparison with (A.10) it is seen that the integral (3.20) represents the expected value of $I[\ ]$. If now $\hat{\mathbf{x}}_j$ represents the jth vector of random observations from $f_{\mathbf{x}}(\ )$, then it follows directly from sample statistics that

$$p_f \approx J_1 = \frac{1}{N} \sum_{j=1}^{N} I[G(\hat{\mathbf{x}}_j \leqslant 0)] \qquad (3.21)$$

is an unbiased estimator of J and hence of p_f. Hence expression (3.21) provides a direct estimate of p_f. This is what was done in section 3.3.1.

Three matters are now of interest: how to extract most information from the simulation points; how many simulation points are needed for a given accuracy; conversely, how to improve the sampling technique to obtain greater accuracy for the same or fewer sample points. These matters will be considered below.

One way that the results of the above sampling technique may be represented is as a cumulative frequency function $F_G(g)$ (see Fig 3.6). The

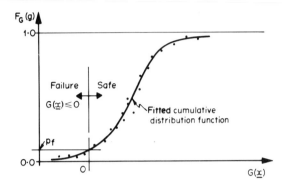

Fig. 3.6 — Use of fitted cumulative distribution function to estimate p_f.

region for which $G(\mathbf{x}) \leq 0$ is shown. Obviously, much of the range of $F_G(g)$ is of little interest, since the structure or member is (usually) amply safe.

The estimate of p_f in equation (3.21) may be improved by fitting an appropriate distribution function through the points obtained as the cumulative frequencies for $G(\mathbf{x}) \leq 0$, i.e. the left-hand tail in Fig. 3.6. However, the choice of distribution function may be difficult. Distributions which could be of use for this purpose are the Johnson and Pearson distribution systems [Elderton and Johnson, 1969]. However, it is possible that the parameters of the distribution function are rather sensitive to extreme results in the region $G(\mathbf{x}) \leq 0$, and in that case the choice of distribution function parameters may not stabilize until N is quite large [Moses and Kinser, 1967].

Rather than fitting a single distribution to the sample points, it is possible to fit a sequence

$$\hat{F}_G(g) = \sum_{i=1}^{n} p_i F_i(g) \tag{3.22}$$

where

$$\sum_{i=1}^{n} p_i = 1 \text{ and } p_i \geq 0 \text{ for all } i$$

Criteria for fitting the sequence might be that each $F_i(\ )$ has the same mean and standard deviation as estimated from the sample points and that the error with respect to one or more higher moments is minimized [Grigoriu, 1983].

An estimate of the number of simulations required for a given confidence level may be made as follows. Since $G(\mathbf{X})$ is a random variable in $\mathbf{X}$, the indicator function $I(G \leqslant 0)$ is also a random variable, albeit with only two possible outcomes. It follows from the central limit theorem that the distribution of J_1 given by the sum of independent sample functions (3.21) approaches a normal distribution as $N \to \infty$. The mean of this distribution is (cf. (A.160))

$$E(J_1) = \sum_{i=1}^{N} \frac{1}{N} E[I(G \leqslant 0)] = E[I(G \leqslant 0)] \tag{3.23a}$$

which is equal to J (see (3.21)), while the variance is given by (cf. (A.161))

$$\sigma_{J_1}^2 = \sum_{i=1}^{N} \frac{1}{N^2} \text{var}[I(G \leqslant 0)] = \frac{\sigma_{I(G \leqslant 0)}^2}{N} \tag{3.23b}$$

This shows that the standard deviation of J_1 and hence of the Monte Carlo estimate (3.21) varies directly with the standard deviation of $I(\ )$ and inversely with $N^{1/2}$. These observations are important in determining the number of simulations required for a particular level of confidence. To actually calculate confidence levels, an estimate of $\sigma_I(\ )$ is required.

Using (A.11), the variance is given by

$$\text{var}[I(\ )] = \int \ldots \int [I(G \leqslant 0)]^2 \, d\mathbf{x} - J^2 \tag{3.24a}$$

so that the sample variance is given by:

$$S_{I(G \leqslant 0)}^2 = \frac{1}{N-1} \left(\left\{ \sum_{j=1}^{N} I^2[G(\hat{\mathbf{x}}_j) \leqslant 0] \right\} - \right.$$

$$\left. - N \left\{ \frac{1}{N} \sum_{j=1}^{N} I[G(\hat{\mathbf{x}}_j) \leqslant 0] \right\}^2 \right) \tag{3.24b}$$

On the basis that the central limit thereom applies, the following confidence statement can be given for the number (J_1) of trails in which "failure" occurs:

$$P(-k\sigma < J_1 - \mu < +k\sigma) = C \tag{3.25}$$

where μ is the expected value of J_1 given by (3.23a) and σ is given by (3.23b). For confidence interval $C = 95\%$, $k = 1.96$, as can be verified from standard normal tables (see Appendix C). As σ is not known, it may be estimated from (3.24b). However, this is not very helpful at the beginning of a Monte Carlo study.

Shooman (1968) has suggested that σ and μ in (3.25) can be approximated by the binomial parameters $\sigma = (Nqp)^{1/2}$ and $\mu = Np$, with $q = 1 - p$, (cf. (A.20) and (A.21)) provided that $Np \geq 5$ when $p \leq 0.5$. If these are substituted in (3.25) there is obtained

$$P[-k(Nqp)^{1/2} < J_1 - np < +k(Nqp)^{1/2}] = C \qquad (3.26)$$

If the error between the actual value of J_1 and the observed is denoted by $\varepsilon = (J_1 - Np)/Np$ and this is substituted into (3.26), it follows easily that $\varepsilon = k[(1-p)/Np]^{\frac{1}{2}}$. Thus, if $N = 100\ 000$ samples, and $p = p_f = 10^{-3}$ (expected), the error in J_1 and hence p_f will be less than 20% with 95% confidence (as then $k = 1.96$).

Broding *et al.* (1964) suggested that a first estimate of the number N of simulations for a given confidence level C in the failure probability p_f can be obtained from

$$N > \frac{-\ln(1 - C)}{p_f} \qquad (3.27)$$

Thus, for a 95% confidence level and $p_f = 10^{-3}$, the required number of simulations is more than 3000. The actual number of variates to be calculated is, of course, N times the number of independent basic variables. Others have suggested that the number of simulations may need to be of the order of 10 000–20 000 for approximately 95% confidence limit, depending on the function being evaluated [Mann *et al.*, 1974].

Example 3.3
The stress resultant S acting on a member in tension has statistical properties estimated to be $N(10.0, 1.25)$. The resistance R is estimated to be $N(13.0, 1.5)$. What probability of failure is estimated using the "crude" Monte Carlo method, if R and S are independent?

Using integral (1.16), both R and S need to be simulated. For convenience, only ten variates $\hat{R}$ and another ten variates $\hat{S}$ will be given below, but generally more samples will be required.

From a table of random numbers (e.g. Appendix D), select a random number, say, $\hat{r}_1 = 0.9311$ for R, say. Then following (3.18) and Fig. 3.5, $\hat{r}_1 = \Phi^{-1}(\hat{r}_1) = +1.49$ from standardized normal tables such as given in Appendix C. The sample value $\hat{R}$ then follows directly (cf. section A.5.7):

$$\hat{r}_1 = 1.49 = \frac{\hat{R} - \mu_R}{\sigma_R} = \frac{\hat{R} - 10}{1.25} \text{ or } \hat{R}_1 = 15.24$$

Similar calculations for another nine $\hat{R}$ values and for ten $\hat{S}$ can be made using the next 19 values in Appendix D. $\hat{S}$ follows from $\hat{s} = (\hat{S} - \mu_S)/\sigma_S$. In particular, using the eleventh random number in Appendix D, $\hat{S}_1$ can be found to be 10.53. Thus $\hat{R} > \hat{S}_1$.

It will be left to the reader to show that, in this case, only one of the sample pairs $\hat{R}_i$, $\hat{S}_i$) led to a failure (i.e. $\hat{R}_i \leqslant \hat{S}_i$) so that $p_f \approx 0.1$. Obviously more sampling is required. Since both R and S are normally distributed, the exact result is directly obtainable from expression (1.22):

$$p_f = \Phi\left[\frac{-(13-10)}{(1.5^2+1.25^2)^{1/2}}\right] = \Phi(-1.54) = 0.0618$$

Example 3.4

If the single integral form (1.18) is to be used in a Monte Carlo analysis, it may be seen from comparing (3.20) with (1.18) that in essence $I[\ ]$ is replaced by $F_R(x)$. This means that the value for $F_R(\ )$, averaged over all samples, is estimated. The samples $\hat{x}_i$ themselves are drawn from the probability distribution given by $f_S(\ )$. This is done using the inverse transform technique (3.18) to give $x_i = F_S^{-1}(r_i)$ where r_i is a random number.

With data as in Example 3.3, the calculation procedure is as follows. Select a random number from Appendix D, $r_1 = 0.9311$, say. Then the sample value $\hat{x}_1$ is obtained from $\hat{X}_i = F_S^{-1}(r_i)$, or in this case, since $S = N(\ ,\ )$, from $\hat{x}_1 = $ standardized normal $N(0,1) = \Phi^{-1}(r_1) = +1.49$. Since $\hat{x}_1 = (\hat{X}_1 - \mu_S)/\sigma_S$ it follows that $\hat{X}_1 = 11.86$. With this sample value, $F_R(\hat{X}_1)$ is evaluated, first by calculating $\hat{r}_1 = (\hat{X}_1 - \mu_R)/\sigma_R = -0.76$, and then $F_R(\hat{X}_1) = \Phi(\hat{r}_1) = 0.2237$. Similar results can be obtained for nine more samples, using the next nine random numbers in Appendix D, say. These will be found to be (0.0631, 0.0188, 0.0918, 0.1075, 0.0075, 0.0136, 0.0005, 0.0021, 0.0036).

The sum of the ten $F_R(x_i)$ samples is 0.532, so that $p_f \approx (1/10) \times 0.532 = 0.0532$. This is 14% in error compared with the correct result of 0.0618, a reasonable result considering that only ten samples were used. Clearly the size of the error will depend on the selection of random numbers, and the number of samples.

3.4.5 Variance reduction

From expression (3.23b) it is seen that the variance σ_I^2 directly affects the variance of J and that the number of samples N does so inversely. This means that the standard deviation of $\hat{J}$ and hence of the Monte Carlo estimate (3.21) decreases in proportion to $N^{-1/2}$. By comparison, for one-dimensional integration, the error in standard deviation reduces as N^{-2} for the trapezoidal rule and N^{-4} for Simpson's rule [Dahlquist and Björck, 1974]. Obviously the slow convergence is a severe penalty for the "crude" Monte Carlo method and has led to so-called "variance reduction" techniques. The essential procedure is to find a way to reduce σ_I^2.

Variance reduction can only be achieved by using additional (*a priori*) information about the problem to be solved. For example, from Fig 1.10 it is evident that only sampling in the region of overlap between f_R and f_S is likely to prove interesting, so that this information could be used to reduce the variance in the estimate. This observation, generalized, forms the basis for "importance sampling" to be considered below.

A number of other variance reduction techniques with roughly similar strategies exist. In each case information about the problem is used to limit the simulation to "interesting" regions. A good overview of the various strategies for variance reduction is given by Rubinstein (1981). Comparison of earlier works [eg Kahn, 1956; Hammersley and Handscomb, 1964] indicates some differences in nomenclature for similar techniques. Warner and Kabaila (1968) reviewed some of the techniques in relation to structural reliability calculations.

3.4.6   Importance sampling
3.4.6.1   *Theory of importance sampling*
The multiple integral (1.31) can be written as in (3.20) using the indicator function $I[\ ]$, or, equivalently, as

$$J = \int \cdots \int I[G(\mathbf{x}) \leq 0] \frac{f_{\mathbf{X}}(\mathbf{x})}{l_{\mathbf{V}}(\mathbf{x})} l_{\mathbf{V}}(\mathbf{x})\ d\mathbf{x} \qquad (3.28)$$

where $l_{\mathbf{V}}(\ )$ is termed the "importance-sampling" probability density function whose definition will be considered in detail below. Again by comparison with (A.10), expression (3.28) can be written as an expected value:

$$J = E\left\{ I[G(\mathbf{v}) \leq 0] \frac{f_{\mathbf{X}}(\mathbf{v})}{l_{\mathbf{V}}(\mathbf{v})} \right\} = E\left(\frac{If}{l}\right),\ \text{say} \qquad (3.29)$$

where $\mathbf{V}$ is any random vector with probability density function $l_{\mathbf{V}}$. Obviously it is required that $l_{\mathbf{V}}(\mathbf{v})$ exists for all valid $\mathbf{v}$ and that $f_{\mathbf{X}}(\mathbf{v}) \neq 0$. Comparison with section 3.4.4 shows that $I[\ ]$ there is here replaced by $I[\ ]f/l$.

An unbiased estimate of J is given (cf. (3.21)) by

$$p_f \approx J_2 = \frac{1}{N} \sum_{j=1}^{N} \left\{ I[G(\hat{\mathbf{v}}_j) \leq 0] \frac{f_{\mathbf{X}}(\hat{\mathbf{v}}_j)}{l_{\mathbf{V}}(\hat{\mathbf{v}}_j)} \right\} \qquad (3.30)$$

where $\hat{\mathbf{v}}_j$ is a vector of sample values taken from the importance sampling function $l_{\mathbf{V}}(\ )$.

It is evident that $l_{\mathbf{V}}(\ )$ governs the distribution of the samples. How this distribution should be chosen is quite important. By direct analogy with (3.23b) and (3.24a), the variance of J_2 is given by

$$\mathrm{var}(J_2) = \mathrm{var}\left(\frac{If}{l}\right)\Big/N$$

where

$$\mathrm{var}\left(\frac{If}{l}\right) = \int \cdots \int \left\{I[\ ]\frac{f_{\mathbf{x}}(\mathbf{x})}{l_{\mathbf{v}}(\mathbf{x})}\right\}^2 l_{\mathbf{v}}(\mathbf{x})\ d\mathbf{x} - J^2 \qquad (3.31)$$

Clearly (3.31) should be minimized to minimize $\mathrm{var}(J_2)$. If the function $l_{\mathbf{v}}(\ )$ is selected as

$$l_{\mathbf{v}}(\mathbf{v}) = \frac{|I[\ ]f_{\mathbf{x}}(\mathbf{v})|}{\int \cdots \int |I[\ ]f_{\mathbf{x}}(\mathbf{v})|\ d\mathbf{v}} \qquad (3.32)$$

then upon substitution into (3.31) it is easily found that

$$\mathrm{var}\left(\frac{If}{l}\right) = \left(\int \cdots \int |I[\ ]f_{\mathbf{x}}(\mathbf{v})|\ d\mathbf{v}\right)^2 - J^2 \qquad (3.33)$$

If the integrand $I[\ ]f_{\mathbf{x}}(\mathbf{v})$ does not change sign, the multiple integral is identical with J and $\mathrm{var}(If/l) = 0$. In this case the optimal function $l_{\mathbf{v}}(\ )$ is obtained from (3.32) as

$$l_{\mathbf{v}}(\mathbf{v}) = \frac{I[G(\mathbf{v}) \le 0]f_{\mathbf{x}}(\mathbf{x})}{J} \qquad (3.34)$$

but evidently this is not very helpful since the very integral to be determined, J, is needed. However, it is clear that, even if J is only approximately evaluated, the variance of If/l can be reduced. Thus, to reduce the variance (3.31), it is desirable that $l_{\mathbf{v}}(\mathbf{v})$ should have the form of the integrand of (3.20), divided by the estimate for J. Equivalently, the variance is reduced if $l_{\mathbf{v}}(\mathbf{v})/(I[\ ]f_{\mathbf{x}}(\mathbf{v}) \approx \text{constant} < 1$ [Shreider, 1966].

If the integrand does change sign, then the addition of a sufficiently large constant may adapt it to the above form. A weaker result when the integrand does change sign can also be given [Kahn, 1956; Rubinstein, 1981].

From (3.33) it follows that a good choice of $l_{\mathbf{v}}(\ )$ can actually produce zero variance in the estimate for J for the special case of non-negative integrand. This may seem to be a surprising result. In essence, it demonstrates that, the more effort is put into obtaining a close initial estimate of J in equation (3.34), the better the Monte Carlo result will actually be; there is thus no limit to discourage such a search. This is in contrast with ordinary numerical analysis. Conversely, the variance can actually be increased by using a very poor choice of $l_{\mathbf{v}}$ [Kahn, 1956].

3.4.6.2 *Importance sampling functions*

While in general the derivation of optimal l_V functions is difficult, appropriate functions may often be selected on *a priori* grounds. Thus, in the n-dimensional reliability problem, the region of most interest is the hyperzone $G(\mathbf{x}) \leq 0$ and, more particularly, the region of greatest probability density within that zone. This is the region just to the right of the point $\mathbf{x}^*$ shown for a two-variable problem in Fig. 3.7. In general the region cannot be

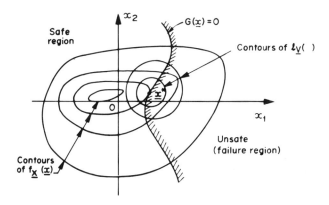

Fig. 3.7 — Importance sampling function $l_V(\ )$ in x space.

easily identified as there are no simple criteria with which to work. However, a surrogate for identifying the region of interest is to identify the point $\mathbf{x}^*$. In this case the point $\mathbf{x}^*$ is clearly the point having the largest ordinate $f_{\mathbf{X}}(\mathbf{x})$. For most probability density functions $f_{\mathbf{X}}(\mathbf{x})$, the point $\mathbf{x}^*$ corresponds to a maximum value of $f_{\mathbf{X}}(\mathbf{x})$ and may be found by the direct application of numerical maximization techniques. However, if $f_{\mathbf{X}}(\mathbf{x})$ is, say, a rectangular distribution, no unique maximum point $\mathbf{x}^*$ can be identified. This, and similar cases, is not as critical as it may seem at first, since p_f is usually quite small, so that the region of $f_{\mathbf{X}}(\mathbf{x})$ satisfying $g(\mathbf{X}) \leq 0$ is also quite small. Therefore the choice of an appropriate point $\mathbf{x}^*$ is quite restricted and may be made rather arbitrarily within $G(\mathbf{X}) \leq 0$, although a point at the maximum mass location of $f_{\mathbf{X}}(\mathbf{x})$ would be preferable.

Once $\mathbf{x}^*$ is identified, one approach for choosing l_V is simply to use the distribution $f_{\mathbf{X}}$ shifted so that its mean is at $\mathbf{x}^*$ [Harbitz, 1983]. However, a more appropriate distribution for l_V is considered to be $l_V(\ ) = \Phi_n(\mathbf{v}, \mathbf{C}_V)$, where $\mathbf{C}_V$ is a strictly diagonal matrix of σ_i^2 and with the mean of $\mathbf{V}$ placed at $\mathbf{x}^*$. Such a distribution will produce sample points unbiased with respect to each variable. It will also give greater sampling density in the neighbourhood of the region where it is of most significance. The choice of $\mathbf{x}^*$ as the locating point for l_V has the obvious advantage that it gives a systematic procedure for selection of the location of l_V.

As can be seen from Fig. 3.7 for two-variable problems, the shape or format of the limit state function $G(\mathbf{x}) = 0$ is of little importance, since with l_V centred at $\mathbf{x}^*$, a wide region is included in the sample space. Thus non-linear regions are handled with ease, a feature not found with the techniques described in Chapter 4. Similarly, the format of $f_{\mathbf{X}}(\mathbf{x})$ is of little significance; dependence between the $\mathbf{X}_i$, such as indicated schematically in Fig. 3.7, does not affect the procedure.

It is also seen that, unless $G(\mathbf{x}) = 0$ is very non-linear, the "success" rate for sample points selected from $l_V(\ )$ is about 50%, i.e. there is an approximately equal likelihood of any sample point falling in either the safe or the unsafe region. It follows directly, that for a given level of confidence, far fewer sample points are required using $l_V(\ )$ as shown than using the "crude" Monte Carlo method with $f_{\mathbf{X}}(\mathbf{x})$ as sampling distribution.

An approximate measure of the relative efficiencies of the two methods can be obtained by considering the number of sample points required to obtain a given number of points (say 100) in the "failure" region. Assuming a "success" rate of 50%, for a linear limit state function with $l_V(\ )$ centred on the "design" point $\mathbf{x}^*$, on average 200 points need to be generated with importance sampling. By contrast, if p is the probability of a sample point based on $f_{\mathbf{X}}(\mathbf{x})$ falling in the "failure" region, then $100/p$ sample points, on average, need to be generated. Hence, if p is of the order of 10^{-3}–10^5, as may be typical, 10^5–10^7 sample points are required for the "crude" Monte Carlo method. Similar results can be obtained for confidence levels, for example, using (3.26) with $p \approx 0.5$ being the "success rate".

An important observation is that, since $l_V(\ )$ can be selected by the analyst, it can have independent components $\mathbf{V}$. Unlike the "crude" Monte Carlo technique, for which the generation of correlated or dependent random variables $\mathbf{X}$ may be required (since $f_{\mathbf{X}}(\mathbf{x})$ is the sampling function!), no such requirement exists for importance sampling. Because of the complexity of obtaining dependent random variables (see Appendix B), this is sufficient reason in itself to use importance sampling instead of "crude" Monte Carlo.

Example 3.5

Example 3.4 will now be repeated using importance sampling. With R being $N(13, 1.5)$ and S being $N(10, 1.25)$, a reasonable choice for the importance-sampling function is $N(11.5, 1.3)$. Hence, the random numbers r_i (identical with those used in the earlier examples, to enable comparison) are transformed to sample values $\hat{v}_i$ through the inverse normal transformation. The first four lines of calculation are shown in Table 3.1. The sampling function is then evaluated as $\phi(\hat{y}_i)$, where $\hat{y}_i = (\hat{v}_i - \mu_v)\sigma_V$. The function $F_R f_S / l_V$ takes the place of $I[\ ]f_X / l_V$ in equation (3.47). For ten samples, it will be found that $p_f \approx (1/10) \times 0.478 = 0.0478$.

Experimentation with sample size and l_V will show that sensible selection

Table 3.1 — Importance sampling: Example 3.5

$\hat{r}_i$	$\hat{y}_i = F_I^{-1}(\hat{r}_i)$ $= \Phi^{-1}(\hat{r}_i)$	$\hat{v}_i = 1.3\hat{y}_i + 11.5$	$l_V(\hat{v}_i) = \phi(\hat{y}_i)$	$F_R(\hat{v}_i)$		$f_S(\hat{v}_i)$		$\dfrac{F_R(\hat{v}_i)f_S(\hat{v}_i)}{l_V(\hat{v}_i)}$
				$\hat{r}_i = \dfrac{\hat{v}_i - 13}{1.5}$	$F_R(\hat{v}_i) = \phi(\hat{r}_i)$	$\hat{s}_i = \dfrac{\hat{v}_i - 10}{1.25}$	$f_S(\hat{v}_i) = \dfrac{1}{\sigma_s}\phi(\hat{s}_i)$[a]	
0.9311	+ 1.49	13.44	0.1315	0.29	0.6141	2.75	0.0073	0.0340
0.7163	+ 0.57	12.24	0.3391	− 0.51	0.3050	1.79	0.6432	0.0578
0.4626	− 0.09	11.38	0.3973	− 1.08	0.1401	1.10	0.1743	0.0614
0.7895	+ 0.80	12.54	0.2897	− 0.31	0.3783	2.03	0.0406	0.0530

[a]Obtained from $\phi(s) = [1/(2\pi)]^{\frac{1}{2}} \exp(-\frac{1}{2}s^2)$ or from a table of standardized normal probability distribution functions.

of l_V is required if rapid convergence is to be achieved. As a general rule, the standard deviation σ_{Vi} for variable X_i should increase the further x_i^* lies from μ_{Xi} [Melchers, 1984].

3.4.7 Conditional expectation

In the special case where the variable with the largest uncertainty, X_1, say, is also an independent variable, the order of integration in the probability integral (1.31) may be reduced by one:

$$p_f = \int_{G(\mathbf{X}) \leqslant 0} \cdots \int_n f_{\mathbf{X}}(\mathbf{x}) \ d\mathbf{x} = \int_{G(\mathbf{X}') \leqslant 0} \cdots \int_{n-1} F_{X_1}(\mathbf{u}) f_{\mathbf{X}'}(\mathbf{u}) \ d\mathbf{u}$$

(3.35)

where X' represents the reduced vector. This form may be used as a basis for a Monte Carlo technique known as "conditional expectation".

The unbiased estimator for (3.35) is given by

$$p_f \approx \frac{1}{N} \sum_{j=1}^{N} F_{X_1}(\hat{x}_{1j})$$

(3.36)

where $\hat{x}_{1j}$ is the jth value of X_1 obtained from the limit state equation $G(\mathbf{x}) = 0$ solved for x_1 when all other X_i are generated sample values $\hat{x}_j$ for the jth trial. Hence, the simulation procedure is to generate the sample $\hat{x}'_j$, to satisfy the limit state equation and to determine an estimate $\hat{p}_{fj} = F_{X_1}(\hat{x}_{1j})$. This process is repeated N times to estimate p_f through (3.36). Expression (3.36) can be shown to have a smaller variance than the equivalent "crude" formulation [Ayyab and Haldar, 1984].

Some additional variance reduction may be achieved through "antithetic" variates in which variates in sample sequences are given negative correlation. This variance reduction technique, as are several others, is discussed in the literature [Rubinstein, 1981].

3.4.8 Practical aspects of Monte Carlo simulation
3.4.8.1 Systematic selection of random variables
It is important to note that random sampling is not always required for every variable. If, as would be usual for structural reliability, the samples can be taken sequentially, say $\hat{x}_1$, $\hat{x}_2$, $\hat{x}_2$, etc., for $G(\mathbf{x})$, then the sampling for X_1, say, could be done systematically, rather than randomly. For example, the interval over which the sampling would be done for X_1 could be divided into n equal intervals. For each interval represented by the point i, (which might be taken as the centre of the interval), the probability density function would be simply $f_{X1}(x_i)$. In a conventional random sampling of X_1, using N samples, it would be expected that N/n values of X_1 would occur in the ith interval. Hence N/n values, each equal to $f_{X1}(x_i)$ could be employed. The results are not biased if the first N/n values for X_1 are systematically set at $f_{X1}(x_i)$, the next N/n at $f_{X1}(x_2)$, etc., with $x_1 < x_2 < x_3$, etc. [Kahn, 1956].

3.4.8.2 Sensitivity studies

If the effect of changing one (or more) variables on the failure probability is required to be evaluated, running two Monte Carlo assessments, with and without the change, is unlikely to be very helpful. This is because the change is obtained by subtracting the outputs, while the variance for the change is obtained by adding the two variances, so that the result sought might have a very large uncertainty. The problem can be circumvented by employing the same set of random numbers for both probability estimates. If this is done, the total variance is significantly less than for independent calculations, since the probability estimates are now (highly) correlated. This type of sampling is sometimes known as "correlated" sampling [Rubinstein, 1981].

If the limit state function is analytic, the differentials $\partial G/\partial X_i$ will give the sensitivity of G to a (unit) change in X_i, but values so obtained will not be comparable unless the variances of the X_i are closely similar and the X_i are independent random variables. For independent variables, an approximate comparable measure of the sensitivity of $G(\mathbf{X})$ to a change in X_i is given by $c_i \approx (\sigma_{X_i}^{-1})\partial G/\partial X_i$. This might be compared with the sensitivity coefficients α_i to be discussed in section 4.3.2.

3.4.8.3 Generalized limit state function

Considerable computational problems may arise if $G(\mathbf{X})$ is at all complex to calculate. This might occur, for example, with a finite element analysis or for iterative non-linear material or structural behaviour. If the number of basic variables is relatively small, some typical calculations for $G(\ )$ might be performed so as to cover the range of values of $\mathbf{X}$ likely to be obtained by simulation. These values may then be used as interpolation points for $G(\ )$ each time that a sample vector $\hat{\mathbf{x}}$ is generated. This procedure is only feasible if $G(\ )$ is reasonably regular. It is sometimes referred to as the "response surface" technique [e.g. Augusti et al., 1984].

If the limit state is defined by a "ramp" or other general function (see section 1.4.8) rather than by the indicator function (3.20) there is in principle little difficulty in applying the crude and importance-sampling techniques, at the expense of very little extra computation. Such an extension is not necessarily straightforward with other methods of probability integration.

3.4.8.4 Error estimates

For every numerical method of integral solution the error involved in application of the method is of importance, since it provides a measure of the degree of confidence that can be attached to the results obtained.

In principle, the sample variance, and hence the sample standard deviation measure of error is adequate to indicate the error inherent in Monte Carlo simulation (see equation (3.25) and analogously for importance sampling). However, it seems that, in practice, such error measures are only seldom employed.

3.4.8.5 Calculation times

A common criticism of the Monte Carlo method is that it is a crude and time-consuming method of solution. Unfortunately for many practical applications where Monte Carlo techniques have been used these criticisms are often well founded, but that need not be the case in general.

The applications noted below show that refinements such as importance sampling are not widely used, despite its considerable potential to reduce the number of samples and hence the required computing time. The importance-sampling technique in particular offers a unique opportunity for the analyst to influence the method and efficiency of calculation. Unfortunately it appears that this opportunity has not thus far been widely appreciated.

3.4.8.6 *Typical applications*
Applications of the Monte Carlo technique to a large variety of problems have been described, for example, by Kahn (1956), Hammersley and Handscomb (1964) and Rubinstein (1981). Some typical structural engineering applications have been described by Moses and Kinser (1967), and Knappe *et al.* (1975). As noted, Warner and Kabaila (1968) reported on various ways of applying the Monte Carlo technique (but not importance sampling) to an idealized reinforced concrete member. Other typical applications include probabilistic dynamics [Ellyin and Chandresekhar, 1977], fatigue reliability [Oswald and Schuëller, 1983] and low-cycle fatigue reliability [Harrington and Melchers, 1985].

3.5 CONCLUSION

Much of this chapter has been devoted to the use of the Monte Carlo technique to enable integration of the convolution integral. Both the "crude" and the "importance sampling" methods were considered in detail and other possibilities as well as some practical matters noted.

The early part of the chapter gave exact results and bounds for the bivariate normal integral and the multivariate normal integral. These are the only multidimensional integral forms available in the literature and form an important basis for the material to be discussed in Chapter 4.

4

Second-moment and transformation methods

4.1 INTRODUCTION

Rather than use approximate integration methods to evaluate (1.18) or (1.31), in this chapter the probability density functions themselves will be simplified. In particular, the first part of this chapter will consider the special case of reliability estimation in which each variable is represented only by its first two moments, i.e., by its mean and standard deviation. This is known as the 'second-moment' level of representation. Higher moments, which might describe skew and flatness of the distribution, are ignored. As noted earlier, a convenient way in which the second-moment representation might be interpreted is that each random variable is represented by the normal distribution. (This is the only continuous probability distribution completely described by its first two moments.) As was seen in section 1.4.3 when both R and S are normally distributed, integration of (1.18) is trivial; further, as already noted in sections 3.2 and 3.3, the multiple integral (1.31) may also be tractable when $f_X(\mathbf{x})$ is multinormal.

Because of their inherent simplicity, the so-called 'second-moment' methods have become very popular, particularly in design code calibration work (Chapter 9). Early works by Mayer (1926), Freudenthal (1956), Rzhanitzyn (1957) and Basler (1961) contained second-moment concepts. Not until the late 1960s, however, was the time ripe for the ideas to gain a measure of acceptance [Cornell, 1969].

From the original second-moment ideas have grown a number of extensions and refinements: principally that it is possible, with iterative techniques, to approximate the actual probability distributions in f_X with normal probability distributions and still to obtain good approximations for the original failure probability.

4.2 SECOND-MOMENT CONCEPTS

From section 1.4.3, with resistance R and load effect S each second-moment random variables (i.e. having normal distributions), the limit state equation is the 'safety margin' $Z = R - S$ and the probability of failure p_f is

$$p_f = \Phi(-\beta), \quad \beta = \frac{\mu_Z}{\sigma_Z} \qquad (4.1)$$

where β is the 'safety index' and Φ the standard normal distribution function. The function Φ is tabulated in Appendix C and approximate expressions are given in section A.5.7.

Evidently, equation (4.1) yields the exact probability of failure p_f when both R and S are normally distributed. However, p_f defined in this way is only a "nominal" failure probability for other distributions of R and S. Conceptually it is probably better in this case not to refer to probabilities at all, but simply to β, the "safety index".

As shown in Fig 1.11, β is simply a measure (in standard deviation σ_Z units) of the distance that the mean μ_Z is away from the origin $z = 0$. This point marks the boundary to the "failure" region. Hence β is a direct (if imprecise) measure of the safety of the structural element and greater β represents greater safety, or lower nominal failure probability p_{fN}. For convenience in what follows, the more general probability interpretation p_{fN} will be used to refer to any probability calculated using one or more conceptual approximations.

If the limit state function $G(\mathbf{X})$ consists of more than two basic random variables, the above ideas are readily extended. If $G(\mathbf{X})$ is a function $Z(\mathbf{X})$ given by

$$Z(\mathbf{X}) = a_0 + a_1 X_1 + a_2 X_2 + \ldots + a_n X_n \qquad (4.2)$$

then $Z(\mathbf{X})$ is normally distributed and (A.160) and (A.161) provide μ_Z and σ_Z from which β and p_{fN} can again be evaluated using (4.1).

In general, however, $G(\mathbf{X})$ is not linear. Thus, with all the random variables X_i implicitly normally distributed, $G(\mathbf{X})$ will not be normally distributed. The only way to obtain the first two moments of $G(\mathbf{X})$ (i.e. its mean and variance) is to linearize $G(\mathbf{X})$. This can be done by using expressions (A.178) and (A.179) which then give approximate moments by expanding $G(\mathbf{X})$ as a first-order Taylor series expansion about some point $\mathbf{x}^*$. Approximations which linearize $G(\mathbf{X})$ will be denoted "first-order" methods.

The first-order Taylor series expansion which linearizes $G(\mathbf{X})$ at $\mathbf{x}^*$ might be denoted $G_L(\mathbf{x}) = 0$ (Fig. 4.1). Expansion about the means μ_X is common in probability theory, but there is no rationale for doing so in general; in fact it will be seen later that there is indeed a better choice for the present

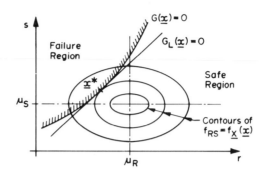

Fig. 4.1 — Limit state surface in the space of the basic variables.

problem. At this stage, it is sufficient to note that the choice of $\mathbf{x}^*$ directly affects the estimate of β. This is demonstrated in the following example.

Example 4.1
The first two moments of Z linearized about $\mathbf{x}^*$ rather that μ_X are given by (cf. (.A178) and (A.179))

$$\mu_Z \approx G(\mathbf{x}^*) \tag{A.178}$$

and

$$\sigma_Z^2 \approx \sum \left(\frac{\partial G}{\partial X_i} \right)^2 \Bigg|_{\mathbf{x}^*} \sigma_{X_i}^2 \tag{A.179}$$

Now, if $G(\mathbf{X})$ is defined as $G(\mathbf{X}) = X_1 X_2 - X_3$ with the random variables X_i being independent and with $\sigma_{X_1} = \sigma_{X_2} = \sigma_{X_3} = \sigma_X$, then $\partial G/\partial X_1 = X_2$, $\partial G/\partial X_2 = X_1$ and $\partial G/\partial X_3 = -1$.
Then (A.178) and (A.179) evaluated at the means μ_{X_i} become

$$\mu_Z \Bigg|_{\mu_{X_i}} \approx \mu_{X1} \mu_{X2} - \mu_{X3}$$

$$\sigma_Z^2 \approx \mu_{X2}^2 \sigma_{X1}^2 + \mu_{X1}^2 \sigma_{X2}^2 + \sigma_{X3}^2 = \sigma_X^2 (\mu_{X1}^2 + \mu_{X2}^2 + 1)$$

However, if, for example, the expansion point $\mathbf{x}^*$ was taken as $x_1^* = \mu_{X_1}/2$, $x_2^* = \mu_{X_2}$, $x_3^* = \mu_{X_3}$, then the corresponding terms become

$$= \frac{\mu_{X1}}{2} \mu_{X2} - \mu_{X3}$$

and

$$\sigma_Z^2 = \left(\frac{\mu_{X2}}{2}\right)^2 \sigma_{X1}^2 + \mu_{X1}^2 \sigma_{X2}^2 + \sigma_{X3}^2 = \sigma_X^2 \left(\frac{\mu_{X1}^2}{4} + \mu_{X2}^2 + 1\right)$$

It follows readily using (4.1) that:

$$\beta\bigg|_{\mu_X} = \frac{\mu_{X1}\mu_{X2} - \mu_{X3}}{\sigma_X(\mu_{X1}^2 + \mu_{X2}^2 + 1)^{1/2}}$$

and

$$\beta\bigg|_{x^*} = \frac{\frac{1}{2}\mu_{X1}\mu_{X2} - \mu_{X3}}{\sigma_X(\frac{1}{4}\mu_{X1}^2 + \mu_{X2}^2 + 1)^{1/2}}$$

These are clearly not equivalent, thus demonstrating the dependence of β on the selection of the expansion point x^*.

4.3   FIRST-ORDER SECOND-MOMENT THEORY

4.3.1   The Hasofer–Lind transformation

The desirability that a safety measure is invariant was noted in Chapter 1. How this can be achieved for the reliability index β will be discussed below. The first step is to transform all variables to their standardized form $N(0, 1)$ (normal distribution with zero mean and unit variance) using the well-known transformation

$$Y_i = \frac{X_i - \mu_{X_i}}{\sigma_{X_i}} \tag{4.3}$$

where Y_i has $\mu_{Y_i} = 0$, $\sigma_{Y_i} = 1$. As a result of applying this transformation to each basic variable X_i, the unit of measurement in any direction in the Y_i hyperspace will be σ_{Y_i} ($= 1$) in any direction. In the y space, the joint probability density function $f_Y(y)$ is thus the standardized multivariate normal (see section 3.3); thus many well-known properties of the multivariate normal distribution can be immediately applied [Hasofer and Lind, 1974]. The limit state function must, of course, also be transformed and is given by $g(Y) = 0$.

The transformation (4.3) can be performed without complication if the random variables X_i are all uncorrelated (i.e. linearly independent) random variables. If this is not the case, an intermediate step is required to find a random variable set X' which is uncorrelated, and this new set can then be transformed according to (4.3). The procedure for finding the uncorrelated set X' (including means and variances) from the correlated set X is essentially that of finding the eigen-values and eigen-vectors. Details are given in

Appendix B. As an approximation, weak correlation (e.g. <0.2) can usually be ignored and the variables treated as independent, whereas strong correlations (e.g. >0.8) can usually be treated as fully dependent, with one of two correlated variables replaced by the other.

4.3.2 Linear limit state function

When $G(\mathbf{X}) = X_1 - X_2$, as in the simple $R - S$ problem, the limit state function is linear, and shown as $G_L(\mathbf{X})$ in Fig. 4.1. The transformation (4.3) changes the joint probability density function $f_\mathbf{X}(\mathbf{x})$ shown in Fig. 4.1 to $f_\mathbf{Y}(\mathbf{y})$ shown in Fig. 4.2 in the transformed (y_1, y_2) space. The joint probability

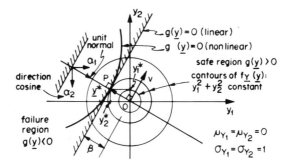

Fig. 4.2 — Limit state surface in the space of the standardized normal variables.

function $f_\mathbf{Y}(\mathbf{y})$ **is now a bivariate normal distribution** $\Phi_2(\mathbf{y})$, symmetrical about the origin. As noted in section 1.4.2, the probability of failure is given by the integral of $\Phi_2(\mathbf{y})$ over the transformed failure region $g(\mathbf{Y}) < 0$. This can be obtained from the results for the standardized bivariate normal given in section 3.2.1 or in Appendix A. It is given more directly, however, by integrating in the direction v ($-\infty < v < +\infty$) to obtain the marginal distribution, shown in Fig. 4.3. By well-known properties of the bivariate normal distribution the marginal distribution is also normal, and hence the shaded area in Fig. 4.3 represents the failure probability $p_f = \Phi(-\beta)$, where β is as shown (note that $\sigma = 1$ in the β direction since the normalized $\mathbf{y}$ space is being used). The direct correspondence between Fig. 4.3 and Fig. 1.11 should be noted.

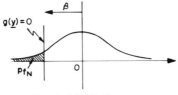

Marginal distribution

Fig. 4.3 — Marginal distribution in the space of the standardized normal variables.

The distance β as shown in Fig. 4.2 is perpendicular to the v axis and hence is perpendicular to $g(\mathbf{y}) = 0$. It clearly corresponds to the shortest distance from the origin in the $\mathbf{y}$ space to the limit state surface $g(\mathbf{y}) = 0$.

The above concepts carry over directly to an n-dimensional space with a (hyper)plane limit state. The shortest distance is then

$$\beta = \min\left(\sum_{i=1}^{n} y_i^2\right)^{1/2} = \min(\mathbf{y}^{\mathrm{T}} \cdot \mathbf{y})^{1/2} \tag{4.4}$$

if the y_i represent the coordinates of any point on the limit state surface.

The particular point for which (4.4) is satisfied, i.e. the point on the limit state surface perpendicular to β, in n-dimensional space, is variously called the "checking" or "design" point $\mathbf{y}^*$. It is evident that $\mathbf{y}^*$ is the projection of the origin on the limit state function.

A direct relationship between the checking point $\mathbf{y}^*$ and β can be established as follows. From the geometry of surfaces [e.g. Sokolnikoff and Redheffer, 1958, p. 309] the outward normal vector to a hyperplane given by $g(\mathbf{y}) = 0$, has components given by

$$c_i = \lambda \frac{\partial g}{\partial y_i} \tag{4.5a}$$

where λ is an arbitrary constant. The total length of the outward normal is

$$l = \left(\sum_i c_i^2\right)^{1/2} \tag{4.5b}$$

and the direction cosines α_i of the unit outward normal are then

$$\alpha_i = \frac{c_i}{l} \tag{4.5c}$$

With α_i known, it follows that the coordinates of the checking point are

$$y_i = -\alpha_i \beta \tag{4.6}$$

where the negative sign is because the α_i are components of the outward normal as defined in conventional mathematical notation (i.e. positive with increasing $g(\ )$). Figure 4.2 shows the geometry for the two-dimensional case (y_1, y_2).

For a linear limit state function (i.e. a hyperplane) the direction cosines α_i do not change with position along the limit state function, so that it is easy to find a set of co-ordinates $\mathbf{y}^*$ satisfying both equation (4.4) and equation (4.6). This will be demonstrated in Example 4.3.

The equation for the hyperplane $g(\mathbf{y}) = 0$ can be written as

$$g(\mathbf{y}) = \beta + \sum_{i=1}^{n} \alpha_i y_i = 0 \tag{4.7}$$

The validity of (4.7) can be verified by applying (4.5); it can also be deduced directly from Fig 4.2 for the two-dimensional case.

The function corresponding to (4.7) in **X** space is obtained by applying (4.3)

$$G(\mathbf{x}) = \beta - \sum_{i=1}^{n} \frac{\alpha_i}{\sigma_{X_i}} \mu_{X_i} + \sum_{i=1}^{n} \frac{\alpha_i}{\sigma_{X_i}} x_i \tag{4.8a}$$

$$= b_0 + \sum_{i=1}^{n} b_i x_i \tag{4.8b}$$

which is again a linear function.

The direction cosines α_i (4.5c) represent the sensitivity of the standardized limit state function $g(\mathbf{y})$ at $\mathbf{y}^*$ to changes in $\mathbf{y}$. The corresponding expressions in the original **x** space, for **X** independent, are obtained by using (4.3) in (4.5a):

$$c_i = \lambda \sigma_i \frac{\partial G}{\partial x_i} \tag{4.9}$$

with (4.5b) and (4.5c) as before. For those X_i which are not independent, the direction cosines α_i have no direct physical meaning. This is because of the transformation to independent standardized **y** space, even if the **X** are (partly) dependent.

Finally, it follows directly from (4.7) that

$$\beta = - \sum_{i=1}^{n} y_i^* \alpha_i = - \mathbf{y}^{*T} \cdot \boldsymbol{\alpha} \tag{4.10}$$

which allows β to be determined from the checking point co-ordinates, y_i^* should this be required.

Example 4.2
For a linear limit state function, $Z(\mathbf{X}) = G(\mathbf{X})$, such as given by (4.8b), β may be estimated in **x** space using (4.1) and (4.8a):

$$\mu_G = b_0 + \sum_{i=1}^{n} b_i \mu_{X_i} = \beta$$

and

$$\sigma_G = \left(\sum_{i=1}^{n} b_i^2 \sigma_{X_i}^2 \right)^{1/2} = \left(\sum_{i=1}^{n} \left(\frac{\alpha_i}{\sigma_{X_i}} \right)^2 \sigma_{X_i}^2 \right)^{1/2} = 1$$

so that $\beta_x = \mu_G/\sigma_G = \beta$. Using (4.7) it follows easily that $\mu_g = \mu_G = \beta$ and $\sigma_g = \sigma_G = 1$. Hence β does not depend on the space used. This is true if (and only if) $G(\mathbf{X})$ is a linear function.

Example 4.3
Let the limit state function $g(\mathbf{y})$ in two-dimensional normalized space (y_1, y_2) be given by $g(\mathbf{y}) = y_1 - 2y_2 + 10 = 0$. This meets the requirement that $g(\mathbf{y}) > 0$ for the safe state and $g(\mathbf{y}) < 0$ for the failure state.

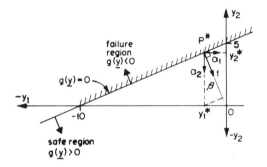

Fig. 4.4 — Linear limit state function.

In Fig. 4.4 the "checking point" is evidently P^*, the projection of the origin onto $g(\mathbf{y}) = 0$, with perpendicular OP^*. From (4.5)

$$c_1 = \lambda \frac{\partial g}{\partial y_1} = \lambda(1) = \lambda$$

$$c_2 = \lambda(-2) = -2\lambda$$

$$l = [\lambda^2 + (-2\lambda)^2]^{1/2} = \sqrt{5}\lambda$$

so that $\alpha_1 = 1/\sqrt{5}$ and $\alpha_2 = -2/\sqrt{5}$. Evidently, this represents the normal anywhere along the linear limit state function (see Fig. 4.4). Note again that the direction of the normal is governed by the definition of $g(\mathbf{y}) = 0$ and the mathematics of planar surfaces.

The checking point is now given by (4.6) as $\mathbf{y}^* = (-\beta/\sqrt{5}, 2\beta/\sqrt{5})$. Also $\mathbf{y}^*$ must satisfy $g(\mathbf{y}^*) = 0$, so that

$$-\frac{\beta}{\sqrt{5}} - 2\left(\frac{2\beta}{\sqrt{5}}\right) + 10 = 0$$

from which $\beta = 2\sqrt{5}$, as is easily verified on Fig. 4.4.

For this special linear problem, of course, it would have been just as correct simply to put $\beta = \mu_g/\sigma_g = [(0) - (2)(0) + 10]/[1^2 + (2)^2(1)^2]^{\frac{1}{2}} = 10/\sqrt{5} = 2\sqrt{5}$.

4.3.3 Non-linear limit state function—general case

As was noted in section 4.2, when the limit state equation is non-linear, the first two moments of $G(\mathbf{X})$ in the x space, and therefore the first two moments of $g(\mathbf{Y})$ in the y space, can no longer be obtained exactly. This is because non-linear combination of the implicit (standardized) normal distributions does not lead to a normal distribution for $G()$ or $g()$ (see Appendix A). The approach suggested in section 4.2 was to linearize $G(\mathbf{X})$ using (A.178) and (A.179) although, as was shown in Example 4.1, β depends then on the choice of the linearization or expansion point used in the Taylor series expansion.

Because of rotational symmetry and the Hasofer–Lind transformation, the "checking point" in the y space (e.g. point $\mathbf{y}^*$ in Fig 4.2) represents the point of greatest probability density or maximum likelihood. It therefore makes the most significant contribution to the nominal failure probability p_{fN} ($= \Phi(-\beta)$).

An intuitively appealing linearization point is therefore the checking point $\mathbf{y}^*$, and indeed this is shown in Fig 4.2. It follows that the concepts and method used for the case of a linear limit state equation $g(\mathbf{y})$ can also be used when $g(\mathbf{Y})$ is non-linear; again the shortest distance β from the origin to $\mathbf{y}^*$ must be found, subject to $g(\mathbf{y}^*) = 0$. How well a linear limit state function $g(\mathbf{y}) = 0$ approximates a non-linear function $g(\mathbf{Y})$ in terms of the nominal probability p_{fN} of failure depends on the shape of $g(\mathbf{y}) = 0$; if it is concave towards the origin, p_{fN} is underestimated by the hyperplane approximation. Similarly, a convex function implies overestimation, as in Fig. 4.2.

Since the point $\mathbf{y}^*$ is not known *a priori*, the problem of finding the shortest distance β in y space, subject to $g(\mathbf{y}) = 0$ is strictly a minimization problem [Flint *et al.*, 1981; Shinozuka, 1983]. There are therefore several ways in which a solution may be found; the classical calculus of variations method will be considered in this section. Its application reveals some useful properties.

The minimization problem (4.4) is subject to the constraint $g(\mathbf{y}) = 0$. Introducing a Lagrangian multiplier λ, the problem becomes [Shinozuka, 1983]

$$\min(\Delta) = (\mathbf{y}^{\mathrm{T}} \cdot \mathbf{y})^{1/2} + \lambda g(\mathbf{y}) \tag{4.11}$$

For a stationary point, $\partial\Delta/\partial y_i = 0$ for all i, and $\partial\Delta/\partial\lambda = 0$:

$$\frac{\partial\Delta}{\partial y_1} = y_1(\mathbf{y}^T \cdot \mathbf{y})^{-1/2} + \lambda\frac{\partial g}{\partial y_1} = 0$$

$$\frac{\partial\Delta}{\partial y_2} = y_2(\mathbf{y}^T \cdot \mathbf{y})^{-1/2} + \lambda\frac{\partial g}{\partial y_2} = 0 \qquad (4.12)$$

$$\vdots$$

and $$\frac{\partial\Delta}{\partial\lambda} = g(\mathbf{y}) = 0$$

which may be written compactly as

$$0 = \delta^{-1}\mathbf{y} + \lambda\mathbf{g}_Y \qquad (4.13a)$$

$$0 = g(\mathbf{y}) \qquad (4.13b)$$

where $\mathbf{g}_Y = (\partial g/\partial y_1, \partial g/\partial y_2, \ldots)$ and the distance δ (cf. (4.4)) is given by $\delta = (\mathbf{y}^T \cdot \mathbf{y})^{1/2}$. Expression (4.13b) is satisfied by definition, and (4.13a) immediately produces the co-ordinates for the stationary point $\mathbf{y}^s$ as [Horne & Price, 1977]

$$\mathbf{y}^s = -\lambda\mathbf{g}_Y\delta \qquad (4.14)$$

Whether the point represents a minimum, maximum or a "saddle point" depends on the nature of $g(\mathbf{y})$. No general statement can be made, although for a given function $g(\mathbf{y})$ standard tests can be applied [Ditlevsen, 1981a, 1981b]. If $g(\mathbf{y})$ is linear, or regular and convex towards the origin, the stationary point is clearly a minimum point [Lind, 1979]. Since many limit state functions depart only slightly from linearity, it will be assumed in what follows that the vector $\mathbf{y}^s$ does indeed locate the minimum point.

Let the minimum distance δ corresponding to $\mathbf{y}^s$, be δ^s. Substituting (4.14) into equation (4.4) with $\mathbf{y}^s$ replacing $\mathbf{y}$ and δ^s replacing β, it is easily shown that $\lambda = \pm(\mathbf{g}_Y^T \cdot \mathbf{g}_Y)^{-1/2}$ and hence, from (4.14), that

$$\delta^s = \frac{-(\mathbf{y}^s)^T \cdot \mathbf{g}_Y}{(\mathbf{g}_Y^T \cdot \mathbf{g}_Y)^{1/2}} = \frac{-\displaystyle\sum_{i=1}^{n} y_i^s\,(\partial g/\partial y_i)}{\left[\displaystyle\sum_{i=1}^{n} (\partial g/\partial y_i)^2\right]^{1/2}} \qquad (4.15)$$

It will now be shown that δ^s is equal to β and that β is therefore the minimum distance from the origin to the limit state $g(\mathbf{y}) = 0$, provided that β is measured to the checking point $\mathbf{y}^s = \mathbf{y}^*$ and that $g(\mathbf{y})$ is linearized at $\mathbf{y}^*$ [Lind, 1979; Shinozuka, 1983].

Let the limit state function $g(\mathbf{y})$ be linearized at $\mathbf{y}^*$ by means of a Taylor series expansion (i.e. in first-order terms only). This approximation provides a tangent (hyper)plane $g_L(\mathbf{y})$ at $\mathbf{y}^*$ to the function $g(\mathbf{y})$:

$$g_L(\mathbf{y}) \approx g(\mathbf{y}^*) + \sum_{i=1}^{n} (y_i - y_i^*) \frac{\partial g}{\partial y_i} = 0 \qquad (4.16)$$

but, since $\mathbf{y}^*$ is on the limit state, $g(\mathbf{y}^*) = 0$. Further, using first- and second-moment rules for added functions ((A.160) and (A.161)), and remembering that $\mu_{Y_i} = 0$, $\sigma_{Y_i} = 1$ by definition,

$$\mu_{g_L}(\mathbf{y}) = -\sum_{i=1}^{n} y_i^* \frac{\partial g}{\partial y_i} = -\mathbf{y}^{*T} \cdot \mathbf{g}_Y \qquad (4.17)$$

and

$$\sigma_{g_L}^2(\mathbf{y}) = \sum_{i=1}^{n} \left(\frac{\partial g}{\partial y_i}\right)^2 = \mathbf{g}_Y^T \cdot \mathbf{g}_y \qquad (4.18)$$

from which, using the usual definition of β (4.1) given by

$$\beta = \frac{\mu_{g_L}}{\sigma_{g_L}} = \frac{-\displaystyle\sum_{i=1}^{n} y_i^* \, (\partial g/\partial y_i)}{\left[\displaystyle\sum_{i=1}^{n} (\partial g/\partial y_i)^2\right]^{1/2}} = \frac{-\mathbf{y}^{*T} \cdot \mathbf{g}_Y}{(\mathbf{g}_Y^T \cdot \mathbf{g}_Y)^{1/2}} \qquad (4.19)$$

which is identical with (4.10), $\beta = -\mathbf{y}^{*T} \cdot \boldsymbol{\alpha}$, as might be expected. It is immediately evident that expression (4.19) is equivalent to δ^s, given by expression (4.15), if $\mathbf{y}^* = \mathbf{y}^s$. Hence $\mathbf{y}^*$ is the point which minimizes the distance from the origin to the non-linear limit state function $g(\mathbf{y})$, and is also its expansion (linearization) point.

It follows immediately that once the point $\mathbf{y}^*$ has been located, the non-linear limit state function may be replaced by its linearized equivalent and the method of section 4.3.2 applied.

Finally, since β is the minimum distance from the origin of the rotationally symmetric standardized normal probability density function $f_{\mathbf{Y}}(\ )$ in y space, it is easily verified that $\mathbf{y}^*$ represents the point of maximum likelihood. This might be compared with the choice $\mathbf{x}^*$ in section 3.3.6.2.

Example 4.4

For the two-dimensional problem of Fig 4.5, the limit state equation is

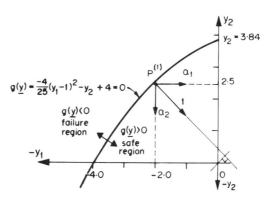

Fig. 4.5 — Non-linear limit state function.

$$g(\mathbf{y}) = -\frac{4}{25}(y_1 - 1)^2 - y_2 + 4 = 0$$

The distance to be minimized is $\delta = (y_1^2 + y_2^2)^{1/2}$ subject to $g(\mathbf{y}) = 0$. The modified function Δ is then (cf. (4.11)

$$\min(\Delta) = (y_1^2 + y_2^2)^{1/2} + \lambda\left[-\frac{4}{25}(y_1 - 1)^2 - y_2 + 4 \right]$$

for which

$$\frac{\partial\Delta}{\partial y_1} = y_1(y_1^2 + y_2^2)^{-1/2} - \lambda\frac{8}{25}(y_1 - 1) = 0 \tag{4.20a}$$

$$\frac{\partial\Delta}{\partial y_2} = y_2(y_1^2 + y_2^2)^{-1/2} - \lambda = 0 \tag{4.20b}$$

$$\frac{\partial\Delta}{\partial\lambda} = -\frac{4}{25}(y_1 - 1)^2 - y_2 + 4 = 0 \tag{4.20c}$$

Eliminating λ from (4.20a) and (4.2b) gives

$$y_2 = \frac{25y_1}{8(y_1 - 1)} \tag{4.20d}$$

which together with (4.20c) leaves, upon eliminating y_2,

$$-\frac{4}{25}(y_1 - 1)^2 + 4 = \frac{25y_1}{8(y_1 - 1)} \tag{4.20e}$$

This equation is cubic in y_1. A trial and error solution shows that $y_1 = 2.36$ satisfies (4.20e) and that $y_2 = +2.19$ from (4.20d). β is then estimated as $(y_1^2 + y_2^2)^{1/2} = 3.22$, and $\mathbf{y}^*$ is given by $(-2.36, 2.19)^T$. It is readily verified that the stationary point $\mathbf{y}^*$ does indeed give a minimum value of β. (Note that $P^{(1)}$ in Fig. 4.5 refers to Example 4.5.)

For problems with rather more basic variables it is apparent that the solution of the system of simultaneous equations requires numerical evaluation and that a more formal verification that the calculated β is a minimum is required.

4.3.4   Non-linear limit state function - numerical solution
For some situations, principally those with a large number of basic variables, and for complex limit state equations, a numerical approach may be preferred. When the constraint is linear, the problem of finding β, given by (4.4), is essentially a quadratic programming problem, for which efficient computer algorithms exist.

If the constraint $g(\mathbf{y})$ is not a linear function, but a vector $\mathbf{y}^{(1)}$ can be found which satisfied $g(\mathbf{y}^{(1)}) = 0$, then there are relatively few efficient algorithms available. In principle Lagrangian multipliers could be used to convert the problem into an unconstrained optimization problem as was done in the previous section and then solved by one of the many efficient algorithms available for this purpose. However, it is probably better, in general, to employ a method such as the modified gradient projection method.

If a vector $\mathbf{y}^{(1)}$ to satisfy $g(\mathbf{y}^{(1)}) = 0$ cannot be easily found, the problem (4.4) is one of non-linear minimization under non-linear inequality constraints. Again, a range of algorithms is available for solution [Beveridge and Schechter, 1970; Schittkowski, 1980]. At the time of writing, no comprehensive comparison between the previous method, numerical solutions and the iteration technique of the next section appears to have been reported.

4.3.5   Non-linear limit state function - iterative solution scheme
In the iterative solution scheme, a trial checking point $\mathbf{y}^{(1)}$ is chosen and equation (4.4) and the limit state expression checked. If $\mathbf{y}^{(1)}$ is poorly chosen, the condition of perpendicularity between the tangent (hyper)plane at $\mathbf{y}^{(1)}$ and the β direction will not be satisfied. How the new trial checking point $\mathbf{y}^{(1)}$ should be chosen is the topic of this section.

Let $\mathbf{y}^{(m)}$ be the mth approximation to the vector representing the local perpendicular to $g(\mathbf{y}) = 0$ from the origin. A better approximation $\mathbf{y}^{(m+1)}$ is sought. The relationship between $\mathbf{y}^{(m+1)}$ and $\mathbf{y}^{(m)}$ can be obtained from a first-order Taylor series expansion of $g(\mathbf{y}^{(m+1)}) = 0$ about $\mathbf{y}^m$ (i.e. a linear approximation) [Hasofer and Lind, 1974]. Using index notation,

$$g_L(y_1^{(m+1)}, \ldots, y_n^{(m+1)}) \approx g(y_1^{(m)}, \ldots, y_m^{(m)})$$

$$+ \sum_{i=1}^{n} (y_i^{(m+1)} - y_i^{(m)}) \frac{\partial g(y_1^{(m)}, \ldots, y_n^{(m)})}{\partial y_i} = 0 \qquad (4.21)$$

or, in matrix (cf. expression (4.16)),

$$g_L(\mathbf{y}^{(m+1)}) \approx g(\mathbf{y}^{(m)}) + (\mathbf{y}^{(m+1)} - \mathbf{y}^{(m)})^{\mathrm{T}} \cdot \mathbf{g}_Y = 0 \qquad (4.22)$$

This expression represents a hyperplane $g_L(\ ) = 0$ approximating the hypersurface $g(\ ) = 0$ in y space, for which, at point $m + 1$, the linearized limit state function must be satisfied, i.e. $g_L(\mathbf{y}^{(m+1)} = 0$. However, for the earlier trial point $\mathbf{y}^{(m)}$, the direction cosines are $\alpha^{(m)}$ which together with the trial value of $\beta^{(m)}$ are given by (4.6):

$$\mathbf{y}^{(m)} = -\alpha^{(m)}\beta^{(m)} \qquad (4.23)$$

where from (4.5)

$$\alpha^{(m)} = \frac{\mathbf{g}_Y^{(m)}}{l} \qquad (4.5c)$$

and

$$l = (\mathbf{g}_Y^{(m)\mathrm{T}} \cdot \mathbf{g}_Y^{(m)})^{1/2} \qquad (4.5b)$$

Substituting (4.23) for $\mathbf{y}^{(m)}$ in (4.22), using (4.5) and rearranging produces the following recurrence relationship:

$$\mathbf{y}^{(m+1)} = -\alpha^{(m)} \left[\beta^{(m)} + \frac{g(\mathbf{y}^{(m)})}{l} \right] \qquad (4.24)$$

In practice the iteration proceeds by assuming a starting point, to give $\mathbf{y}^{(1)}$, evaluating the gradients $g_{Y_i} = \partial g(\mathbf{y}^{(m)})/\partial y_i$ and also evaluating $\beta^{(m)} = \left[\sum_{i=1}^{n} (y_i^{(m)})^2 \right]^{1/2}$ and then substituting into (4.24) to obtain a new trial checking point and β estimate. This is continued until convergence on $y_i^{(m)}$ or β is reached.

Comparison of (4.24) and (4.6) shows the essential similarity between the procedures and that the term in the square brackets of (4.24) is essentially a correction term to allow for the fact that $g(\mathbf{y})$ is not zero.

In parallel to the discussion in section 4.3.3, there is no guarantee that the repeated use of (4.24) does indeed converge to a minimum value of β; again either an appeal to intuition or a more formal check is required.

The above may be formalized to the following algorithm.

(a) Standardize basic random variables $\mathbf{X}$ to the independent standardized normal variables $\mathbf{Y}$, using (4.3) and Appendix B if necessary.
(b) Transform $G(\mathbf{x}) = 0$ to $g(\mathbf{y}) = 0$.
(c) Select initial checking point $\mathbf{x}^{(1)}$, $\mathbf{y}^{(1)}$.
(d) Compute $\beta^{(1)} = (\mathbf{y}^{(1)T} \cdot \mathbf{y}^{(1)})^{1/2}$; let $m = 1$.
(e) Compute direction cosines $\boldsymbol{\alpha}^{(m)}$ using (4.5).
(f) Compute $g(\mathbf{y}^{(m)})$.
(g) Compute $\mathbf{y}^{(m+1)}$ using (4.24).
(h) Compute $\beta^{(m+1)} = (\mathbf{y}^{(m+1)T} \cdot \mathbf{y}^{(m+1)})^{1/2}$.
(i) Check whether $\mathbf{y}^{(m+1)}$ and/or $\beta^{(m+1)}$ have stabilized; if not go to (e) and increase m by unity.

The above algorithm has been programmed, sometimes using slightly different logic, first by Fiessler *et al.* (1976) [see also Fiessler 1979] and later by others [e.g. Ellingwood *et al.*, 1980].

It is possible to transform the recurrence relationship (4.24) into the space of the original variables X_i; this has the advantage that transformation to the $\mathbf{y}$ space is not required. Substituting $X_i = y_i \sigma_{X_i} - \mu_{X_i}$ into equation (4.24) yields [Parkinson, 1980]

$$\mathbf{X}^{(m+1)} - \boldsymbol{\mu}_X = -\mathbf{CG}_X^{(m)} \frac{(\mathbf{X}^{(m)} - \boldsymbol{\mu}_X)^T \cdot \mathbf{G}_X^{(m)}}{\mathbf{G}_X^{(m)T} \mathbf{CG}_X^{(m)}} \qquad (4.25)$$

where

$$\mathbf{G}_X^{(m)} = \left(\frac{\partial G}{\partial x_1}, \frac{\partial G}{\partial x_2}, \dots, \frac{\partial G}{\partial x_n} \right)^T \Bigg|_{\mathbf{x} = \boldsymbol{\mu}_X^{(m)}}$$

$\mathbf{X}$ is the vector of random variables in original x space and $\mathbf{C}$ is the covariance matrix for $\mathbf{X}$. (If the X_i are independent, $c_{ii} = \sigma_i^2$, $c_{ij} = 0$ for $i \neq j$; see section A.11.1.)

Example 4.5
Example 4.4 will now be reworked, this time using the above algorithm. The geometry of the problem is shown in Fig 4.5.

Let the trial solution for the checking point be $P^{(1)}$ given by $y^{(1)} = (-2.0, 2.5)^T$ which does not quite satisfy $g(y) = 0$ (this would require $y_2^{(1)}$ to be 2.56, as is easily verified).

$$\frac{\partial g}{\partial y_1} = \frac{-4}{25} 2(y_1 - 1) = \frac{24}{25} = +0.96$$

$$\frac{\partial g}{\partial y_2} = -1 \quad \text{and} \quad \alpha_i = \frac{\partial g}{\partial y_i} \bigg/ l$$

where

$$l = [(0.96)^2 + (-1)^2]^{1/2} = 1.386$$

$$\beta = (y_1^2 + y_2^2)^{1/2} = (2^2 + 2.5^2)^{1/2} = 3.20$$

$$g(y) = \frac{-4}{25} (-2 - 1)^2 - (2.5) + 4 = +0.06$$

so that

$$\begin{bmatrix} y_1 \\ y_2 \end{bmatrix}^{(2)} = \frac{-\begin{bmatrix} +0.96 \\ -1 \end{bmatrix}}{1.386} \left(3.20 + \frac{+0.06}{1.386}\right) = \begin{bmatrix} -2.22 \\ 2.309 \end{bmatrix}$$

which, with two iterations, will converge very closely to the results given in Example 4.4.

4.3.6 Geometric interpretation of iterative solution scheme

It is useful to study the iteration scheme a little more closely so that its limitations can be revealed. Firstly, expression (4.24) can be written also as

$$y^{(m+1)} = -\alpha^{(m)} \beta^{(m+1)} \tag{4.26}$$

where the $\beta^{(m+1)}$ denotes the term [ ] in (4.24) and consists of $\beta^{(m)}$ plus a correction term.

Consider now the two-dimensional space (y_1, y_2) given in Fig. 4.6 [after Fiessler, 1979], with the limit state surface as shown, together with a contour in the safe domain $g(y) = c$. Assume the initial trial $y^{(1)}$ values are given by point $P^{(1)}$, for which $\beta = \beta^{(1)}$. The linear approximation to $g(y)$ at $P^{(1)}$ is given by line AA. The direction cosines α_i evaluated at $P^{(1)}$ give the direction of the vector $P^{(1)}T$ normal to AA; the gradient $\partial g/\partial y$ gives the slope in the direction of the vector, as shown by section BB. The true slope of the vector $P^{(1)}T$ is given by

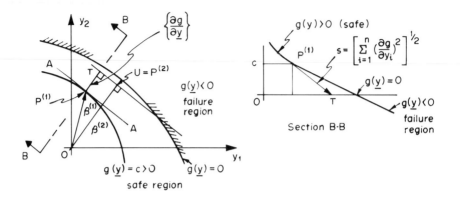

$$s = \left[\sum_{i=1}^{n} \left(\frac{\partial g}{\partial y_i} \right)^2 \right]^{1/2}$$

and hence the horizontal projection in section BB is given by c/s, which is the second (i.e. the correction) term in the [ ] in equation (4.24). This term then gives the vector length $\beta^{(1)} + P^{(1)}T$ and multiplication by $\alpha_i^{(1)}$ fixes the new trial $y^{(2)}$ as OU parallel to $P^{(1)}T$, in Fig. 4.6.

The iteration procedure can fail in certain circumstances. One case is illustrated in Fig. 4.7. For a highly non-linear limit state function it is

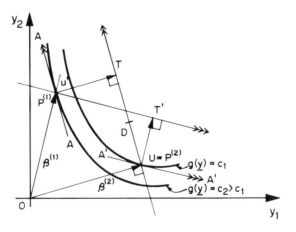

Fig. 4.7 — Breakdown of iteration through oscillation.

possible to alternate between successive approximation points i and $i+1$ [Fiessler, 1979]. Thus, starting at $P^{(1)}$, the local gradient perpendicular to the tangent AA at $g(\mathbf{y}^{(1)}) = c_2$ leads to point T (given by $g(\mathbf{y}) = 0$). The new checking point U is then defined by the parallel TU to AA and the perpendicular $\beta^{(2)}$ from O. This is point $P^{(2)} \equiv U$. However, for the situation shown, starting at $P^{(2)}$ for the next iteration, the new point $P^{(3)}$ is found to be at $P^{(1)}$. Obviously a breakdown situation exists because of the non-linear nature of $g(\mathbf{y}) = 0$. The difficulty is easily overcome. Rather than selecting $P^{(2)}$ on TU at U, some lesser correction, such as point D, might be chosen for $P^{(2)}$.

A second breakdown case is when the trial checking point lies close to a stationary point which is not a minimum. As noted in section 4.3.5, the iteration procedure can only search for local stationary points and cannot distinguish between maxima, minima or saddle points [Ditlevsen and Madsen, 1980]. The problem can only be overcome by selecting different starting points and commonsense appraisal of results.

4.3.7 Interpretation of first-order second-moment theory

The theory discussed so far is a "first-order" approximation to the correct reliability problem when the limit state function is non-linear. As was seen, the probability content of the failure region is estimated at the "checking" point. For this reason the above theory is also sometimes referred to as a "single-check-point" method.

Unfortunately, an ambiguity of interpretation of the probability expressed by the safety index β arises when the limit state function is non-linear (see Fig. 4.8). For the linear limit state bb, containing P_1 as checking

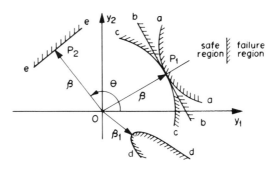

Fig. 4.8 — Inconsistency between β and p_{fN} for different forms of limit state functions.

point, the failure probability for normal variables is given exactly by $p_f = \Phi(-\beta)$. However, the point P_1 is also the checking point for the non-linear limit state functions aa and cc. In terms of first-order theory, each of these limit states has an identical value of β, and hence an identical nominal failure probability $p_{fN} = \Phi(-\beta)$; yet it is quite clear from Fig. 4.8 that the actual probability contents of the respective failure regions are not identical. Similarly, the limit state dd represents what is probably a lower failure probability still; yet its safety margin β_1 is less than β. Evidently β as defined so far lacks a sense of "comparativeness" or an "ordering property" with respect to the implied probability content for non-linear limit states [Ditlevsen, 1979a].

A further point of interest is that no limitation has been placed on the direction of β in **y** space so that, for some other checking point P_2, the probability content for the linear limit state ee should be identical with that implied by bb when both have the same distance β from the origin.

A measure of comparativeness can be introduced by defining a formal probability density function $f_Y(y)$ in the reduced variable space. The probability content associated with each limit state can then formally be calculated and compared. It is not difficult to see that such a formal density function must give greater reliability with greater β, that it must allow for the shape of the limit state function and that it must be rotationally symmetric. It appears that the only function that can satisfy all requirements is the n-dimensional standardized normal density function with independent variables:

$$\phi_n(\mathbf{y}) = \prod_{i=1}^{n} \left[\frac{1}{(2\pi)^{\frac{1}{2}}} \exp(-\tfrac{1}{2}y_i^2) \right] \tag{4.27}$$

The safety index β associated with a given limit state (hyper)surface with safe domain D_k, say, is then obtained by integrating $\phi_n(\mathbf{y})$ over the domain D_k, to give the β value for the safe state as:

$$\Phi[\beta(k)] = \int_{D_k} \phi_n(\mathbf{y}) \, d\mathbf{y} \tag{4.28}$$

where Φ is the standardized normal distribution function. It follows that

$$\beta(k) = \Phi^{-1} \int_{D_k} \phi_n(\mathbf{y}) \, d\mathbf{y} \tag{4.29}$$

is a more general definition of the safety index, now a function of the shape of the limit state function for safe domain D_k. As known from Chapter 3, the calculation of the integral (4.29) is generally not straightforward, owing to the nature of the shape of the domain D_k of integration. Some specific cases are considered in Section 4.3.8 below.

4.3.8 Probability calculation for general limit state functions
4.3.8.1 Bounds

With the function ϕ_n in expression (4.29) taken as the n-dimensional standardized normal distribution, estimates of p_f can be obtained as a function of $\beta(k)$ for a given non-linear limit state function k. This allows the development of bounds on the reliability implied by a given value of β.

If the limit state function $g(\mathbf{y})$ is linear, it follows directly from section 4.3.2 that $p_f = \Phi(-\beta)$. This result forms an upper bound on p_f if $g(\mathbf{y})$ is actually convex to the origin and a lower bound if it is actually concave. More generally, a lower bound on p_f is given by a (hyper)spherical limit state function with radius β, centred at the origin of the standardized normal variable space $\mathbf{y}$:

$$g(\mathbf{y}) = \beta^2 - \sum_{i=1}^{n} y_i^2 = 0 \qquad (4.30)$$

Since Y_i are normally distributed and independent, $\sum_{i}^{n} y_i^2$ has a chi-squared distribution χ_n with n degrees of freedom [Benjamin and Cornell, 1970]. The probability of failure is then $p_f = 1 - \chi_n(\beta^2)$ which is shown in Fig. 4.9 as a function of n [Lind, 1977]. When $n = 1$ the probability content is twice that given by a single linear limit state surface, as is readily verified.

A number of more complex approximations for general limit state functions have been proposed; these include the use of piecewise spherical sectors [Veneziano, 1974], the use of quadratic expressions centred about the checking point [Fiessler, et al. 1979; Horne and Price, 1977] and the use of (linear) tangents at various predefined locations on $g(\mathbf{y}) = 0$ (the so-called "polyhedral approximation") [Berthellemy and Rackwitz, 1979]. Piecewise linear limit state functions may be evaluated using the system bounds of Chapter 5 and replaced by a single linear equivalent limit state function [Gollwitzer and Rackwitz, 1983]. For obvious reasons, the above methods are also known as "multiple-check-point methods".

The above methods are also all second- and higher-order approximations to the actual probability content when the limit state functions are of general (non-linear) form. Fortunately, in many practical situations, the relevant limit state function is not highly non-linear and it is usually adequate to use first-order theory only.

4.3.8.2 Multiple linear limit states*

A special and important case arises when there is only one limit state function for each standard normal (but not necessarily independent) basic

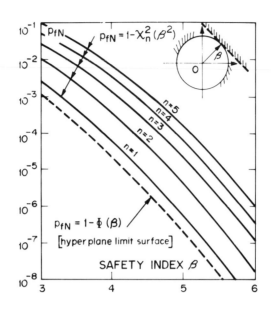

Fig. 4.9 — Safety index β and p_{fN} for hyperspherical limit state surface.

variable, i.e. $g_i(y_i) = \beta_i$ for each Y_i. This represents essentially the standard multinormal distribution function (cf. (3.10))

$$P\left(\bigcap_{i=1}^{n} Y_i \leqslant -\beta_i \right) = \Phi_n(-\boldsymbol{\beta}; \mathbf{R}) \qquad (4.31)$$

where $\mathbf{Y}$ is the standard (correlated) normal vector, $\mathbf{R}$ is the (positive definite) correlation matrix for $\mathbf{Y}$ and β_i is the safety index in the Y_i direction.

Expression (4.31) was considered in section 3.3.1 with the particular value y_i rather than $-\beta_i$, but the latter will be retained here if only to emphasize that the probability that Y_i is greater than β_i is sought (intersected for all directions). The result to be presented here approximates (4.31) using first order second-moment (FOSM) concepts and the Rosenblatt transformation described in Appendix B. The result is of particular use in system reliability (see Chapter 5).

The original normal random variables $\mathbf{X}$ may be reduced to $\mathbf{Y}$ using the usual transformation (4.3), and the correlated $\mathbf{Y}$ may be transformed to the uncorrelated standard normal $\mathbf{U}$ by means of the inverse Rosenblatt transformation (see Appendix B):

$$\mathbf{Y} = \mathbf{BU} \text{ or } Y_i = \sum_{j=1}^{i} b_{ij} U_j \qquad (4.32)$$

where the (lower triangular) elements b_{ij} are given by (B.33).

The dimension of (4.31) may be reduced by one by first substituting (4.32) into (4.31) and then conditioning on $U_1 \leq -\beta_1$ as follows [Hohenbichler and Rackwitz, 1983a]:

$$\Phi_n(\;) = P\left(\bigcap_{i=1}^{n} Y_i \leq -\beta_i\right)$$

$$= P\left[\bigcap_{i=1}^{n} \left(\sum_{j=1}^{i} b_{ij}U_j + \beta_i\right) \leq 0\right]$$

$$= P\left[\bigcap_{i=2}^{n} \left(\sum_{j=1}^{i} b_{ij}U_j + \beta_i \leq 0\right) \;\middle|\; U_1 \leq -\beta_1\right] P(U_1 \leq -\beta_1) \quad (4.33)$$

It is now desired to replace the first probability statement in (4.33) by one unconditioned on U_1;

$$P\left[\bigcap_{i=2}^{n} \left(b_{i1}\tilde{U}_1 + \sum_{j=2}^{i} b_{ij}U_j + \beta_i \leq 0\right)\right] \quad (4.34)$$

where now $\tilde{U}_1$ is required to be constrained to $\tilde{U}_1 \leq -\beta_1$. Since all variables U_i are independent of each other, the conditioning on U_1 does not affect U_2, U_3, ...

The conditional distribution function of U_1 given that $U_1 \leq -\beta_i$ is

$$F_1(u) = P(U_1 \leq u \;\middle|\; U_1 \leq -\beta_1)$$

$$= \frac{\Phi(u)}{\Phi(-\beta_1)}, \qquad \text{if } U \leq -\beta$$

$$= 1, \qquad \text{if } U > -\beta_1 \quad (4.35)$$

so that

$$\tilde{U}_1 = \Phi^{-1}[\Phi(-\beta_1)F_1(\tilde{U}_1)], \qquad \tilde{U}_1 \leq -\beta_1 \quad (4.36)$$

In equation (4.36), almost any continuous distribution function, say $F_X(x)$, may be equated to $F_1(\tilde{U}_1)$, without really affecting the outcome of $\tilde{U}_1$:

$$F_1(\tilde{U}_1) = F_X(x) \quad (4.37)$$

For first-order reliability (FOR) theory to be applicable, equation (4.34) must contain only normal variables. Hence, replacing $F_1(\tilde{U}_1)$ by a normal distribution function $\Phi(U_1)$ rather than by $F_X(x)$, it follows that (4.36) becomes

$$\tilde{U}_1 = \Phi^{-1}[\Phi(-\beta_1)\Phi(U_1)] \qquad (4.38)$$

Here the new standard normal variable U_1 is not conditional, as is the original U_1 in equation (4.33). The reuse of the U_1 is permissible since all U_i are independent (Tang and Melchers, 1984a]. Noting that $P(U_1 \leqslant -\beta_1) = \Phi(-\beta_1)$, and using (4.34) and (4.38), expression (4.33) can now be written as

$$\Phi_n(\ ) = P\left(\bigcap_{i=2}^{n} \left\{ b_{i1}\Phi^{-1}[\Phi(-\beta_1)\Phi(U_1)] + \sum_{j=2}^{i} b_{ij}U_j + \beta_i \leqslant 0 \right\} \right)\Phi(-\beta_1) \qquad (4.39)$$

$$= P\left\{ \bigcap_{i=2}^{n} [g_i(\mathbf{U}) \leqslant 0] \right\}\Phi(-\beta_1) \qquad (4.40)$$

where $g_i(\mathbf{U}) \leqslant 0$ is the non-linear limit state term in { } in (4.39). According to (4.7) in section 4.3.2, the approximating hyperplane, and hence the linearized limit state function, is

$$g_i(\mathbf{U}) \approx g_{Li}(\mathbf{U}) = \beta_i^{(2)} + \sum_{j=1}^{i} \alpha_{ij}U_j$$

Hence expression (4.40) is linearized as

$$\Phi_n(\ ) \approx P\left\{ \bigcap_{i=2}^{n} \left(\sum_{j=1}^{i} \alpha_{ij}U_j + \beta_i^{(2)} \leqslant 0 \right) \right\}\Phi(-\beta_1) \qquad (4.41)$$

where α_{ij} are the direction cosines $\mathbf{U}^*$ (a "checking point") and $\beta_i^{(2)}$ is the shortest distance to the approximating plane

$$g_{Li}(\mathbf{U}) = \beta_i^{(2)} + \sum_{j=1}^{i} \alpha_{ij}U_j$$

The intersection of the $n-1$ approximating hyperplanes in $\mathbf{U}$ space can now be recast in the original correlated $\mathbf{Y}$ space:

$$\Phi_n(\ ) \approx P\left(\bigcap_{i=2}^{n} Y_i^{(2)} \leq -\beta_i^{(2)}\right)\Phi(-\beta_1) \tag{4.42}$$

$$\approx \Phi_{n-1}(-\beta^{(2)}; \mathbf{R}^{(2)})\Phi(-\beta_1) \tag{4.43}$$

where $\mathbf{R}^{(2)}$ now represents the matrix of correlation coefficients for the linear approximating hyperplanes.

Repetition of the whole process in equation (4.43) over and over again eventually produces

$$\Phi_n(-\boldsymbol{\beta}; \mathbf{R}) \approx \Phi(-\beta_1)\Phi_2^{(2)})...\Phi(-\beta_n^{(n)}) \tag{4.44}$$

which has been found to produce excellent accuracy even for quite high n [Hohenbichler and Rackwitz, 1983a; Tang and Melchers, 1984a], provided that the linearization of limit states $g_i(\mathbf{U})$ in (4.40) can be efficiently carried out.

Fortunately, each limit state $g_i(\ )$ in equation (4.40) is only non-linear in U_1. If the linear space $(U_2, U_3, ..., U_i)$ is condensed into V space, an equivalent limit state function in two-dimensional (U_1, V) space may be written as

$$g_i(U_1, V) = b_{i1}\Phi^{-1}[\Phi(-\beta_1)\Phi(U_1)] + b_2V + \beta_i = 0 \tag{4.45}$$

where $b_2 = (1 - b_{i1}^2)^{1/2}$ and $V = \sum_{j=2}^{i} b_{ij}U_j/b_2$, $i = 2, 3, ..., n$. The b_{ij} are again defined by (B.33).

This non-linear limit state function can be used directly to determine $\beta_i^{(2)}$ and α_{ij} in (4.41). For this purpose, the iteration routine of section 4.3.5 may be used. The approach is to approximate $g_i(U_1, V)$ with a hyperplane, as in (4.40), but with each hyperplane translated such that the enclosed probability content is equal to the more accurate estimate of $\beta_i^{(2)}$. Details are given by Hohenbichler and Rackwitz (1983a) (see also section 5.5.4).

A better estimate of $\beta_i^{(2)}$ may be obtained by applying a reverse Rosenblatt transformation to each conditional probability content in (4.33) [Tang and Melchers, 1984a]. The conditional probability P_i for each

$$\sum_{j=1}^{i} b_{ij}U_j + \beta_i \leq 0 \text{ in (4.33) is}$$

$$P_i = P\left(\sum_{j=1}^{i} b_{ij}U_j + \beta_i \leq 0 \middle| U_1 \leq -\beta_1\right) \tag{4.46}$$

$$= P(Y_i + \beta_i \leq 0 | Y_1 \leq -\beta_1) \tag{4.47}$$

since $Y_1 = U_1$ and $Y_i = \sum_{j=1}^{i} b_{ij} U_j$ from (4.32) and b_{ij} as defined in Appendix B.

Using (A.3) this becomes

$$P_i = \frac{P(Y_i \leqslant -\beta_i \cap Y_1 \leqslant -\beta_1)}{P(Y_1 \leqslant -\beta_1)} \tag{4.48}$$

$$= \frac{\Phi_2(-\beta_1, -\beta_i; \rho_{i1})}{\Phi(-\beta_1)}, \qquad i = 2, 3, \ldots, n \tag{4.49}$$

$\Phi_2(\ )$ may be evaluated from (A.140b) or (3.7). Equation (4.49) then becomes

$$P_i = \Phi(-\beta_i) + \frac{1}{\Phi(-\beta_1)2\pi} \int_0^{\rho_{i1}} \frac{1}{(1-u^2)^{1/2}}$$

$$\exp\left[\frac{-(\beta_1^2 + \beta_i^2 - 2\beta_1\beta_i u)}{2(1-u^2)}\right] du \tag{4.50}$$

and $\beta_i^{(2)} = -\Phi^{-1}(P_i)$ would be a better estimate for use in expressions (4.41) and (4.42) than the previous $\beta_i^{(2)}$.

Both the original method and its extension have been programmed using direct minimization of β rather than iteration [Tang and Melchers, 1984a].

4.4   THE TRANSFORMATION OR FIRST-ORDER RELIABILITY METHOD

4.4.1   Simple transformations

Thus far, only the first two moments of each random variable have been considered in the probability calculation. However, it may be that information about probability distributions is available for some of or all the basic variables and it would be unfortunate simply to disregard or ignore this. Such extra information might be incorporated in the reliability analysis by transforming non-normal distributions into equivalent normal distributions [Paloheimo and Hannus, 1974]. For example, if X has a lognormal distribution with mean μ_X and variance σ_X^2, the transformation to an equivalent variable U is given by $U = \ln X$, with $\mu_X \approx \ln \mu_X$ and $\sigma_V^2 \approx V_X^2$ for $V_X < 0.3$. If U is then converted to the standardized normal random variable Y, then $Y = (U - \mu_U)/\sigma_U$ which is approximated by $[\ln(X/\mu_X)]/V_X$. It follows that the original variable X can be represented in terms of the standard normal variable Y and the first two moments μ_X and $\sigma_X = \mu_X V_X$ as $X \approx \mu_U \exp(YV_X)$. This can be done for other non-normal random variables as well. In addition, the limit state function $G(\mathbf{X}) = 0$ must be transformed to the **y** space; generally the transformation yields a non-linear $g(\mathbf{y}) = 0$ function.

For this simple example it is seen that the non-normally distributed variables can be directly transformed to equivalent standardized normal variables which can now be used in the second-moment calculation procedures. The disadvantage is that the limit state function becomes more complex and usually more non-linear.

In the discussion to follow in this section, a general approach for transforming independent non-normal basic variables to equivalent normal variables will be described and it will be shown that the transformation is best made about the "checking point" already introduced in FOSM theory. Additional requirements for dealing with dependent variables will then be described before the algorithm for the transformation method is outlined. An example is also given.

Historically the transformation approach was developed as an extension of the FOSM method, and is therefore sometimes called the "advanced" or "extended" FOSM method. It is probably less confusing to call it the FOR method, since limit state functions are still linearized (hence of "first order") but probability distributions are no longer approximated by their first and second moments.

Exercise
Show that, if X has an extreme type I distribution, X can be given in terms of the standardized normal random variable Y as

$$x = \frac{-\ln\{-\ln[\Phi(y)]\} - 0.5772}{1.2825} \sigma_X + \mu_X$$

4.4.2   The normal tail transformation
The transformation of an independent basic random variable X of non-normal distribution to an equivalent standardized normally distributed random variable Y is shown schematically in Fig. 4.10 and mathematically

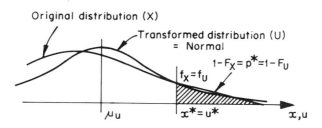

Fig. 4.10 — Original and transformed probability density functions.

can be expressed as

$$p = F_X(x) = \Phi(y) \text{ or } y = \Phi^{-1}[F_X(x)] \tag{4.51}$$

where p is some probability content associated with x, and hence with y; $F_X(\ )$ is the marginal cumulative distribution function of X and Φ is the cumulative distribution function for the standardized normal random variable Y. The transformation is shown by the lines abcde in Fig. 4.11.

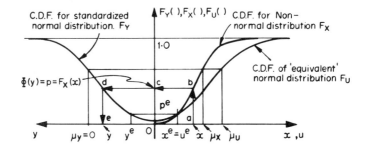

Fig. 4.11 — Relationships between original non-normal X, standardized normal Y and equivalent normal U cumulative distribution functions.

As in section 4.4.1, an equivalent normal variable U with cumulative distribution function $F_U(\ )$ might be introduced to represent X; (see Fig. 4.11). Evidently, many choices for U can be made, depending on the selection of μ_U and σ_U. What constitutes an appropriate choice will now be discussed.

Consider the first-order Taylor series expansion of (4.51) about some point x^e:

$$y \approx \Phi^{-1}[F_X(x^e)] + \frac{\partial}{\partial x}\{\Phi^{-1}[F_X(x)]\}|_{x^e}(x - x^e) \tag{4.52}$$

By letting $\Phi^{-1}[F_X(x)] = T$, it can be shown that the term $(\partial/\partial x)\{\ \}$ in (4.52) is

$$\frac{\partial}{\partial x}\{\ \} = \frac{\partial T}{\partial x} = \frac{f_X(x)}{\phi(T)}$$

or

$$= \frac{f_X(x)}{\phi\{\Phi^{-1}[F_X(x)]\}} \tag{4.53}$$

Upon substituting (4.53) into (4.51) and rearranging,

$$y \approx \frac{x - \{x^e - \Phi^{-1}[F_X(x^e)]\ \phi\{\Phi^{-1}[F_X(x^e)]\}/f_X(x^e)\}}{\phi\{\Phi^{-1}[F_X(x^e)]\}/f_X(x^e)} \tag{4.54a}$$

which may be written as

$$y = \frac{u - \mu_U}{\sigma_U} \tag{4.54b}$$

if

$$u = x \tag{4.54c}$$

with

$$\mu_U = x^e - y^e \sigma_U \tag{4.54d}$$

$$\sigma_U = \frac{\phi(y^e)}{f_X(x^e)} \tag{4.54e}$$

and

$$y^e = \Phi^{-1}[F_X(x^e)] \tag{4.54f}$$

This shows that it is possible to express the transformation (4.51) in terms of a new random variable U, normally distributed, with mean and standard deviation given by (4.54d) and (4.54e) above. Using (4.51), (4.54b) and (4.54c) it follows easily that $F_X(x^e) = F_U(x^e)$ and using (4.54e) with (4.54b), (4.54c) and (A.146) that $f_X(x^e) = f_U(x^e)$. Thus the probability density and the cumulative distribution function (i.e. the tail probability) for the equivalent normal random variable U should both be set equal to those of the original non-normal random variable X. This is shown in Figs 4.10 and 4.11. It is the so-called "normal tail approximation" [Ditlevsen, 1981a]. It may be carried out for each random variable X_i separately, provided that the expansion point x^e is known.

The fact that the expansion point x^e is identical with the checking point x^* is shown as follows. Since $\beta = \min(y^T \cdot y)^{1/2}$ from (4.4) and since each component of y is given by (4.51), it follows that β is a function of the location of the expansion point x^e. For β to be stationary,

$$0 = \frac{\partial \beta}{\partial x_i^e} = \frac{\partial}{\partial x_i^e} \left\{ \sum_{i=1}^{n} [y_i(x_i^e)]^2 \right\}^{\frac{1}{2}}, \quad \text{for all } i \tag{4.55}$$

Noting that the Y_i are independent, it is convenient to drop the subscript i from y_i, x_i and x_i^e. Hence (4.55) becomes

$$0 = \frac{1}{\beta} \, y(x^e) \, \frac{\partial y(x^e)}{\partial x^e}$$

from which $\partial y(x^e)\partial x^e = 0$ since, in general, $y(x^e) \neq 0$. Using y obtained from expression (4.52) and differentiating it by parts,

$$
\begin{aligned}
0 = \frac{\partial y(\ )}{\partial x^e} = \\
= \frac{\partial}{\partial x^e} \{\Phi^{-1}[F_X(x^e)]\} + \frac{\partial^2}{\partial(x^e)^2} \{\Phi^{-1}[F_X(x)]\}(x - x^e) - \\
- \frac{\partial}{\partial x^e} \{\Phi^{-1}[F_X(x)]\}
\end{aligned}
\qquad (4.56)
$$

Since neither the first nor the second derivative terms are generally zero, it follows that, for β to be a stationary value, $x_i = x_i^e$. Furthermore, in section 4.3.3 it was shown that a stationary value of β is achieved if $x_i = x_i^*$ (the "checking point") so that the expansion point considered thus far is identical with the checking point; $x_i^e = x_i^*$ for all i [cf. Lind, 1977].

Because each non-normal random variable is individually approximated by a normal distribution at the checking point, the latter may not correspond exactly to the point of maximum joint probability density [Horne and Price, 1977]. Any resulting error in β or p_f, is thought to be small.

A refinement of the above approach, which appears to improve computational efficiency, is to relax the relationships (4.54d) and (4.54e) to introduce the additional requirement that the slopes of the probability density functions at the design point must be the same [Chen and Lind, 1983]. It is evident that, provided that the approximating distributions are valid probability distributions, there is scope for a variety of normal tail approximations.

The above considerations immediately suggest an iterative procedure for finding the checking point x^* and hence the safety index β [Rackwitz and Fiessler, 1978]. In practical problems the basic variables may not be independent as assumed here, so that attention will first be given to the use of the Rosenblatt transformation (see Appendix B) before considering an algorithm in detail.

4.4.3   Dependent basic variables
4.4.3.1   *Transformation to independent random variables*
Assuming the appropriate data are available, the uniformly distributed random variables **R** of the Rosenblatt transformation (B.1) for n-dimensional space may be equated to the cumulative distribution functions of standardized normal random variables **Y**:

$$\Phi(y_1) = r_1 = F_1(x_1)$$
$$\Phi(y_2) = r_2 = F_2(x_2|x_1)$$

.

. (4.57)

.

$$\Phi(y_n) = r_n = F_n(x_n|x_1, \ldots, x_{n-1})$$

where $\Phi(\ )$ is the standard normal cumulative distribution function for **Y** and F_i the conditional cumulative distribution function for the random variable X as given by (B.4). The y_i can be obtained from (4.57) by successive inversion:

$$y_1 = \Phi^{-1}[F_1(x_1)]$$
$$y_2 = \Phi^{-1}[F_2(x_2|x_1)]$$

.

. (4.58)

.

$$y_n = \Phi^{-1}[F_n(x_n|x_1, \ldots, x_{n-1})]$$

In general the valuation will need to be done numerically. Similarly, the inverse transformation is obtained from

$$x_1 = F_1^{-1}[\Phi(y_1)]$$
$$x_2 = F_2^{-1}[\Phi(y_2)|x_1]$$

.

. (4.59)

.

$$x_n = F_n^{-1}[\Phi(y_n)|x_1, \ldots, x_{n-1}]$$

If the X_i are independent random variables, all the conditions in (4.57) to (4.59) disappear, and the transformation is essentially identical with that discussed in the previous section and given by (4.51). This means that, for the present case also, the expansion point is identical with the checking point.

Before the above transformations can be incorporated into an iteration algorithm, it is necessary to determine the transformation of the limit state function to $g(\mathbf{X})$ to $g(\mathbf{Y})$. A probability density function defined in the x space is transformed to the y space according to identity (A.150). This transformation also holds for the relationship between any (continuous) functions in $\mathbf{X}$ and $\mathbf{Y}$; in particular it also holds for $G(\mathbf{X})$:

$$G(\mathbf{x}) = g(\mathbf{y})|\mathbf{J}| \qquad (4.60)$$

where the Jacobian **J** has elements $j_{ij} = \partial y_i / \partial x_j$ (see A.151). The differential may be evaluated by substituting for y_i using (4.58), and then noting that, from (4.57), $\partial y_i = [\phi(y_i)]^{-1} \partial F_i(x_1|x_1, ..., x_{i-1})$, so that

$$\frac{\partial y_i}{\partial x_j} = \frac{1}{\phi(y_i)} \frac{\partial F_i(x_i|x_1, ..., x_{i-1})}{\partial x_j} \tag{4.61}$$

Evidently, if $i < j$, $\partial F_i / \partial x_j = 0$, so that **J** is lower triangular. This allows the inverse $\mathbf{J}^{-1}$ to be easily obtained from **J** by backsubstitution.

The components α_i of the gradient of $g(\mathbf{Y})$ are given by (4.5), but this is not easily evaluated since no explicit expression for $g(\mathbf{Y})$ is usually available. However, it follows easily that

$$\frac{\partial G(\mathbf{x})}{\partial x_j} = \sum_{i=1}^{n} \frac{\partial g(\mathbf{y})}{\partial y_i} \frac{\partial y_i}{\partial x_j} \tag{4.62}$$

so that

$$\begin{bmatrix} \dfrac{\partial G(\mathbf{x})}{\partial x_1} \\ \vdots \\ \dfrac{\partial G(\mathbf{x})}{\partial x_n} \end{bmatrix} = [J] \begin{bmatrix} \dfrac{\partial g(\mathbf{y})}{\partial y_1} \\ \vdots \\ \dfrac{\partial g(\mathbf{y})}{\partial y_n} \end{bmatrix} \tag{4.63}$$

from which the components of the gradient $\partial g / \partial y_i$ can be computed by matrix inversion. This allows the direction cosines α_i to be evaluated. However, the earlier interpretation of α_i as sensitivity factors (see section 4.3.2) is not now necessarily valid. This is because the y_i have, in general, no direct physical meaning (unless the basic variables are independent).

4.4.3.2 Algorithm for first-order reliability method
The algorithm of section 4.3.5 can now be generalized [Hohenbichler and Rackwitz, 1981].

(1) Select an initial checking point vector $\mathbf{x}^* = \mathbf{x}^{(1)}$ where $\mathbf{x}^{(1)}$ might be $\boldsymbol{\mu}_X$.
(2) Use the transformation (4.58) to obtain $\mathbf{y}^{(1)}$; this transformation will be of the simple form (4.51) for any random variables **X** which are independent.
(3) Use expressions (4.60) and (4.61) to obtain the Jacobian **J** and its inverse $\mathbf{J}^{-1}$.
(4) Compute the direction cosines α_i according to equations (4.5) and (4.63), the current value of $g(\mathbf{y}^{(1)})$, and the current estimate of β according to $\beta = -\mathbf{y}^{*T} \cdot \boldsymbol{\alpha}$ (equation (4.10)).
(5) A new estimate of the co-ordinates of the checking point in **y** space is then given by expression (4.24) using the current β value.

(6) The co-ordinates in $\mathbf{x}$ space of the current estimate of the checking point are then given by the reverse transformation (4.59).

Repeat steps (2)–(6) until $\mathbf{x}^*$ (or $\mathbf{y}^*$) and β stabilize in value.

It is evident that this algorithm is essentially a generalization of that given in section 4.3.5 with the more complex transformation (4.58) and the inverse (4.59) used in steps (2) and (6).

Example 4.6

The following example is adapted from Dolinsky (1983) and Hohenbichler and Rackwitz (1981). It is one of the few cases which can be treated wholly analytically.

Consider the problem in which the limit state function is $G(\mathbf{x}) = 6 - 2x_1 - x_2 = 0$ (e.g. X_1, X_2 are actions) with $\mathbf{X}$ being rather highly correlated and having the joint probability density function

$$f_{\mathbf{X}}(\mathbf{x}) = (ab - 1 + ax_1 + bx_2 + x_1x_2)\, \exp(- ax_1 - bx_2 - x_1x_2), \text{ for } (x_1, x_2) \geqslant 0$$
$$= 0, \quad \text{otherwise}$$

By appropriate integration (see section A.6), it follows that the marginal probability density functions for X_1 and X_2 are

$$f_{X_1}(x_1) = a\, \exp(- ax_1), \qquad \text{for } x_1 \geqslant 0$$
$$f_{X_2}(x_2) = b\, \exp(- bx_2), \qquad \text{for } x_2 \geqslant 0$$

while the joint cumulative distribution function is

$$F_{\mathbf{X}}(\mathbf{x}) = 1 - \exp(- ax_1) - \exp(- bx_2) + \exp(- ax_1 - bx_2 - x_1x_2),$$
$$\text{for } (x_1, x_2) \geqslant 0$$
$$= 0, \quad \text{otherwise}$$

Substituting into the first and second lines of the Rosenblatt transformation (4.57) produces

$$\Phi(y_1) = F_1(x_1) = \int_0^{x_1} f_{X_1}(t)\, dt = 1 - \exp(- ax_1), \text{ for } x_1 \geqslant 0 \tag{4.64a}$$

$$\Phi(y_2) = F_2(x_2|x_1) = \frac{\int_0^{x_2} f_{\mathbf{X}}(x_1, t)\, dt}{f_{X_1}(x_1)} =$$

$$= 1 - \left(1 + \frac{x_2}{a}\right)\exp(- bx_2 - x_1x_2) \tag{4.64b}$$

The limit state function $G(\mathbf{X})$ becomes

$$6 - 2F_1^{-1}[\Phi(y_1)] - F_2^{-1}[\Phi(y_2|x_1)] = 0$$

which can only be evaluated numerically. For example, for the case $a = 1$, $b = 2$ and $x_1 = 1$, then (4.64a) becomes $\Phi(y_1) = 1 - \exp(-1) = 0.6321$ or $y_1 = 0.34$. Then, in (4.64b), $\Phi(y_2) = 1 - (1 + x_2) \exp(-2x_2 - x_1 x_2)$, but $x_2 = 4$ for $G(\mathbf{x}) = 0$, so that $\Phi(y_2) = 1 - 5 \exp(-12) = 0.999\ 969$ or $y_2 \approx 4.01$. This marks point A on Fig. 4.12(a). The complete curve can be

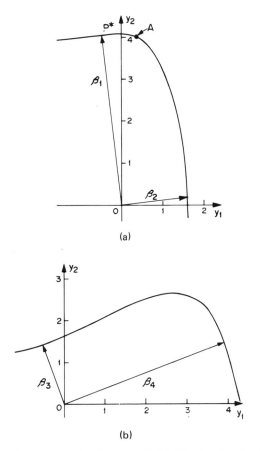

(a)

(b)

Fig. 4.12 — Local stationary points for Example 4.6; (a) original order of dependent basic variables; (b) interchanged order.

constructed in a similar manner by taking other values of x_1.
    With the use of (4.61), the Jacobian (A.151) becomes

$$\mathbf{J} = \begin{bmatrix} \dfrac{1}{\phi(y_1)}\dfrac{a}{\exp(ax_1)} & 0 \\[4mm] \dfrac{1}{\phi(y_2)}\dfrac{(1+x_2/a)x_2}{\exp(bx_2+x_1x_2)} & \dfrac{1}{\phi(y_2)}\dfrac{-1/a+(b+x_1)(1+x_2/a)}{\exp(bx_2+x_1x_2)} \end{bmatrix}$$

$$(4.64c)$$

from which it follows by back-substitution that

$$\mathbf{J}^{-1} = \begin{bmatrix} \phi(y_1)\dfrac{\exp(ax_1)}{a} & 0 \\[4mm] -\Phi(y_1)\dfrac{(1+x_2/a)x_2\exp(ax_1)}{1/a-(b+x_1)(1+x_2/a)a} & \text{or} \quad -\phi(y_2)\dfrac{\exp(bx_2+x_1x_2)}{1/a-(b+x_1)(1+x_2/a)} \end{bmatrix}$$

$$(4.64d)$$

With $a=1$, $b=2$, the algorithm can now be used as follows.

(1) Let a trial design point be, arbitrarily, $x_1=1$, $x_2=4$, which it has already been seen corresponds to the point A (Fig. 4.12(a)).

(2) From (4.64a), $\Phi(y_1)=1-\exp(-1)$, or $y_1=0.34$ and, from (4.64b), $\Phi(y_2)=1-(1+4)\exp(-12)$ or $y_2=4.01$. Hence $\mathbf{y}=\{0.34, 4.01\}^\mathrm{T}$.

(3) From (4.64c) and (4.64d)

$$\mathbf{J} = \begin{bmatrix} 0.977 & 0 \\ 0.917 & 0.642 \end{bmatrix}$$

$$\mathbf{J}^{-1} = \begin{bmatrix} 1.024 & 0 \\ 1.462 & 31.289 \end{bmatrix}$$

(4) (a) direction cosines (see 4.5) and 4.63))

$$\begin{bmatrix} c_1 \\ c_2 \end{bmatrix} = \dfrac{\dfrac{\partial g}{\partial y_1}}{\dfrac{\partial g}{\partial y_2}} = \mathbf{J}^{-1} \begin{bmatrix} \dfrac{\partial G(\mathbf{x})}{\partial x_1} \\ \vdots \\ \dfrac{\partial G(\mathbf{x})}{\partial x_n} \end{bmatrix} = \begin{bmatrix} 1.024 & 0 \\ 1.462 & 31.829 \end{bmatrix} \begin{bmatrix} -2 \\ -1 \end{bmatrix} = $$

$$\begin{bmatrix} -2.048 \\ -34.21 \end{bmatrix}$$

so that $l \approx 34.27$, $\alpha_1 = -0.060$, $\alpha_2 = -0.998$.

(b) By original choice of x_1 and x_2 the limit state function is identically satisfied for this cycle, i.e.

$$g(\mathbf{y}) = G(\mathbf{x}) = 0$$

(c) Calculate β:

$$\beta = (\mathbf{y}^T \cdot \mathbf{y})^{1/2} = (0.34^2 + 4.01^2)^{1/2} = 4.02$$

(5) From (4.23) the new estimates for $\mathbf{y}$ are

$$\begin{bmatrix} y_1 \\ y_2 \end{bmatrix} = -\begin{bmatrix} -0.060 \\ -0.998 \end{bmatrix} (4.02 - 0) = \begin{bmatrix} 0.24 \\ 4.02 \end{bmatrix}$$

which by inspection of Fig. 4.12(a) is obviously a better estimate of the (apparent) design point P*.

(6) From (4.64a) inverted according to (4.59), the second estimate for x_1 is

$$x_1 = F_1^{-1}[\Phi(y_1)] = -\ln[1 - \Phi(0.24)] = 0.903$$

and from (4.64b)

$$1 - \Phi(y_2) = (1 + x_2) \exp[-(2 + x_1)x_2]$$

i.e.

$$0.262 \times 10^{-4} = (1 + x_2) \exp[-2.903x_2]$$

or $x_2 = 4.2$ obtained by trial and error.

The algorithm should now be repeated until $\mathbf{y}$ or $\mathbf{x}$ stabilizes. This should then give β_1 as shown in Fig. 4.12(a).

4.4.3.3 Observations

Without actually using the algorithm, it is evident that, if a rather different starting point was chosen in the above example, say $\mathbf{y} = (1.5, 1.3)^T$, the algorithm would converge to β_2 in Fig. 4.12(a), with $\beta_2 < \beta_1$! This shows again that the algorithm provides only stationary points and that, for each of these, the respective β (and hence p_f) must be checked [Dolinsky, 1983].

This difficulty does not arise if a numerical algorithm for seeking the point of maximum likelihood is used instead of the iteration algorithm (which identifies only stationary points). Such an approach was outlined in section 4.3.4 for a second-moment problem with non-linear limit state function. The present problem is essentially similar but has an unusually highly non-linear limit state function, owing to the format of the joint cumulative distribution function used. The example represents an unusually severe test for the FOR method; in most practical situations highly correlated exponential distributions, as used here, would not occur.

A further difficulty can arise. The Rosenblatt transformation (4.57) can be given in $n!$ different ways, depending on how the basic variables are arranged. The importance of this can be demonstrated as follows. If the order x_1, x_2 in (4.57) is interchanged, then

$$\Phi(y_1) = F_2(x_2)$$
$$\Phi(y_2) = F_1(x_1|x_2) \qquad\qquad (4.65)$$

and for $a = 1$, $b = 2$ as before

$$\Phi(y_1) = 1 - \exp(-2x_2)$$

$$\Phi(y_2) = 1 - \left(1 + \frac{x_1}{2}\right)\exp(-x_1 - x_1 x_2)$$

Using a similar procedure as before, the limit state function $G(\mathbf{X})$ is transformed to that shown in Fig. 4.12(b). Again two stationary points exist. Let the respective distances from the origin be marked as β_3 and β_4. Then it is easily verified that $\beta_3 < \beta_2 \ll \beta_4$. This demonstrates that in principle *all n!* possible combinations of arrangement of $\mathbf{X}$ in (4.57) must be considered if the critical (i.e. lowest) is to be identified [Dolinsky, 1983]. In practice some prior knowledge of the problem may be advantageous in selecting an appropriate ordering of variables, and again the severity of the problem is less likely to be as great as indicated here for a less extreme choice of joint cumulative distribution function $F_\mathbf{X}(\ )$.

4.5   APPLICATION OF FIRST-ORDER SECOND-MOMENT AND FIRST-ORDER RELIABILITY METHODS

The FOSM method in particular has proved very popular with those wishing to obtain probability statements about particular problems. The procedure to determine the safety index β is very straightforward even for non-linear limit state functions. Once experience has been gained with the FOSM method, its "extended" or "advanced" version, here called the FOR method, is a natural step to incorporate distributional probability information. As was discussed in relation to Example 4.6 it is only in rather extreme situations that difficulties arise or that the linearization of the transformed limit state equation leads to grossly inaccurate results.

Both methods require the use of derivatives of the limit state function. For the simple examples described in the preceding sections, explicit expressions could be obtained for derivatives. More generally, however, such as when the limit state function is a complex one dependent on structural behaviour or analysis, resort may have to be made to numerical procedures. This will have a significant effect on the amount of computation necessary as the number of basic variables increases.

The failure probabilities calculated from β for the two methods are, of course, nominal values, except in the special case of a linear limit function and normal basic variables. Even in that case, the basic variables must adequately describe all parameters of importance (see Chapter 2).

It is important to note that the necessary data to allow $f_X(\ )$ to be described may not be available in practice, particularly for dependent variables. In some cases, only correlation data may be available. In such situations one approach is to transform dependent variables X' to independent X using the orthogonal transformation (see Appendix B). This is strictly valid only for normally distributed variables (i.e. second-moment representation) for which correlation is a sufficient descriptor of dependence. The normal tail approximation of section 4.4.2 can then be applied to the uncorrelated (but non-normal) variables. The normal tail approximation is, of course, the special case of the Rosenblatt transformation (4.57) with independent variables. For this reason the note in section 4.4.2 that the checking point does not necessarily coincide exactly with the point of maximum probability density will also apply for the Rosenblatt transformation. More generally use can be made of partial information by employing the ideas of der Kiureghian and Liu (1985).

It is possible to extend the concept of using a second-moment representation to approximate a non-normal distribution function by using also approximating higher moments. One such approach is to use a weighted system of second-moment cumulative distribution functions to represent the non-normal distribution, and to modify the limit state function $G(X)$ accordingly [Grigoriu, 1982]. Such methods are beyond the scope of this book.

Finally, it is possible also to specify safety indices other than β. A useful overview of most of these has been given by Turkstra and Daly (1978). None has found wide acceptance.

4.6 CONCLUSION

In this chapter the FOSM method has been developed. It was shown that for linear limit state functions the failure probability can be obtained directly from the safety or reliability index β. If the limit state function is non-linear, β can still be defined, but only with respect to a tangent (hyper)plane to approximate the non-linear limit state function. β is in each case the shortest distance from the (hyper)plane and therefore perpendicular to it in standardized normal space. The corresponding point on the hyperplane was termed the checking or design point. It is also the point of greatest probability density within the space encompassed by the failure region.

The checking point may be found by direct minimization using a

Lagrangian multiplier, by numerical maximization or by iteration to find a saddle point.

When more than second-moment information is available for some of or all the basic variables, the FOSM method of determining a nominal failure probability, or β, may still be used, provided that each basic variable is first transformed to an equivalent normal random variable. The procedure for doing this has been termed the FOR method or the transformation method. It is based on Rosenblatt's transformation (see Appendix B). When all basic variables are independent, the transformation degenerates into transformation of each variable independently. When the full Rosenblatt transformation must be used, it is possible in certain highly correlated or highly non-linear problems to obtain a non-critical β value with iteration routines, since only stationary (rather than maximum) points are found. This difficulty may be overcome by using physical insight or by applying a direct maximization routine.

5

Reliability of structural systems

5.1 INTRODUCTION

This chapter will be concerned with structures for which more than one limit state must be considered. Even in simple structures composed of just one element, various limit states such as bending action, shear, buckling, axial stress, deflection, etc., may all need to considered. Most structures are, in addition, composed of many members or elements. Such a composition will be referred to as a "structural system". A system may be subject to many loads, either singly or in various combinations. In this chapter, only time-invariant random loading will be considered.

The reliability of a structural system will be a function of the reliability of its members for the following reasons:

(a) Load effects (stress resultants) on different members are obtained from one or more common loads.
(b) Loads and resistances may not be independent (e.g. dead loads may be related to member size, and strength may be related to previous applied loadings).
(c) Correlation of member strength properties may exist between different locations in the structure.
(d) Construction practices may influence member strength for a group of members.

Further factors include the configuration of the structure, and the possible existence of limit states for the structure as a whole rather than its elements (e.g. overall deflection, foundation settlement, residual stiffness).

The reliability assessment of structural systems, therefore, will involve consideration of multiple, perhaps correlated, limit states. Methods to deal with such problems will be the subject of this chapter. The problem will be formulated in the next section. Monte Carlo solutions are considered in section 5.3 and bounding methods, using the First-order second-moment (FOSM) or First-order reliability (FOR) methods, in section 5.4. The analysis of large complex structures is discussed in section 5.5.

5.2 SYSTEMS RELIABILITY FORMULATION

Discussion in this chapter will be confined to framed structures such as trusses and rigid frames. These are essentially "one-dimensional systems". Two-dimensional systems such as plates, slabs and shells will not be considered, nor will three-dimensional continua such as earth embankments and dams. For these more complex structural systems the principles given here are also valid and are being used to develop calculation techniques appropriate, for example, for finite element analysis [e.g. der Kiureghian and Taylor, 1983].

5.2.1 Member and structural failure criteria

Member or material behaviour in structural engineering is usually idealized to one of the forms of strength–deformation relationships shown in Fig. 5.1.

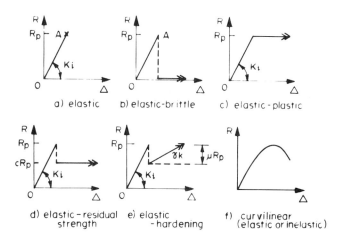

Fig. 5.1 — Various strength–deformation (R–Δ) relationships.

Elastic behaviour (Fig. 5.1(a)) corresponds to the maximum permissible stress concept of section 1.2.1. With this idealization, failure at any one location within the structure, or of any one element, is considered to be identical with structural failure. Although clearly unrealistic for most structures, it is nevertheless a convenient idealization.

When one load system acts on a structure, the location of the peak stress or stress resultant can be identified from an elastic stress analysis. For such an analysis, use of deterministic elastic properties and dimensions is often adequate owing to the very low coefficient of variation associated with these variables (see Chapter 8). When more than one load system acts, the location of the peak stress (resultant) will depend on the relative magnitudes of the load systems, and several candidate locations or members may need to be considered. For large structures, such identification may not be easy by inspection alone.

Brittle failure of a member does not always imply structural failure, owing to redundancy of the structure. The actual member behaviour can therefore be better idealized as "elastic–brittle" indicating that deformation at zero capacity is possible for a member, even after the peak capacity has been reached (Fig. 5.1(b)).

Elastic–plastic member behaviour (Fig. 5.1(c)) allows individual members or particular regions within the structure to sustain the maximum stress resultant as deformation occurs. When the elastic member stiffness K_i approaches infinity this behaviour is known as idealized rigid–plastic behaviour. A generalization of both elastic–brittle and elastic–plastic behaviour is elastic–residual strength behaviour (Fig. 5.1(d)) and a further generalization is elastic–hardening (or softening) behaviour (Fig. 5.1(e)). The latter may be seen as an approximation to general behaviour including post-buckling effects. Even without introducing reliability concepts, the analysis of structures with these latter behaviours is complex. Of course, general non-linear (curvilinear) strength–deformation relationships (Fig. 5.1.(f)) present even more difficulty.

Structural failure (as distinct from individual member or material failure) may be defined in a number of different ways, including:

(a) maximum permissible stress reached anywhere ($\sigma(\mathbf{x}) = \sigma_{max}$);
(b) (plastic) collapse mechanism formed (i.e. zero structural stiffness attained: $|K| = 0$);
(c) limiting structural stiffness attained ($|K| = K_{limit}$);
(d) maximum deflection attained ($\Delta = \Delta_{limit}$);
(e) total accumulated damage reaches a limit (e.g. as in fatigue).

Structural failure modes which consist of the combined effects of two or more member or material failure events, such as for statically indeterminate structures, are of particular interest in the determination of structural systems reliability. If all failure modes for the system have been identified, then the various events contributing to these failure modes may be systematically enumerated using the "fault-tree" concept. An example of a fault-tree is shown in Fig 5.2(b) for the elementary structure of Fig 5.2(a). The procedure is to take each failure event and to decompose it into contributing subevents, which are themselves decomposed in turn. The lowest subevents in the tree correspond, for structures, to member or material failure. At this level (if not earlier) limit state equations can be written. Fault-tree methodology has found most application in general, rather than structural systems, reliability analysis [Henley and Kumamoto, 1981].

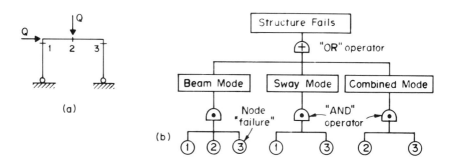

Fig. 5.2 — Fault-tree represenatation.

5.2.2 Solution approaches
For the reliability analysis of multimember structures (or structures which can be idealized as such), two complementary approaches can be adopted [Bennett and Ang, 1983]:

(1) failure modes approach;
(2) survival modes approach.

The failure modes approach is based on the identification of all possible failure modes for the structure. A common example is the collapse mechanism technique for ideal plastic structures. Each mode of failure for the structure will normally consist of a sequence of member "failures" (i.e. reaching of an appropriate member limit state) sufficient to cause the structure as a whole to reach a limit state such as (a)–(e) above. The possible ways in which this might occur can be represented by an "event tree" (Fig. 5.3) or as a "failure graph" (Fig. 5.4). Each branch of the failure graph

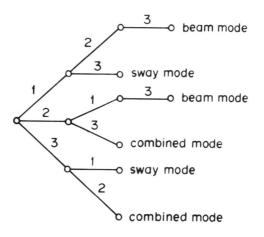

Fig. 5.3 — Event-tree representation for structure of Fig. 5.2(a).

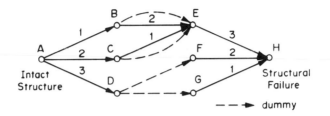

Fig. 5.4 — Failure-graph representation for structure of Fig. 5.2(a).

represents the failure of a member of the structure, and any complete forward path through the branches starting from the "intact structure" node and concluding at the "failure" node represents a possible sequence of member failures. This information is also conveyed in the event tree [Henley and Kumamoto, 1981].

Since failure through any one failure path implies failure of the structure, the event "structural failure" (F_S) is the union of all m possible failure modes:

$$p_f = P(F_S) = P(F_1 \cup F_2 \cup \ldots \cup F_m) \tag{5.1}$$

where F_i is the event "failure in the ith mode". For each such mode, a sufficient number of members (or structural "nodes") must fail; thus

$$P(F_i) = P(F_{1i} \cap F_{2i} \cap \ldots \cap F_{ni}) \tag{5.2}$$

where F_{ji} is the event "failure of the jth member in the ith failure mode" and n_i represent the number of members required to form the ith failure mode. For the simple example of Fig. 5.2(a), there are $m = 3$ failure modes, and $n_1 = 3$, $n_2 = n_3 = 2$.

The survival modes approach is based on identifying various states (or modes) under which the structure survives. For the structure of Fig. 5.2(a), each of A, B, C, D, E, F, G (but not H!) in the failure graph Fig. 5.4 represents such a state (see also Fig. 5.3). For each survival mode the structure has partially failed but is still capable of supporting the load (i.e. it is still statically and geometrically stable).

Survival of the structure requires survival in at least one survival mode, or

$$p_s = P(S_S) = P(S_1 \cup S_2 \cup \ldots \cap S_k) \tag{5.3}$$

where S_S is the event "structural survival", and S_i the event "structural survival in mode i", $i = 1, \ldots, k$, with k not equal to the final node index. From (A.5) it follows that

$$p_f = P(\bar{S}_1 \cap \bar{S}_2 \cap \ldots \cap \bar{S}_k) \qquad (5.4)$$

where $\bar{S}_i$ is the event "structure does not survive in survival mode i". Clearly, to attain survival in any particular survival mode, all the members contributing to that survival mode must survive. For example, survival in mode B of Fig. 5.4 requires member 3 to survive.

It follows that failure to survive in a given survival mode is equivalent to failure of a sufficient number of the contributing members, or

$$P(\bar{S}_i) = P(F_{1i} \cup F_{2i} \cup \ldots \cup F_{\ell_i i}) \qquad (5.5)$$

(where F_{ji} is the event "failure of the jth member in the ith survival mode" and where ℓ_i represent the number of members required to ensure survival of the ith survival mode.

Owing to difficulties in conceptualization of survival modes and in formulating the limit state equations, the survival mode approach has not received much attention. For this reason, it will not be explored herein, although some results exist for structural systems composed of ideal rigid–plastic members, [Bennet and Ang, 1983; Ditlevsen and Bjerager, 1983; Melchers, 1983a].

It follows directly from (5.1) and (5.5) that any estimate of the probability of structural failure based on failure modes will be unconservative (i.e. underestimate p_f) unless all possible failure modes have been included in the analysis; conversely, the failure probability based on the survival mode approach (5.4) will be conservative (i.e. overestimate p_f) unless all possible survival modes have been incorporated in the analysis.

It is important to note that both approaches are unconservative with respect to member failure. If the possibility of member failure (or one or more of the member failure modes) is ignored, the failure probability of the structure will usually be underestimated [Bennett and Ang, 1983].

5.2.3 Structural system idealizations
5.2.3.1 Series systems
In a series system, typified by a chain, and also called a "weakest link" system, attainment of any one element limit state constitutes failure of the structure (Fig. 5.5). For this idealization the precise material properties of

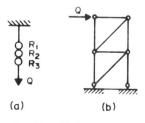

Fig. 5.5 — Series systems.

the elements or members do not matter. If the members are brittle, failure is caused by member fracture; if the members have a plastic deformation capacity, failure is by excessive yielding. It is evident that a statically determinate structure is a series system since the failure of any one of its members implies failure of the structure. Each member is therefore a possible failure mode. It follows that the system failure probability for a weakest link structure composed of m members is [Freudenthal, 1961; Freudenthal *et al.*, 1966]

$$p_f = P(F_1 \cup F_2 \cup F_3 \cup \ldots \cup F_m) \tag{5.6}$$

Comparison with (5.1) shows that the series systems formulation (5.6) is of the "failure mode" type.

If each failure mode F_i ($i = 1, m$) is represented by a limit state equation $G_i(\mathbf{x}) = 0$ in basic variable space, the direct extension of the fundamental reliability problem (1.31) is

$$p_f = \int_{D \in \mathbf{X}} \cdots \int f_\mathbf{X}(\mathbf{x}) d\mathbf{x} \tag{5.7}$$

where $\mathbf{X}$ represents the vector of all basic random variables (loads, strengths of members, member properties, sizes, etc.) and D (and D_1) is the domain in $\mathbf{X}$ defining failure of the system. This is defined in terms of the various failure modes as $G_i(\mathbf{X}) \leq 0$. In two-dimensional $\mathbf{x}$ space, expression (5.7) is defined in Fig. 5.6 with D and $G_i(\mathbf{X}) \leq 0$ shown shaded.

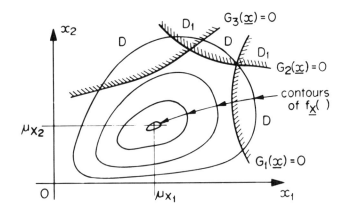

Fig. 5.6 — Basic structural reliability problem in two dimensions.

The safe region in Fig. 5.6 is given by

$$\overline{D} : \overline{F}_1 \cap \overline{F}_2 \cap \ldots \cap \overline{F}_m \tag{5.8}$$

where $\overline{F}_i$ is defined as "survival in mode i" or $G_i(\mathbf{X}) > 0$. The probability of survival is then

$$p_s = P\left(\bigcap_{i=1}^{m}\overline{F}_i\right) = \int_{D}\cdots\int f_{\mathbf{X}}(\mathbf{x})d\mathbf{x} \tag{5.9}$$

clearly indicating that this formulation is equivalent to the "survival mode" approach of Section 5.2.2.

A particularly simple result can be derived immediately without recourse to the integrations required by (5.7) or (5.9) above. For the chain of Fig. 5.5 the load effect S in each link of the chain is identical with the load Q. If $F_{R_i}(r)$ is the cumulative distribution function for the strength of the ith link, then the cumulative distribution function $F_R(\ )$ for the chain as a whole is given by

$$\begin{aligned} F_r(r) &= P(R \leqslant r) = 1 - P(R > r) \\ &= 1 - P(R_1 > r_1 \cap R_2 > r_2 \cap \ldots \cap R_m > r_m) \end{aligned}$$

which, for independent strength properties, becomes

$$\begin{aligned} F_r(r) = P(R \leqslant r) &= 1 - [1 - F_{R_1}(r_1)][1 - F_{R_2}(r_2)]\ldots \\ &= 1 - \prod_{i=1}^{m}[1 - F_{R_i}(r_i)] \end{aligned} \tag{5.10}$$

This expression forms the basis for the probability distribution of the mechanical resistance of brittle materials [Weibull, 1939]. When each R_i is identically and normally distributed the distribution of R as m approaches infinity is given by the type III extreme value distribution for the smallest value (see section A.5.13).

5.2.3.2 *Parallel systems*

When the elements in a structural system (or subsystem) behave in such a way or are so interconnected that the reaching of the limit state in any one or more elements does not necessarily mean failure of the whole system the reliability problem becomes one of a "parallel" or "redundant" system analysis. Parallel systems can be modelled as in Fig. 5.7.

Redundancy in systems may be of two types "Active redundancy" occurs when the redundant member(s) actively participates in structural behaviour even at low loading. "Passive (or stand-by) redundancy" occurs when the redundant member(s) does not come into play until the structure has suffered a sufficient degree of degradation or failure of its members. Passive redundancy, or "fail-safe" design, implies the availability of a reserve capacity. It increases the reliability of a system as is easily demonstrated.

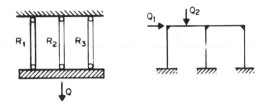

Fig. 5.7 — Parallel systems.

However, whether active redundancy is beneficial depends on the behaviour characteristics of the members or elements. As might be expected, for ideal plastic systems, the "static theorem" guarantees that active redundancy cannot reduce the reliability of a structural system [cf. Augusti and Barratta, 1973].

With active redundancy, the failure probability of an n-component parallel system (or subsystem) is

$$p_f = P(F_S) = P(F_1 \cap F_2 \cap \ldots \cap F_n) \tag{5.11}$$

where F_i is the event "failure of the ith component". It follows immediately that (5.11) is equivalent to (5.4), and can be represented in x space by

$$p_f = \int_{D_1 \in \mathbf{X}} \cdots \int f_{\mathbf{X}}(\mathbf{x}) d\mathbf{x} \tag{5.12}$$

as shown in Fig. 5.6 for the intersection domains D_1.

Since a parallel system can only fail when all its contributory components have reached their limit states, it follows that the behaviour characteristics of the components are of considerable importance in defining "system failure". This is in contrast with the situation for series systems.

In parallel systems of low redundancy and with brittle elements, failure of one element is usually sufficient to cause failure of the system. Unless the failed element contributed very little to the system strength immediately prior to its failure, the load redistribution caused by it usually leads to overloading of other elements in sequence (causing so-called "progressive collapse"). This has led to the common assumption that, with brittle members, failure of the most highly stressed member is tantamount to failure of the system.

The situation is completely different for ideal plastic structures, such as rigid frames, for which each collapse or failure mode (i.e. limit state) can be represented by an equation of the following type [Moses, 1974]:

$$\sum_i Q_i \Delta_i - \sum_j M_j \theta_j = 0 \tag{5.13}$$

where Q_i is the external load i, Δ_i are the deflections corresponding to Q_i (a function of θ_j and dimensions), M_j is the plastic moment resistance at section j and θ_j is the plastic rotation at section j. Expression (5.13) is clearly of the "parallel" type since each resistance M_j must be mobilized to develop the total resistance against the loading Q_i.

Collectively, a set of failure mode equations, each like (5.13), constitutes a series system since the structure will fail when any one failure mode occurs. It follows that plastic moment capacities M_j may occur in more than one failure mode expression. This means that the structural capacities obtained from different failure modes may be correlated. (Note that this is quite distinct from any correlation which may exist between individual M_j values.)

Example 5.1
Consider the idealized parallel system shown in Fig. 5.8. If the elements are

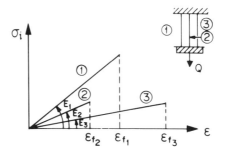

Fig. 5.8 — Brittle material behaviour in parallel system.

all brittle, with different fracture strains ε_f the maximum load Q which can be supported at any particular strain level ε is

$$R_{sys} = \max_\varepsilon [R_1(\varepsilon) + R_2(\varepsilon) + R_3(\varepsilon)] \tag{5.14a}$$

where $R_i = A_i \sigma_i(\varepsilon)$ for $i = 1$ to 3. Here A_i represents cross-sectional area and σ_i stress.

Since each resistance R_i ($i = 1, 2, 3$) is a random variable, expression (5.14a) is not easy to apply since each possible state ε_{f1}, ε_{f2} and ε_{f3} must be considered as a possible state of maximum capacity. Thus, all possible combinations of failed and surviving members must be considered; each such combination will be a "parallel" subsystem; thus

$$R_{sys} = \max\{[R_1(\varepsilon_{f2}) + R_2(\varepsilon_{f2}) + R_3(\varepsilon_{f2})], [R_1\varepsilon_{f1}) + R_3(\varepsilon_{f1})], R_3(\varepsilon_{f3})\}$$

and   (5.14b)

$$p_f = P(R_{sys} - Q < 0) \ .$$

The cumulative probability distribution function for (5.14b) is given by [Hohenbichler and Rackwitz, 1983b]:

$$F_{R_{sys}}(r) = P\left\{\max_{\varepsilon_{fi}}\left[\sum_{i=1}^{n} R_i(\varepsilon_{fi})\right] \leq r\right\}$$

$$= P\left\{\bigcap_{\varepsilon_{fi}}\left[\sum_{i=1}^{n} R_i(\varepsilon_{fi}) - r \leq 0\right]\right\}$$ (5.14c)

which can be evaluated by the methods to be discussed subsequently. The material properties can be generalized to include curvilinear and unloading behaviours. For the particular case when the elastic modulii E_i in Fig. 5.8 are identical then $F_{R_{sys}}(\ )$ approaches a normal distribution as n approaches infinity [Daniels, 1945].

Example 5.2
For the rigid–plastic portal frame shown in Fig. 5.9, the collapse mode

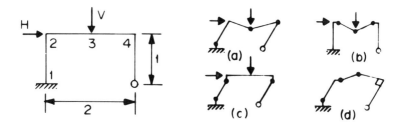

Fig. 5.9 — Rigid frame and collapse modes: Example 5.2.

equations (limit state equations) for the four collapse modes shown are

mode a:	M_1		$+2M_3$	$+2M_4$	$-H$	$-V$	$= 0$	(5.15a)
mode b:		M_2	$+2M_3$	$+2M_4$		$-V$	$= 0$	(5.15b)
mode c:	M_1	$+M_2$		$+2M_4$	$-H$		$= 0$	(5.15c)
mode d:	M_1	$+2M_2$	$+2M_3$	$+$	$-H$	$+V$	$= 0$	(5.15d)

Let each random variable $X_i = (M_1, M_2, \ldots, H, V)$ be normally distributed, with properties $\mu_{X_i} = (1.0, 1.0, 1.0, 1.0, 1.0, 1.0)$ and $\sigma_{X_i} = (0.15, 0.15, 0.15, 0.15, 0.17, 0.50)$. The β index for each collapse mode can be obtained using the FOSM concepts of Chapter 4.

For mode a,

$$G(\mathbf{X}) = M_1 + 2M_3 + 2M_4 - H - V$$

so that

$$\mu_G = 1 + 2 + 2 - 1 - 1 = 3$$

and

$$\sigma_G^2 = (0.15)^2 + 2^2(0.15)^2 + 2^2(0.15)^2 + (0.17)^2 + (0.5)^2 = 0.4814$$

Thus

$$\beta_a = \frac{\mu_G}{\sigma_G} = \frac{3}{\sqrt{0.4814}} = 4.32$$

The β indices for modes b, c and d may be similarly determined; $\beta_b = 4.83$, $\beta_c = 6.44$ and $\beta_d = 7.21$. How these results for individual failure modes can be combined will be considered in Example 5.4.

5.2.3.3 *Conditional and combined systems*

Any real structural system usually requires a combination of both series and parallel descriptions for its subsystems. It may also require the use of conditioned events. The latter arises if the failure of one independent element or group of elements affects the likelihood of failure of other elements or element groups. For example, in Fig. 5.10, if the upper beam

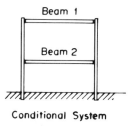

Conditional System

Fig. 5.10 — Conditional system.

collapses, it will affect the performance and reliability of the lower beam (since it may be damaged and will be subject to extra load) [Benjamin, 1970].

In this situation, the failure probabilities of the structural elements are dependent on the behaviour of the structure under extreme events. Provided that the sequence of events can be enumerated, it is evident that a structure with conditional events can be reduced to one containing both "series" and "parallel" element groups.

Example 5.3
Consider a single-span three-girder bridge. It will fail if any two adjacent girders fail, or if any deck panel fails or if either (or both) of its abutments fail. The abutments could fail through failure of its two bored piles or through overturning. All this can be expressed as

$$p_f = P[(G_1 \cap G_2) \cup (G_2 \cap G_3) \cup D \cup A_1 \cup A_2]$$

where

$$D = \bigcup_i D_i$$

and

$$A_j = (P_{1j} \cap P_{2j}) \cup O_j$$

with obvious notation for events.

5.3 MONTE CARLO TECHNIQUES FOR SYSTEMS

5.3.1 Series systems
The Monte Carlo techniques of Chapter 3 can be directly extended to deal with the calculation of systems reliability. For a series system (section 5.2.3.1), the probability of failure represented by (5.7) can be rewritten as

$$p_f = \int \cdots \int I[\ ] f_X(x) dx \qquad (5.16)$$

where now the indicator function $I[\ ]$ is generalized from (1.33) and (3.20) to

$$I\left[\bigcup_{i=1}^{m} G_i(x) \leq 0\right] = 1, \qquad \text{if } [\ ] \text{ is true}$$

$$= 0, \qquad \text{if } [\ ] \text{ is false}$$

where $G_i(\mathbf{x}) = 0$ represents the ith (known) limit state function.

For a two-dimensional $\mathbf{x}$ space, $I[\ ]$ represents the integration domain abcd of Fig. 5.11(a), i.e. a sample point $\hat{\mathbf{x}}$ lies in the failure region if any one

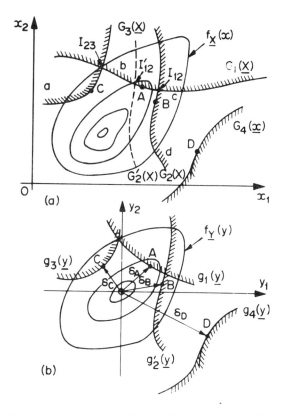

Fig. 5.11 — Two variable system problems: (a) original space; (b) hypothetical standardized space.

$G_i(\mathbf{x}) \leq 0$. This formulation is directly applicable if the "crude" Monte Carlo approach is used.

For importance sampling with m different limit state functions, an appropriate sampling function $\ell_\mathbf{v}(\ )$ can be defined as [Melchers, 1984]:

$$\ell_\mathbf{v}(\ ) = a_1 \ell_{1\mathbf{v}}(\ ) + a_2 \ell_{2\mathbf{v}}(\ ) + a_3 \ell_{3\mathbf{v}}(\ ) + \ldots + a_m \ell_{m\mathbf{v}}(\ ) \quad (5.17)$$

with

$$\sum_{i}^{m} a_i = 1$$

where the a_i are weighing coefficients. Each component $\ell_{iv}(\ )$ is selected for the ith limit state in the same way as for an individual limit state, with most interest being the regions contributing the greatest probability density mass for the limit state. Points such as A, B, C and D in Fig. 5.11(a) represent the points of maximum probability density and these may be used as systematically obtainable surrogates for these regions (see section 3.4.6.2).

Normally, not all limit states will be of equal importance for a reliability analysis. This can be taken into account by appropriate selection of the weighing coefficients a_i. In particular the calculations will be simplified if those limit states contributing in only a minor way to p_f can be identified. One way in which this can be done is with reference to FOSM concepts. The suggested algorithm runs as follows [Melchers, 1984].

(a) For each limit state i determine $\mathbf{x}_i^*$ the point in n space $\mathbf{x}$ having the highest probability density $f_\mathbf{X}(\ )$ consistent with $G_i(\mathbf{X}) \leq 0$.
(b) For each $\mathbf{x}_i^*$ calculate

$$\delta_i = \left[\sum_{j=1}^{n} (y_j^*)_i^2 \right]^{1/2}$$

with $(\mathbf{y}^*)_i$ given by $y_j^* = (x_j^* - \mu_{Xj})/\sigma_{xj}$ (i.e. a "standardized" space such as shown in Fig. 5.11(b) might be visualized in which the relative importance of each limit state function is considered).
(c) Ignore all limit state functions for which $\delta_i > \delta_L$, where δ_L is some arbitrarily chosen limit. As a first-order approximation, the error in p_f associated with any limit state which is ignored in this way is given by $p_{\text{error}} \approx \Phi(-\delta_L)$.
(d) For the remaining k limit states, use (5.17) as the sampling function in (5.16) with a_i chosen on the basis of the δ_i values.

The total number of sample points required in this approach if M points are to fall in the failure region is $2kM$, which is considerably less than M/p required for crude Monte Carlo, unless k is extremely (and impracticably) large.

When the limit states are closely clustered over all or part of the region of integration, one sampling function $\ell_v(\ )$ may be used for such limit states considered as a group. Such an approach might be invoked when

$$\Delta_{ij} = \left[\left| \sum_{k=1}^{n} (y_k^*)_i^2 - (y_k^*)_j \right| \right]^{1/2}$$

(which represents a measure of the nominal distance between the "checking points" for the ith and jth limit states in the hypothesized "standardized" space) is less than some criterion Δ_C, say. The suggested coordinates for $\ell_v(\ )$ would be the mean of x_i^*, x_j^*. This concept is readily extendable to situations with more than two-limit-state "checking points" in a cluster. A suggested criterion Δ_C might be one standard deviation in **Y** space.

5.3.2 Parallel systems

For parallel systems the probability content associated with the intersection of two or more limit states is given by (5.12), which may be rewritten in the form (5.16) but with $I[\ ]$ defined as

$$I\left[\bigcap_{i=1}^{n} G_i(\mathbf{x}) \leqslant 0\right].$$

Typically the regions of most interest are those bounded by the appropriate limit state functions and near their intersection. Again, appropriate systematically determinable surrogates for these regions are the co-ordinates of the intersection(s) of the appropriate limit state functions, such as points I_{12}, I_{23} in the two-dimensional space of Fig. 5.11(a). It follows readily that the procedure of the previous section applies directly but now with I_{ij}, etc., instead of points A, B, etc. Since the sampling efficiency may be quite low (e.g. at point I_{23} Fig. 5.11(a)), rather more sample points may need to be used.

The point I_{ij} in **x** is obtained, of course, directly by equating limit state functions

$$G_i(\mathbf{x}) = G_j(\mathbf{x}) = 0$$

for a two-dimensional intersection in n (hyper)space. In general, if the dimension of the intersection is less than that of the (hyper)space, I refers to a (hyper)crease in $f_{\mathbf{X}}(\ )$ space, and the "checking" point **x*** will then lie along the (hyper)crease. If the dimension of the intersection equals that of the (hyper)space, **x*** will be at the "vertex" formed by the intersection of the limit state functions. In two dimensions this is equivalent to the points I in Fig. 5.11(a).

It is important to note, however, that **x*** may not lie at I, or along a crease. This may occur where the limit state functions involved in the interaction are significantly different in their contribution to p_f. Consider limit state $G_2(\mathbf{x})$ translated to the left in Fig. 5.11(a), i.e., to the location marked $G_2'(\mathbf{x})$. It is immediately evident that the intersection $G_2'(\mathbf{x}) \cap G_1(\mathbf{x})$ has its greatest probability density at A, and not at I'_{12}. The extension to multidimensional space is obvious; the implication is that the (hyper)cusps and/or (hyper)creases are likely points of greatest probability density, and hence good surrogates for the region of interest, but other locations may need to be checked. This may be accomplished using mathematical programming to locate the points of maximum probability density as indicated in section 3.4.6.

5.3.3   Failure modes identification

By proper accounting of the Monte Carlo results it is possible to identify the most common failure modes for a structure, given that the modes of failure for individual members and the failure criteria for the structure have been specified. (The problem makes no sense otherwise). Limited experience suggests that rather large samples are required to discriminate between all important modes of failure, particularly for structural failure probabilities of realistic magnitude. This is because so few samples tend to fall in the failure regions with the crude Monte Carlo technique.

Direct application of the importance-sampling technique so far discussed is not yet feasible since the limit state functions $G_i(X)$ in Fig. 5.11(a) are not yet known. One approach is to use $f_X(\ )$ (or some simplified form of it) as the sampling distribution ℓ_V but with very much larger coefficients of variation. This tends to spread the region over which sample points are taken and results in rather more samples in the various failure regions [Vrouwenvelder, 1983]. However, the procedure is still quite inefficient. An alternative approach, based on enumeration, is outlined in section 5.5.2.

5.4   SYSTEM RELIABILITY BOUNDS

Rather than proceed with the direct integration of expressions (5.7) and (5.12), an alternative approach is to develop upper and lower bounds on the probability of failure of a structural system. Consider a structural system subject to a sequence of loadings and which may fail in any one (or more) of a number of possible failure modes under any one loading in the loading sequence. The total probability of structural failure may then be expressed in terms of mode failure probabilities as (see section A.1)

$$P(F) = P(F_1) \cup P(F_2 \cap S_1) \cup P(F_3 \cap S_2 \cap S_1)$$
$$\cup P(F_4 \cap S_3 \cap S_2 \cap S_1) \cup \ldots \qquad (5.18)$$

where F_i denotes the event "failure of the structure due to failure in the ith mode, for all loading" and S_i denotes the complementary event "survival of the ith mode under all loading (and hence survival of the structure). Since $P(F_2 \cap S_1) = P(F_2) - P(F_2 \cap F_1), \ldots$, expression (5.18) may be written also as

$$P(F) = P(F_1) + P(F_2) - P(F_1 \cap F_2) + P(F_3) - P(F_1 \cap F_3)$$
$$- P(F_2 \cap F_3) + P(F_1 \cap F_2 \cap F_3) + \ldots \tag{5.19}$$

where $(F_1 \cap F_2)$ is the event that failure occurs in both modes 1 and 2, etc.

5.4.1 First-order series bounds
The probability of failure for the structure can be expressed as $P(F) = 1 - P(S)$, where $P(S)$ is the probability of survival. For independent failure modes, $P(S)$ can be represented by the product of the mode survival probabilities, or, noting that $P(S_i) = 1 - P(F_i)$, by

$$P(F) = 1 - \prod_{i=1}^{m} [1 - P(F_i)] \tag{5.20}$$

where, as before, $P(F_i)$ is the probability of failure in mode i. This result can, by expansion, be shown to be identical with (5.19). It follows directly from (5.19) that, if $P(F_i) \ll 1$, then (5.20) can be approximated by [Freudenthal et al., 1966]

$$P(F) \approx \sum_{i=1}^{m} P(F_i) \tag{5.21}$$

In the case where all failure modes are fully dependent, it follows directly that the weakest failure mode will always be weakest, irrespective of the random nature of the strength. Hence

$$P(F) = \max_{i=1}^{m} [P(F_i)] \tag{5.22}$$

Equations (5.20) or (5.21) and (5.22) can be used to define relatively crude bounds on the failure probability of any structural system of the series type when the failure modes are neither completely independent nor fully dependent [Cornell, 1967]:

$$\max_{i=1}^{m} [P(F_i)] \leq P(F) \leq 1 - \prod_{i=1}^{m} [1 - P(F_i)] \tag{5.23}$$

Unfortunately for many practical structural systems the series bounds (5.23) are too wide to be meaningful [cf. Grimmelt and Schuëller, 1982]. Better bounds can be developed, but at the expense of more computation.

Example 5.4
For the rigid–plastic frame shown in Fig. 5.9 the mode safety indices $\beta_a, \ldots,$ β_d were calculated in Example 5.2 as (4.32, 4.83, 6.44, 7.21). From Appendix C, the corresponding (nominal) failure probabilities are $(0.77 \times 10^{-5}, 0.70 \times 10^{-6}, 0.59 \times 10^{-10}, 0.28 \times 10^{-12})$ and from (5.23) the first-order system failure probability is then bounded by

$$\max_{i=1}^{4}(p_i) \leq p_f \leq 1 - \prod_{i=1}^{4}(1 - p_i) \approx \sum_{i=1}^{4} p_i$$

or

$$0.77 \times 10^{-5} \leq p_f \leq 0.84 \times 10^{-5}$$

It is evident that modes c and d have negligible effect on the failure probability of the structure.

5.4.2 Second-order series bounds
Second-order bounds are obtained by retaining terms such as $P(F_1 \cap F_2)$ in expression (5.19), which for ease of exposition, may be rewritten as

$$
\begin{aligned}
P(F) = P(F_1) & \\
+ P(F_2) & - P(F_1 \cap F_2) \\
+ P(F_3) & - P(F_1 \cap F_3) - P(F_2 \cap F_3) + P(F_1 \cap F_2 \cap F_3) \\
+ P(F_4) & - P(F_1 \cap F_4) - P(F_2 \cap F_4) - P(F_3 \cap F_4) + P(F_1 \cap F_2 \cap F_4) \\
& + P(F_1 \cap F_3 \cap F_4) + P(F_2 \cap F_3 \cap F_4) - P(F_1 \cap F_2 \cap F_3 \cap F_4) \\
+ P(F_5) & - \ldots \\
= \sum_{i=1}^{m} P(F_i) & - \sum\sum_{i<j} P(F_i \cap F_j) + \sum\sum\sum_{i<j<k} P(F_i \cap F_j \cap F_k) - \ldots (5.24)
\end{aligned}
$$

Because of the alternating signs as the order of the terms increases, it is evident that consideration only of first-order terms (i.e. $P(F_i)$) produces an upper bound on $P(F)$, consideration only of first- and second-order terms a lower bound, first-, second- and third-order terms again on upper bound, and so on [Bonferroni, 1936].

It should also be clear that consideration of an additional failure mode cannot reduce the probability of structural failure, so that each complete line in equation (5.24) makes a non-negative contribution to $P(F)$. Noting that $P(F_i \cap F_j) \geq P(F_i \cap F_j \cap F_k), \ldots,$ a lower bound to (5.24) can be obtained if only the terms $P(F_i) - P(F_i \cap F_j)$ are retained, provided the each makes a non-negative contribution [Ditlevsen, 1979b]:

$$P(F) \geqslant P(F_1) + \sum_{i=2}^{m} \max\left\{\left[P(F_i) - \sum_{j=1}^{i-1} P(F_i \cap F_j)\right], 0\right\} \qquad (5.25)$$

An alternative way of using $P(F_i)$ and $P(F_i \cap F_j)$ terms is to select only those combinations of all such terms in (5.24) which give the maximum value (of the lower bound) [Kounias, 1968]:

$$P(F) \geqslant P(F_1) + \max\left\{\sum_{\substack{i=2 \\ j<i}}^{k \leqslant m} [P(F_i) - P(F_i \cap F_j)]\right\} \qquad (5.26)$$

In both formulations the result depends on the order in which the various modes of failure are labelled. Algorithms for optimal ordering of events to obtain the best bounds have been proposed [Dawson and Sankoff, 1967; Hunter, 1977]; a useful rule of thumb is to order the modes in order of decreasing importance. For a given ordering, (5.25) may give a better bound than (5.26); both bounds are equal if all possible orderings are considered [Ramachandran, 1984].

An upper bound may be obtained by simplifying each line in (5.24). As noted, a typical line, such as line 5, makes a non-negative contribution to $P(F)$ and can be written, with P_{ijk} for $P(F_i \cap F_j \cap F_k)$ etc., as

$$\begin{aligned} U_5 = \; & P_5 - P_{15} - P_{25} - P_{35} - P_{45} + P_{125} + P_{135} + P_{145} + P_{235} + \\ & + P_{245} + P_{345} \\ & - P_{1235} - P_{1245} - P_{1345} - P_{2345} + P_{12345} \end{aligned} \qquad (5.27)$$

Apart from the term P_5, the rest of the line can be written as

$$-V_5 = -P(E_{15} \cup E_{25} \cup E_{35} \cup E_{45}) \qquad (5.28)$$

where E_{ij} represents the event ij. For any pair of events A, B, say, it is well known that $P(A \cup B) \geqslant \max[P(A), P(B)]$. It follows readily that

$$V_5 \geqslant \max[P(E_{15}), P(E_{25}), P(E_{35}), P(E_{45})] \qquad (5.29)$$

and, since V_5 makes a negative contribution to U_5, use of the bound (5.29) will increase the right-hand side of (5.27). Hence

$$U_5 \leqslant P_5 - \max_{j<5}(P_{j5}) \qquad (5.30)$$

but, since $P_{j5} \equiv P(F_j \cap F_5)$, and since line 5 was a typical example, it follows that [Kounias, 1968; Vanmarcke, 1973; Hunter, 1976; Ditlevsen, 1979b]

$$P(F) \leqslant \sum_{i=1}^{m} P(F_i) - \sum_{i=2}^{m} \max_{j<i} [P(F_j \cap F_i)] \qquad (5.31)$$

This result may also depend on the ordering of failure events F_i.

Bounds (5.25) and (5.31) have been compared with (crude) Monte Carlo simulation results for a series of rigid-framed, rigid–plastic frames [Grimmelt and Schuëller, 1982]. It was found that for these realistic structures and for a range of distribution types and a range of variances the bounds were mainly quite close to the simulation results. However, the bounds are not necessarily always close [Ditlevsen, 1979b; Ditlevsen, 1982a].

Example 5.5
Consider the three linear limit state functions shown in Fig. 5.12(a) in two-

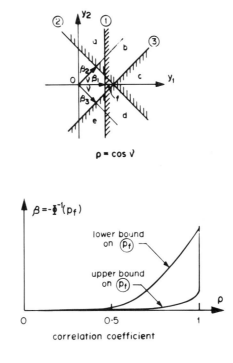

Fig. 5.12(a) — Limit states for Example 5.5; (b) typical bounds for equi-correlated limit states.

dimensional **y** space. For simplicity, the safety indices β_1, β_2 and β_3 are shown of equal length. The probability contents enclosed by each pair of limit state functions have been denoted $a, b, \ldots, e,$ and f, the last bounded by all three limit state functions.

For three limit state functions, the lower bound (5.25) becomes

$$p^- = p_1 + (p_2 - p_{21})^+ + (p_3 - p_{31} - p_{32})^+ \qquad \text{with } (\ )^+ \equiv \max(\ , 0)$$
$$= (b + c + d + f) + [(a + b + c) - (b + c)]^+ + [(c + d + e) - [(d + c) - c]^+$$
$$= a + b + d + e + f$$

while the upperbound (5.31) is

$$p^+ = p_1 + p_2 + p_3 - [\max(p_{21}) + \max(p_{31}, p_{32})]$$
$$= (b + c + d + f) + (a + b + c) + (c + d + e) - [(b + c) + \max(c + d, c)]$$
$$= a + b + c + d + e + f$$

In this case, the upper bound gives the correct result. As the limit states 1, 2, 3 become more correlated, i.e. as the angle v becomes smaller (see section 3.2.3), the probability content of region c will increase and the bounds will diverge further. This is indicated schematically in Fig. 5.12(b). When all three limit states coincide, they are perfectly correlated (dependent) so that (5.22) gives a unique result for p_f. This is responsible for the cut-off effect shown at the right of Fig. 5.12(b).

5.4.3   Second-order series bounds by loading sequences

The series bounds discussed so far consider failure under all loadings, but according to different modes of failure. It may be shown that these bounds are symmetric with respect to loading sequences with the failure event now defined over all modes of failure [Melchers, 1981]. From expression (5.18) the probability of structural failure under a sequence of loading vectors $\mathbf{Q}_1$, $\mathbf{Q}_2, \mathbf{Q}_3, \ldots$, may be expressed as

$$P(F) \;=\; P(F_1) + P(F_2|S_1)P(S_1) + P(F_3|S_2 \cap S_1)P(S_2 \cap S_1) + \ldots \tag{5.32}$$

where F_i denotes the event "structural failure due to the ith loading", and S_i denotes the complementary event "structure survival under the ith loading", $P(F_i) = 1 - P(S_i)$.

In general there will be dependence between failure under the ith loading and survival under previous loadings in the sequence; let this be denoted the "transition" probability

$$p_i \;=\; P(F_i|S_{i-1} \cap S_{i-2} \cap S_{i-3} \cap \ldots) \tag{5.33}$$

Since

$$P(S_3 \cap S_2 \cap S_1) = P(S_3|S_2 \cap S_1)P(S_2 \cap S_1) = P(S_3|S_2 \cap S_1)P(S_2|S_1)P(S_1)$$

and similarly for other corresponding terms in equation (5.32), it follows, upon using equation (5.33) and writing p_1 for $P(F_1)$, that

$$P(F) = p_1 + p_2(1 - p_1) + p_3(1 - p_2)(1 - p_1) + p_4(1 - p_3)(1 - p_2)(1 - p_1) + \ldots \qquad (5.34)$$

or, for n loads in the loading sequence

$$P(F) = \sum_{i=1}^{n} p_i - \sum_{j<i}^{n} \sum p_i p_j + \sum_{k<j<i}^{n} \sum \sum p_i p_j p_k -$$

$$- \sum_{\ell<k<j<i}^{n} \sum \sum \sum p_i p_j p_k p_\ell + \ldots \qquad (5.35)$$

which is identical with equation (5.24) provided that $P(F_i)$ is interpreted as p_i, $P(F_iF_j)$, $j < i$ is interpreted as p_ip_j, $j < i$, etc. Hence, with this interpretation, the bounds (5.25) and (5.31) can be interpreted as bounds on $P(F)$ for structural failure under load sequences.

5.4.4 Series bounds by modes and loading sequences
The series bounds (5.25) and (5.31) for failure modes can be generalized to include loading sequences. For failure in several modes, each of the bounds given by equations (5.25) or (5.31) can be interpreted as the probability $P(F_k)$ of failure under kth loading in a load sequence. To determine the total probability of failure, under the complete loading sequence, equations (5.25) and (5.31) are again applicable, but now interpreted for loading sequences as given in section 5.4.3 above.

In practice, however, it may not always be easy to evaluate the terms $P(F_i \cap F_j)$ since F_i and F_j will usually be correlated (see section 5.2.3 and Example 5.5). One approach is then to use a simpler result obtained by considering two limiting cases for the evaluation of the terms $P(F_i \cap F_j)$: complete independence of (F_i, F_j) or complete dependence. If the events (F_i, F_j) are completely independent, (A.4) applies and (5.19) reduces to (5.20).

Similarly, if F_i, F_j are completely dependent, then all terms of form $P(F_1 \cap F_2 \cap _3)$, $\ldots$, reduce to $\max[P(F_1), P(F_2), P(F_3)]$, $\ldots$, since the critical case will govern. As a result, equation (5.18) reduces to (5.22).

It is now possible to extend the first order bounds (5.23) to consider the effect of both loading sequences and failure modes:

$$\max_{i}^{m} \left\{ \max_{j}^{n} [P(F_{ij})] \right\} \leq P(F) \leq 1 - \prod_{ij}^{mn} [1 - P(F_{ij})] \qquad (5.36)$$

where $P(F_{ij})$ is the probability of failure in the ith mode under the jth load in the loading sequence. The right-hand bound may be replaced by a slightly looser bound, provided that $P(F_{ij}) \ll 1$ [Cornell, 1967]:

$$\max_i \left\{ \max_j [P(F_{ij})] \right\} \leq P(F) \leq \sum_j^m \sum_i^n P(F_{ij}) \tag{5.37}$$

When it is known that either load sequences or failure modes are independent, the appropriate left-hand maximum operator can be replaced by a summation operator to improve the bounds; similarly, if it is known that either load sequences or failure modes are completely dependent, the appropriate right-hand summation operator can be replaced by a maximum operator.

If it is known that the load sequence consists of N successive, mutually exclusive independent loads having the same probability density function, it follows readily that the right-hand bound can be replaced by

$$N \sum_i^m P(F_{ij})$$

[Freudenthal *et al.*, 1966].

Because the first-order bounds (5.37) are based on the extremes of full or no dependence between failure events, they are often rather too wide for practical application. Usually second-order bounds are preferred.

5.4.5 Improved series bounds and parallel system bounds
As demonstrated in Example 5.5, the second-order series bounds generally deteriorate as the correlation between (linear) limit state functions increases. One approach towards obtaining improved series bounds is therefore to attempt to transform the problem to one resulting in lower correlation between limit state functions. This may be achieved using expression (3.12) to obtain a new set of variables [Ditlevsen, 1982b].

Alternatively, improved bounds for series systems may be obtained if, for example, terms of the form $P_{ijk} = P(F_i \cap F_j \cap F_k)$ are retained in expression (5.19) [Hohenbichler and Rackwitz, 1983a; Ramachandran, 1984]. Again some ordering of failure events may be necessary to obtain the best bounds. However, the greatest difficulty is to evaluate the trisection P_{ijk}. When the events F_i are all expressible as linear functions, a non-linear lower bound is given by [Ramachandran, 1984]

$$P_{ijk} \geq \frac{P(F_i \cap F_k)P(F_i \cap F_j)}{P(F_i)} \tag{5.38}$$

provided that $\rho_{kj} > \rho_{ik}\rho_{ij} > 0$ where ρ_{ij} is the correlation coefficient between the linear failure functions for events i and j, etc.

For parallel systems the failure probability is given by (5.2) or (5.11). Good-quality bounds for this expression have not yet been developed, although rather poor bounds can be obtained from applying series bounds to the right-hand side of the identity:

$$P\left(\bigcap_{i=1}^{m} F_i\right) = 1 - P\left(\bigcup_{i=1}^{m} F_i\right) \tag{5.39}$$

The resulting bounds are poor since the second term on the right will be close to unity for high-reliability systems.

In the special case of parallel systems with linear safety functions, a better approach is to apply the result given in section 4.3.8.2 for the multinormal integral with multiple linear limit state functions.

5.4.6   First-order second-moment method in systems reliability

Utilization of the second-order series bounds (5.25) and (5.31) requires evaluation of terms of the form $P(F_i \cap F_j)$ where F_i denotes the event "failure in limit state i". The intersection terms refer to domains such as D_1 shown bounded by the non-linear limit state functions $G_k(\mathbf{x}) = 0$ ($k = 1, 2, 3$) in Fig. 5.6.

It would be possible to use Monte Carlo integration to evaluate the probability content $P(F_i \cap F_j)$ for use with the bounds given above. This is seldom appropriate, however, since a Monte Carlo approximation of the whole system problem is likely to be more efficient than the use of system bounds together with Monte Carlo evaluation of the intersections.

The second-order series bounds are more useful when used in conjunction with linear limit state functions and with the FOSM methods of Chapter 4. The evaluation of the intersections $P(F_1 \cap F_2)$ is then over regions such as D_1 in Fig. 5.13 bounded by, for example, $g_1(\mathbf{y}) = 0$ and $g_2(\mathbf{y}) = 0$. These limit state functions can be linearized, once the checking points $\mathbf{y}_i^*$ have been identified, and the direction cosines $\boldsymbol{\alpha}$ determined (see section 4.3.2). The approximating linearized limit state function $g_L(\ ) = 0$ is then given by (4.7).

For linearized limit states given by $g_{L1}(\ ) = 0$ and $g_{L2}(\ ) = 0$, the probability content enclosed by these limit states is

$$P(F_1 \cap F_2) = P\left[\bigcap_{i=1}^{2} g_{Li}(\mathbf{y}) \leq 0\right] \tag{5.40}$$

which for the standardized normal vector $\mathbf{y}$ can be evaluated exactly using

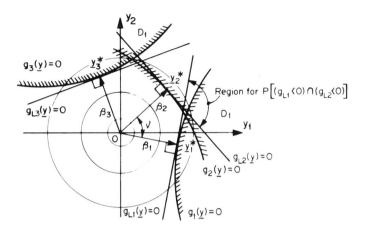

Fig. 5.13 — Linearization of limit states.

the bivariate normal integral $\Phi_2(\ )$ (see (3.7) or (A.140b)] or by means of the bounds (3.8). In each case the correlation coefficient ρ is require to be known. This can be obtained for two intersecting limit states 1 and 2 as follows.

In standardized independent normal **y** space, a linear limit state function is given by (4.7):

$$g_i(\mathbf{y}) \;=\; \beta_i + \sum_{j=1}^{n} \alpha_{ij} y_j \tag{5.41}$$

For g_1 and g_2, the variance and covariance are then (A.162) and (A.163)

$$\sigma_i^2 \;=\; \sum_{j=1}^{n} \alpha_{ij}^2 \;, \qquad i = 1,\, 2$$

$$\mathrm{cov}(g_1, g_2) \;=\; \sum_{j}^{n} \alpha_{1j} \alpha_{2j}$$

and from (A.124)

$$\rho_{12} \;=\; \frac{\mathrm{cov}(g_1, g_2)}{(\sigma_1^2 \sigma_2^2)^{1/2}}$$

$$= \frac{\sum\limits_{j}^{n} \alpha_{1j}\alpha_{2j}}{\left(\sum\limits_{j}^{n} \alpha_{1j}^2 \sum\limits_{j}^{n} \alpha_{2j}^2\right)^{1/2}} \tag{5.42}$$

These expressions can be given a geometric interpretation when attention is restricted to the plane, in **y** space, containing only y_1 and y_2 (see Fig. 5.14).

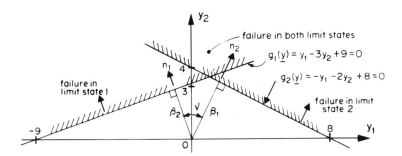

Fig. 5.14 — Intersection of linear limit states: Example 5.6.

From elementary geometry, the unit outward normals $\mathbf{n}_i$ to the hyperplanes g_1 and g_2 shown in Fig. 5.14 are obtained from (5.41) [Ditlevsen, 1979b]

$$\mathbf{n}_1 = \frac{1}{(\Sigma\alpha_{1j})^{1/2}}\left(\sum_{j=1}^{n} \alpha_{1j}\mathbf{e}_j\right) \tag{5.43a}$$

$$\mathbf{n}_2 = \frac{1}{(\Sigma\alpha_{2j})^{1/2}}\left(\sum_{j=1}^{n} \alpha_{2j}\mathbf{e}_j\right) \tag{5.43b}$$

where $\mathbf{e}_i$ represent the unit base vectors and $\mathbf{n}_j$ represent unit normal vectors.

The scalar product of these normals is given by $\mathbf{n}_1 \cdot \mathbf{n}_2 = n_1 n_2 \cos v = \cos v$, where v is the angle between $\mathbf{n}_1$ and $\mathbf{n}_2$ in the plane common to both (see Fig. 5.14). It follows readily from expressions (5.43) that the scalar product is also equal to the right-hand side of expression (5.42) so that (cf. section 3.2)

$$\rho_{12} = \rho(y_1, y_2) = \cos v = \mathbf{n}_1 \cdot \mathbf{n}_2 \tag{5.44}$$

Example 5.6

For the two linear limit state functions $g_1(\mathbf{y}) = 0$ and $g_2(\mathbf{y}) = 0$ shown in
Fig. 5.14, in standardized normal space, it follows from (A.162) and (A.163)
that $\sigma_1^2 = +10$, $\sigma_2^2 = +5$ and $\mathrm{cov}(y_1, y_2) = +5$ from which
$\rho = 5/(10 \times 5)^{1/2} = 1/\sqrt{2} > 0$ so that v arcos$(1/\sqrt{2}) = 45°$. Using (4.1),
$\beta_1 = 9/\sqrt{10} = 2.85$ and $\beta_2 = 8/\sqrt{5} = 3.59$. Applying now expressions (3.8)
and (3.9) it follows that $h = \beta_1$, $k = \beta_2$, $a = (\beta_1 - \rho\beta_2)/(1 - \rho^2)^{1/2}$ and
$b = (\beta_2 - \rho\beta_1)/(1 - \rho^2)^{1/2}$ so that $\Phi(-h) = 0.002\,186$, $\Phi(-k) = 0.000\,165$,
$\Phi(-a) = 0.3264$, $\Phi(-b) = 0.0135$. Substituting these values in (3.8)
yields $0.000\,055 \leqslant P(g_1 \leqslant 0 \cap g_2 \leqslant 0) \leqslant 0.000\,084$.

Let the event $g_1\ 0$ be denoted F_1 and similarly for F_2. Expressions (5.25)
and (5.31) for the failure probability bounds for a two-failure mode structure
then lead to

$$P(F_1) + P(F_2) - P(F_1 \cap F_2)^+ \leqslant P(F) \leqslant P(F_1) + P(F_2) - P(F_1 \cap F_2)^-$$

or $0.002\,26 \leqslant p_f \leqslant 0.002\,29$, where the choices for the intersections have
been made to give the widest bounds.

Example 5.7

The nominal failure probability for the rigid–plastic frame of Example 5.2
will now be determined using second-order system bounds and the FOSM
result for $P(F_i \cap F_j)$. Only the first three collapse modes of Example 5.2 will
be considered. These are

mode 1: M_1 $2M_3 + 2M_4 - H - V = 0$ (combined)

mode 2: $M_2 + 2M_3 + M_4 \quad\ - V = 0$ (beam)

mode 3: $M_1 + M_2$ $M_4 - H \qquad\ \ = 0$ (sway)

Let each random variable be standardized to $N(0, 1)$, such that $x_i = (X_i - \mu_{X_i})/\sigma_{X_i}$. Noting that $\mu_{X_i} = 1.0$ for all $X_i = (M_1, \ldots, M_4, H, V,)$ and
that $\sigma_{X_i} = (0.15, 0.15, 0.15, 0.15, 0.17, 0.50)$ it follows that the limit state
equations in standard normal space are

$$g_1 = 0.15m_1 + 0.30m_3 + 0.30m_4 - 0.17h - 0.5v + 3 = 0$$
$$g_2 = 0.15m_2 + 0.30m_3 + 0.15m_4 \qquad\quad - 0.5v + 3 = 0$$
$$g_3 = 0.15m_1 + 0.15m_2 + 0.15m_4 - 0.17h \qquad + 2 = 0$$

Further, using (A.162) and (A.163)

$$\sigma_{g_1}^2 = (0.15)^2 + (0.3)^2 + (0.3)^2 + (0.17)^2 + (0.5)^2 = 0.481$$
$$\sigma_{g_2}^2 = 0.385$$

$$\sigma_{g_3}^2 = 0.096$$

$$\text{cov}(g_1, g_2) = (0 + 0 + 0.3 \times 0.3 + 0.3 \times 0.15) + 0 + (0.5)^2 = 0.385$$

$$\text{cov}(g_1, g_3) = 0.096$$

$$\text{cov}(g_2, g_3) = 0.045$$

From (5.42) it follows that the correlation coefficients are

$$\rho_{12} = \frac{0.385}{(0.481 \times 0.385)^{1/2}} = 0.895$$

$$\rho_{13} = 0.447$$

$$\rho_{23} = 0.234$$

so that the angles $v = \text{arcos } \rho$ are $v_{12} = 26.5°$, $v_{13} = 63.5°$ and $v_{23} = 76.5°$.

The β indices were calculated already in Example 5.2 and are also given by μ_{g_i}/σ_{g_i} or

$$\beta_1 = \frac{3}{\sqrt{0.481}} = 4.32, \ \beta_2 = 4.83, \ \beta_3 = 6.44$$

since $\mu_{x_i} = 0$. From Appendix C the corresponding nominal probabilities are

$$0.77 \times 10^{-5}, \ 0.70 \times 10^{-6}, \text{ and } 0.59 \times 10^{-10}.$$

It is now possible to calculate the terms $P(F_i \cap F_j)$ needed for the second-order bounds (5.25) and (5.31). This can be done by using the bounds given by (3.8) (see also Example 5.6). Writing p_{ij} for $P(F_i \cap F_j)$, for p_{12},

$$h = \beta_1 = 4.32, \ k = \beta_2 = 4.83$$

$$a = \frac{\beta_1 - \rho_{12}\beta_2}{(1 - \rho_{12}^2)^{1/2}} = \frac{4.32 - 0.895 \times 4.83}{(1 - 0.895^2)^{1/2}} = -0.0064$$

$$b = \frac{\beta_2 - \rho_{12}\beta_1}{(1 - \rho_{12}^2)^{1/2}} = \frac{4.83 - 0.895 \times 4.32}{(1 - 0.895^2)^{1/2}} = 2.160$$

so that, according to (3.8), the bounds are

$$p_{12}^+ = [\Phi(-h)\Phi(-b)^+, \ \Phi(-k)\Phi(-a)]$$
$$= [(0.77 \times 10^{-5})(0.01539)^+, (0.70 \times 10^{-6})(\sim 0.5)]$$
$$= (0.35 \times 10^{-6}, \ 0.47 \times 10^{-6})$$

for p_{13},

$$p_{13}^- = 0.317 \times 10^{-11}, \; p_{13}^+ = 0.479 \times 10^{-11}$$

and, for p_{23}

$$p_{23}^- = 0.189 \times 10^{-13}, \; p_{23}^+ = 0.374 \times 10^{-13}$$

Substituting into (5.25) to obtain the lower bound on the system failure probability,

$$\begin{aligned} p_f^- &= 0.77 \times 10^{-5} + (0.70 \times 10^{-6} - 0.47 \times 10^{-6})^+ + \\ &\quad (0.59 \times 10^{-10} - 0.479 \times 10^{-11} - 0.374 \times 10^{-13})^+ \\ &= 0.79 \times 10^{-5} \end{aligned}$$

The upper bound is given by (5.31):

$$\begin{aligned} p_f^+ &= (0.77 \times 10^{-5} + 0.70 \times 10^{-6} + 0.59 \times 10^{-10}) - 0.35 \times 10^{-6} \\ &\quad - \max(0.317 \times 10^{-5}, 0.189 \times 10^{-13}) \\ &= 0.81 \times 10^{-5} \; . \end{aligned}$$

In each case, the p_{ij} bound giving the worst p_f^+ value has been chosen.

The bounds on the nominal probability of failure for the system coincide within the limit of accuracy of the calculations used here. As expected, the present result is contained within the first-order bounds determined in Example 5.4.

5.4.7 Correlation effects

The above examples show that the failure modes are correlated if some of or all of the resistance and load basic variables are shared between two or more of the limit state functions. Correlation can also arise from dependence between loads and, more commonly, from correlation between structural members and within individual members [Garson, 1980; Melchers, 1983a].

Computationally, correlation of basic variables in second-moment reliability analysis is dealt with by transforming the correlated set to an uncorrelated set (see Appendix B and section 4.4.3).

A set of typical results showing correlation effects for a rigid–plastic portal frame is given in Fig. 5.15 [Frangopol, 1985]. Experimental data on actual strength (and load) correlation are, however, scarce. In practical problems conservative assumptions will need to be made.

5.5 COMPLEX STRUCTURAL SYSTEMS*

5.5.1 Description

For structures which are not highly statically indeterminate, it is usually possible to determine the various limit states by inspection, such as was done

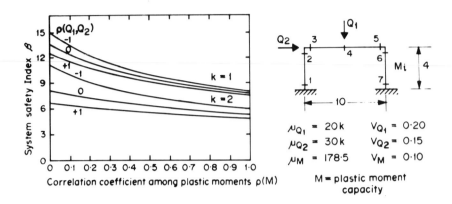

Fig. 5.15 — Effect of correlation on system safety index $\beta = -\Phi^{-1}(p_f)$.

in Example 5.2. For a more complex framed structure (such as an offshore "jacket" structure) the limit state functions which describe the structure failure conditions will depend on the degree of indeterminacy, the loading(s) applied and the member response characteristics. Complex framed structural systems will be taken here as highly statically indeterminate ones. Time-independent or extreme value loading will again be assumed; some ways in which the effect of time may influence the reliability assessment will be considered in Chapter 6. The concepts to be discussed here can be directly extended to cases where time needs to be considered explicitly.

In a reliability analysis using the failure mode approach (see section 5.2.2), all modes of failure making a significant contribution to the failure probability must be identified; if this is not done, the probability calculated will be too low. One way of identifying all structural failure modes is by exhaustive enumeration of all possible combinations of member failure (or failure of local regions within the structure).

In order not to limit discussion to any particular type of structure, members in trusses, or local failure regions in frames (e.g. plastic hinges) will be denoted "nodes"; these may have any appropriate strength–deformation behaviour. Some of or all the node properties may be dependent, so that in general, it would be expected that any failure mode for a structural system will consist of a dependent sequence of nodes. This dependence must be taken into account in evaluating the system reliability.

A simple but primitive procedure for exhaustive enumeration is to consider systematically the sequencing possibilities of node failure. Each node can be considered in turn and added to a sequence of previous node failure. Each new sequence is examined to ascertain whether a valid structural failure mode has been attained; if not, a further node may be added. When structural failure is defined as collapse, the failure mode must be kinematically admissible (i.e. a valid collapse mode must exist); in addition, there must be correspondence everywhere between internal actions and local strains. Equilibrium must, of course, always be satisfied. It is not always easy to ensure that all these requirements are met. When structural failure is defined in some other way, the requirements to be met must again be of an appropriate form.

Continuing the above procedure, when a failure mode has been identified, enumeration of the next failure mode is achieved by "back-tracking" until a new combination or ordering of nodes becomes possible. The concept is shown schematically in the event tree of Fig. 5.16(a). It will evidently

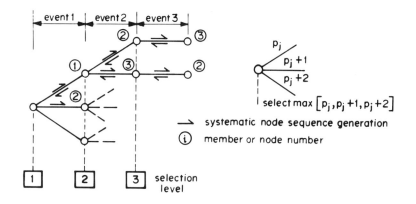

Fig. 5.16 — Systematic enumeration procedure.

produce all failure modes dependent only on node failure, although for complex node behaviour, unloading, stress reversal, loss of stiffness, etc. will also need to be considered as possibilities in the event tree. This will add greatly to computation effort unless some systematic procedure is adopted. For rigid–plastic structures, considerable simplifications are possible since the precise order of node occurrence in a sequence of modes should not matter.

It would be extremely beneficial in terms of computational effort if modes which make little contribution to the structural failure probability can be identified (and hence eliminated) early in the enumeration procedure, preferably before the full calculation of mode failure probability is carried out. A technique which satisfies this requirement is the so-called "truncated enumeration" method [Melchers and Tang, 1984] outlined below.

5.5.2 Truncated enumeration

The truncated enumeration method identifies sequentially the likely occurrences of node sequences forming structural failure modes. The approach used is to increment one of the applied (random) loads acting on the structure and to identify, in turn, the first, second, third, etc., node to reach (local) node failure. Because node strength properties are random, actual load levels at which nodes fail are not identified; only the sequences are developed.

For any one load increment, an elastic stress analysis is used to identify the node which is most likely to have reached its failure condition; the behaviour of this node is then modified. The way in which node modification occurs is described in section 5.5.3 below.

An elastic stress analysis of the structure is used even in the case of rigid–plastic member (node) properties. This is permissible since for rigid–plastic structures the collapse mode and corresponding stress field should be independent of any elastic properties of the structure.

The probability that the kth failure mode, consisting of n nodes, will occur is given by

$$p_{f_k} = P(E_1 \cap E_2 \cap \ldots \cap E_n) \tag{5.45}$$

where E_i denotes the event "resistance of the ith node exceeded". Such a mode is shown in Fig. 5.16 as 1, 2, 3; in general a number of such modes exist. The number n of nodes required to form a structural failure mode depends on the criterion adopted for structural failure (see section 5.2.1). For collapse of the structure, n must be sufficiently large to allow a valid collapse mode to be formed. The actual evaluation of (5.45) depends on the precise definition of the node failure events E_i and will be discussed later.

The number of sequences (5.45) which need to be considered may be reduced if all modes for which $p_{f_k} \leqslant \delta p_f$ are identified (and hence ignored). Here δ is an appropriately valued "truncation criterion" and p_f is the estimated nominal failure probability for the structural system. Exhaustive enumeration will clearly occur if $\delta = 0$.

Unless a subjective estimate is available prior to the analysis, p_f may be approximated by the most significant (i.e. maximum) value of p_f so far estimated in the procedure, e.g. from an earlier estimation of a mode failure probability p_{f_k} (see (5.45)). Let this value be denoted p_f^*.

$$p_f^* = \max_{\ell=1}^{k-1}(p_{f_\ell}) , \qquad k \geqslant 2 \tag{5.46}$$

Clearly $p_f^* \leqslant p_f$ and therefore the use of (5.46) is conservative, since fewer modes will be rejected.

The mode failure probability p_{f_k} given by (5.45) may be bounded, for $q \leqslant n$, using the inequalities

$$p_{f_k} = P(E_1 \cap E_2 \cap \ldots E_n) \leqslant P(E_1 \cap E_2 \cap \ldots \cap E_q) \ldots$$
$$\leqslant P(E_1 \cap E_2) \leqslant P(E_1) \tag{5.47}$$

For any partial set $q \leqslant n$ of (5.47) the truncation criterion $p_{f_k} \leqslant \delta p_f^*$ is conservatively evaluated and any mode which satisfies it may be ignored immediately.

In principle the node sequence selection may be carried out in any order. However, an efficient algorithm will select the most significant contributing nodes as early as possible. This will then improve the estimate p_f^* in (5.46) and thus eliminate insignificant modes earlier in the procedure. A reasonable strategy is to select the next node such that the probability of occurrence of the partial node sequence including the selected node is maximized. Working from the first to the qth node in a sequence, this means that the selections are made as

first node: $P(E_1) = \max_i[P_i(E_1)]$ (5.48a)

second node: $P(E_1 \cap E_2) = \max_i[P_i(E_1 \cap E_2)]$ (5.48b)

$$\vdots$$

or in general

qth node: $P(E_1 \cap E_2 \ldots \cap E_q) = \max_i[P_i(E_1 \cap E_2 \ldots \cap E_q)]$

(5.48c)

where the maximization is over the set of all "eligible nodes" at each selection level (see Fig. 5.16). Thus, at the qth level, the events E_1 to E_{q-1} are held fixed and a decision is made about selection of event E_q from the remaining nodes.

Simplifications in evaluating (5.48) may be desirable to reduce computation times. Approximations for (5.48c) might be the upper bounds [Murotsu $et\ al.$, 1977, 1983]

$$P(E_1 \cap E_2 \ldots \cap E_q) \leq \max_i[P_i(E_q)] , \qquad q \geq 1 \qquad (5.49a)$$

or

$$P(E_1 \cap E_2 \ldots \cap E_q) \leq \max_i[P_i(E_1 \cap E_q)] , \qquad q \geq 2 \qquad (5.49b)$$

These have been called the "one-dimensional" and the "two-dimensional" branching criterion respective. The physical interpretation of (5.49a), for example, is that the node E_q with the greatest probability of failure is selected, rather than searching for the whole sequence $E_1, E_2 \ldots$, E_q with the highest probability of occurrence. The decision is not necessarily the same as that given by (5.48) since there is usually dependence between the events $E_1, \ldots, E_q$ because the applied loading is common to the stress distribution acting at each node.

The mode failure probability P_{f_k}, conservatively estimated (i.e. overestimated) by (5.47) may be further conservatively estimated by (5.49) for any desirable level q. As p_{f_k} is now overestimated, it follows from (5.46) that p_f^* may also be overestimated and hence some modes may be rejected by the truncation criterion $p_{f_k} \leqslant \delta p_f^*$ (unless the approximations are about equal for p_{f_k} and p_f^*). This is only of importance, however, for those modes which have failure probabilities close to δp_f^*. The most significant modes and therefore those contributing most to the estimate of system failure probability are unlikely to be affected [Tang and Melchers, 1984b].

It is important to note that in principle any method may be chosen to obtain the sequences of node failure which result in structural failure modes. If the method is to be useful it must be able to enumerate all possible failure modes (exhaustive enumeration). Hence it is sufficient to demonstrate that the chosen method can do this; clearly the truncated enumeration method can do so, provided that $\delta = 0$. All other discussion was concerned with how the choice of node sequence is to be made; it is clearly desirable that the choice be efficient, but that is not a necessary condition.

The truncated enumeration procedure relies on being able to formulate the sequence of node failure events and to calculate their probabilities. This requires that a stepwise structural stress analysis procedure is available. Ways of evaluating the probabilities associated with the node failure events will be described in the next section.

Example 5.8
To illustrate the truncated enumeration procedure, it will be sufficient to consider only a simple structure. In this way the complexities introduced by structural analysis are reduced.

The three-bar structure shown in Fig. 5.17(a) has members composed of

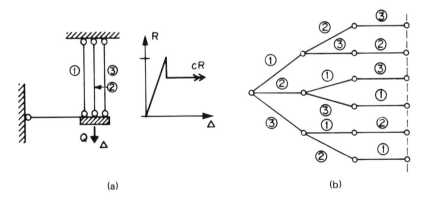

(a)   (b)

Fig. 5.17 — Idealized parallel system and event tree: Example 5.8.

elastic–residual strength behaviour (Fig. 5.1(d)). Each bar has a mean strength $\mu_R = 100$ with a coefficient of variation 0.15. The load Q has a mean value 120 and a coefficient of variation 0.30. For purposes of structural analysis it will be assumed that the elastic properties of the members 1, 2 and 3 are such that the stress distribution between them is in the ratio 10:9:6. The complete event tree for member failure is in this case easily obtained (see Fig. 5.17(b)); in general it need not be known to perform the following analysis.

The decision as to which member is likely to "fail" first is governed by (5.48a). This requires evaluation of the probability $P(E_1)$ for each member. Assuming second-moment calculations for convenience, and noting that the mean load is distributed between the members as 48, 43.2 and 28.8 respectively,

$$\beta_1 = \frac{100 - 48}{[(100 \times 0.15)^2 + 48 \times 0.3)^2]^{1/2}} = 2.50$$

$$P_1(E_1) = 0.621 \times 10^{-2}$$

Similarly

$$\beta_2 = 2.87$$

$$P_2(E_1) = 0.205 \times 10^{-2}$$

and

$$\beta_2 = 4.11$$

$$P_3(E_1) = 0.198 \times 10^{-4}$$

Hence member 1 is likely to fail first. Assuming now that this occurs, the load which it supported must be (partly) redistributed to members 2 and 3. If $c = 0.5$ say, (see Fig. 5.1(d)), only half the load is redistributed; if $c = 0$, all is redistributed. Assume $c = 0.5$ for illustration. Then $cR_1 = 24$ and member 2 gains $(9/15)(1 - c)R_1 = 14.4$, member 3 gains 9.6, in accordance with the relevant stiffnesses of the members.

Let one-dimensional branching (5.49a) be used to approximate the calculation of $P_1(E_1 \cap E_2)$ as would be required to determine which member should be selected next. Then

$$\beta_2 = \frac{100 - (48 + 14.4)}{[(100 \times 0.15)^2 + (48 + 0.3)^2 + (14.4 \times 0.15)^2]^{1/2}} = 1.80$$

$$P_2(E_2) = 0.359 \times 10^{-1}$$

and

$$\beta_3 = 2.37$$
$$P_3(E_2) = 0.889 \times 10^{-2}$$

Hence member 2 would be expected to fail next, followed by member 3. The failure mode so far derived is 1, 2, 3. The truncation criterion δp_f, approximated as δp_f^*, may be used to examine whether other failure modes should be truncated. At this stage the best estimate for p_f is $P_1(E_1) = 0.621 \times 10^{-2}$. The second failure mode, 1, 3, 2 is not easily examined for truncation unless $P_1(E_1 \cap E_3 \cap E_2)$ can be calculated; if (5.49a) is used to estimate it, then $p_{f_2} = P(E_1) = 0.621 \times 10^{-2}$. This is not less than δp_f^* for $\delta \leqslant 1$.

The third and fourth possible failure modes are each estimated similarly by $P_2(E_1)$, which is 0.205×10^{-2}. If $\delta > 0.33$, these modes would be truncated and their contribution to the system failure probability ignored.

The fifth and sixth failure modes are each estimated by $P_3(E_1)$, which is 0.198×10^{-4}. They would usually both be truncated since this crude estimate of their contribution to the system failure probability is less than 0.32% of p_f^*; i.e. δ would need to be less than 0.0032 to retain these modes.

5.5.3 Limit state formulation
Once the various failure modes which are to be considered for the system have been identified, the limit state equation for the structural failure condition must be obtained. Two procedures for doing this have been proposed. These will be discussed in this section. In each case the failure mode of interest is the kth mode for which (in general) the order of the node (or element or member) failure occurrences is known.

5.5.3.1 Node replacement("artificial load") technique
In the early forms of the class of procedures described here [e.g. Murotsu et al., 1977], any member or node that was considered to have failed was replaced by its post-failure resistance. For all further increments of external applied load, this resistance can be treated as a locally applied artificial "load" with random properties identical with those of the post-failure resistance. Thus, if the node has perfect plastic behaviour, the plastic resistance would be applied as an artificial "load"; if the behaviour is elastic–brittle, no artificial "load" would be applied. The applied artificial "load" might need to be modified, however, in the case of hardening behaviour (Fig. 5.1(e)) or if stress reversal occurred at the node.

Evidently, careful accounting is needed of the stress situation at all nodes as the externally applied load is incremented. The node replacement procedure is particularly useful for ideally plastic structures, since stress reversal is then very unlikely to occur. An illustration of the artificial load technique is given in Example 5.9.

5.5.3.2 Incremental load technique
A more systematic technique can be developed from a procedure due to Moses (1982) and Gorman (1979). For elastic–plastic member behaviour

and for a given failure mode k, the failure probability can be determined by comparing the system resistance R_s to the single-parameter loading system Q_1:

$$P_{f_k} = P(R_s - Q_1 \leq 0) \qquad (5.50)$$

The system resistance R_s at collapse of the structure is given by the maximum of all the cumulative load increments:

$$R_s = \max(r_1, r_1 + r_2, r_1 + r_2 + r_3, \ldots, r_1 + r_2 + \ldots + r_n) \qquad (5.51)$$

where r_j is the jth load increment associated with the mode failure event E_j. Thus, for a structure in which three members must fail before structural collapse can occur, r_1 represents the load (from zero load) at which the first member fails, $r_1 + r_2$ the total load at which the second member also fails, and $r_1 + r_2 + r_3$ the total load at which all three members have failed. Since it is possible that members may unload after failure (e.g. Fig. 5.1(b) and Fig. 5.1(d)), the structural resistance R_s is given by the maximum load reached, and hence expression (5.51).

The corresponding failure probability for this (the kth) mode is

$$p_{f_k} = P[\max(r_1 - Q_1 \leq 0, r_1 + r_2 - Q_1 \leq 0, \ldots r_1 + r_2 + \ldots + r_n - Q \leq 0)] \qquad (5.52)$$

which may be rewritten as

$$p_{f_k} = P[(r_1 - Q_1 \leq 0) \cap (r_1 + r_2 - Q_1 \leq 0) \cap \ldots \cap (r_1 + r_2 + \ldots r_n - Q_1 - 0)] \qquad (5.53)$$

The system resistance R_s is governed by the strengths R_i of the nodes. These can be related to the load increments r_j through the matrix equation

$$\mathbf{R} = \mathbf{Ar} \qquad (5.54)$$

where $\mathbf{R} = \{R_i\}$, $i = 1, 2, \ldots, n$, is a vector of nodal resistances and $\mathbf{A} = \{A_{ij}\}$, $j \leq i, j = 1, 2, \ldots, n$, is the so-called "utilization matrix" in which the components A_{ij} represent the stress or stress resultant caused in node number i due to $Q_1 = 1$ at the jth loading increment, i.e. corresponding to the load increment r_j. For a given failure mode, with given node occurrence sequence, the A_{ij} can be determined from conventional structural analysis. Inversion of (5.54) produces the load in increments r_j and hence allows (5.53) to be evaluated:

$$\mathbf{r} = \mathbf{A}^{-1}\mathbf{R} \qquad (5.55)$$

Formulation (5.54) has been extended to cater for more general node behaviour such as elastic–unloading–strain hardening (or post-buckling) (see Fig. 5.1) and for more than one loading system [Melchers and Tang, 1985a, 1985b]:

$$\mathbf{BR} - \mathbf{CQ} - \mathbf{Ar} = 0 \qquad (5.56)$$

where **B** is called "unloading matrix" associated with the unloading portion of load deformation response (Fig. 5.1(b), 5.1(d) and 5.1(e)); **CQ** is the random member action imposed by the other loads (non-incremental) and **Ar** is defined by (5.54). It should be evident that any one loading system may be chosen as the incremental load system for purposes of determining the values of the load increments r_j. The incrementation process has no other purpose since it will be recalled that the sequence of node failure is here taken as given. The sequence itself must be selected from one obtained using a technique such as truncated enumeration method of section 5.5.2.

The incorporation of unloading and strain-hardening (or post-buckling) behaviour in the (incremental) structural analysis for structural reliability brings with it all the limitations and problems of such analysis as known for deterministic analysis. Thus the extent of hardening regions, the possibility of stress reversal and the concomitant change in local stiffness must be accounted for in the incremental analysis; otherwise the eventual structural failure condition will not be properly modelled. Nevertheless, for the reasonably realistic structure shown in Fig. 5.18 a number of analyses with

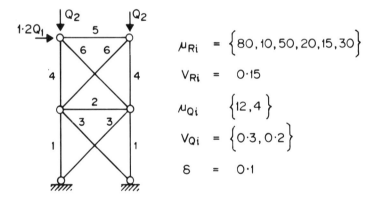

Fig. 5.18 — Pin-jointed truss with unloading and hardening properties.

member behaviour of the type in Fig. 5.1(e) have been performed. The results are tabulated in Table 5.1 [Melchers and Tang, 1985a] in terms of the safety index β_{sys} for the structural system. The parameters μ and γ are defined in Fig. 5.1(e) and the means and coefficient of variations for the members and loads (all normally distributed) are shown in Fig. 5.18.

For this example, structural collapse was assumed to have occurred if $|K|$

Table 5.1 — System bounds for non-linear members

Case	γ	μ	Bounds for β_{sys}
1	0.2	0	3.776–3.883
2	0.1	0	3.656–3.759
3	0	0	3.080–3.098
4	− 0.1	0	2.922–2.989
5	− 0.2	0	2.738–2.810
6	0.1	0.25	3.305–3.368
7	0	0.25	2.654–2.730
8	− 0.1	0.25	2.497–2.620
9	0	0.5	2.534–2.553
10	0	1.0	2.537–2.537

≤ 0, where K is the structural stiffness matrix, or if three or more members had been strained beyond the elastic range.

For case 10, if it were assumed that collapse of the system coincided with the failure of any one member, the bounds for β_{sys} would be 2.491–2.495, which is nearly the same as bounds for elastic–brittle systems. This confirms the well-known result that, for low-redundancy brittle systems, redundant members make little contribution to the safety of the system as a whole.

Example 5.9 [Melchers and Tang, 1983]
For the same structure as in Example 5.8 (see Fig. 5.17), with the same material properties, consider now the formulation of the limit state equation for the failure mode having the mode failure sequence 1, 2, 3.

Using the "artificial load" approach of section 5.5.3.1 and treating failed member resistances as external loads, the first event E_1 equals failure of member 1, given by

$$E_1 = R_1 - 0.4Q < 0 \tag{5.57a}$$

where the 0.4 coefficient is the fraction 10/25 of the load Q in member 1. When member 1 fails, its residual strength is cR_1 which is then applied as a load on the remaining structure; hence

$$E_2 = R_2 - 0.6(Q - cR_1) < 0 \tag{5.57b}$$

where the 0.6 coefficient is the fraction 9/15 of the load in member 2 after member 1 has failed. When member 2 fails the event expression becomes

$$E_3 = R_3 - (Q - cR_2 - cR_1) < 0 \tag{5.57c}$$

Expression (5.45) leads then directly to

$$
\begin{aligned}
P_{f_{1,2,3}} &= P(E_1 \cap E_2 \cap E_3) \\
&= P\{(R_1 - 0.4Q < 0) \cap [(R_2 + 0.6cR_1 - 0.6Q) < 0] \\
&\quad \cap [(R_3 + cR_2 + cR_1 - Q) < 0]\}
\end{aligned}
\tag{5.57d}
$$

Alternatively, the incremental load technique may be used. With members failing in the sequence 1, 2, 3 the progressive load distribution in the members is given in Table 5.2 for unit applied external load. Note that Table 5.2 is the transpose of matrix **A**.

Table 5.2 — Unit load distribution

	Member 1	Member 2	Member 3
Intact	0.40	0.36	0.24
Member 1 failed	—	0.60	0.40
Members 1 and 2 failed	—	—	1.00

Expression (5.56) with $\mathbf{C} = 0$ then becomes, for a three-stage failure mode:

$$
\begin{bmatrix}
1.0 & 0 & 0 \\
-0.60(1-c) & 1.0 & 0 \\
-0.40(1-c) & -1.0(1-c) & 1.0
\end{bmatrix}
\begin{bmatrix}
R_1 \\ R_2 \\ R_3
\end{bmatrix}
=
\begin{bmatrix}
0.40 & 0 & 0 \\
0.36 & 0.60 & 0 \\
0.24 & 0.40 & 1.0
\end{bmatrix}
\begin{bmatrix}
r_1 \\ r_3 \\ r_3
\end{bmatrix}
\tag{5.57e}
$$

From (5.55),

$$
\begin{bmatrix}
r_1 \\ r_2 \\ r_3
\end{bmatrix}
=
\frac{1}{0.24}
\begin{bmatrix}
0.60 & 0.0 & 0 \\
-0.60 + 0.24c & 0.4 & 0 \\
0.0 & -0.4 + 0.24c & 0.24
\end{bmatrix}
\begin{bmatrix}
R_1 \\ R_2 \\ R_3
\end{bmatrix}
\tag{5.57f}
$$

so that

$$
r_1 = 2.5R_1
$$

$$
r_1 + r_2 = cR_1 + \frac{5}{3}R_2
$$

$$
r_1 + r_2 + r_3 = cR_1 + cR_2 + R_3
$$

which can be substituted into equation (5.52):

$$P_{f_{1,2,3}} = P\{\max[(2.5R_1 - Q), (cR_1 + \tfrac{5}{3}R_2 - Q), (cR_1 + cR_2 + R_3 - Q)] \leq 0\}$$
$$(5.57g)$$

which represents the probability of failure in the given sequence 1, 2, 3. Because of the equivalence of (5.52) and (5.53), expressions (5.57d) and (5.57g) are identical.

Example 5.10 [Melchers and Tang, 1983]
In this example a simple case involving stress reversal will be considered. For the structure shown in Fig. 5.19, assume that the resistance properties R_i^+,

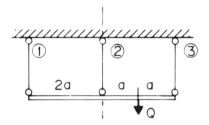

Fig. 5.19 — Structure with stress reversal: Example 5.10.

R_i^-, $c_i R_i^+$ and $c_i R_i^-$ are known ($i = 1, 2, 3$), where the superscripts represent, respectively, tensile and compressive capacity. It will be sufficient to demonstrate the effect of stress reversal in the formulation of one particular failure sequence; the truncated enumeration procedure may be used to obtain other failure sequences. Let the sequence of interest be $1^+ \rightarrow 3^+ \rightarrow 2^+$ where the superscripts indicate tensile failure. Assume that the elastic member properties are such that the ratio of the member forces is 1:3:3. The member forces which are obtained as the structure fails progressively are given in Table 5.3. It has been assumed also that $c_3 R_3^+ = 0$.

Table 5.3 — Progressive member forces

	Member 1	Member 2	Member 3
Stucture intact	$Q/7$	$3Q/7$	$3Q/7$
Member 1 failed	cR_1^+	$(Q - 4cR_1^+)/2$	$(Q + 2cR_1^+)/2$
Member 3 failed	$-Q/2$	$+3Q/2$	0

It is evident that member 1 undergoes stress reversal owing to the failure of member 3, having first attained the tensile strength state. The mode failure expression (5.45) is now

$$P_{f_{1\to 3\to 2}} = P\left[\left(R_1^+ - \frac{Q}{7} \leqslant 0\right) \cap \left(R_3^+ - \frac{1}{2}(Q + 2cR_1^+) \leqslant 0\right)\right.$$

$$\left.\cap \left(R_2^+ - \frac{3Q}{2} \leqslant 0\right)\right]$$

Finally, it may be remarked that for this problem each member has two potential failure events: E_i^+ is failure in tension, and E_i^- is failure in compression. There are thus six nodes to be considered in the event-tree construction.

5.5.4 System probability evaluation
The evaluation of the (nominal) probability of failure for the structural system will involve the calculation of each of the dominant mode failure probabilities given by (5.45) as

$$p_{f_k} = P(E_1 \cap E_2 \cap \ldots \cap E_n) \qquad (5.58)$$

and the series combination of the p_{f_k}.

These calculations will normally be separate from those used to compute probabilities for branching choices and for truncation, unless those probabilities were already determined to a high degree of accuracy (where necessary taking possible correlation effects into account).

If n is smaller than about 5, numerical integration of (5.58) may be possible (see section 3.1), provided that the probability density function for each variable, or, if not independent, the joint probability density function is known. Monte Carlo integration of (5.58) for larger n values is unlikely to be economical compared with a more direct Monte Carlo analysis of the whole systems reliability problem. Alternatively, approximations such as those used by Murotsu et al. (1977) may be employed.

The multinormal integral (4.31) may be advantageously employed to calculate p_{f_k} if all random variables are normal [Melchers and Tang, 1985a, 1985b]. This would automatically account for possible correlation between basic variables.

If the random variables are not normal, the transformation method (FOR) of section 4.4 may be employed, at least in principle, although computation times are expected to be long owing to the need to iterate to find the most appropriate expansion point.

With the modal probabilities p_{f_k} for the dominant modes known, the nominal failure probability for the structural system can be obtained, in principle, from (5.1) as

$$p_f = P(F_1 \cup F_2 \cup F_3 \cup \ldots \cup F_m) \qquad (5.59)$$

where the event F_i ($i = 1, \ldots, m$), represents failure in the ith dominant mode. By considering only the dominant modes, the failure probability is underestimated (see section 5.2.2).

An upper bound estimate for the error in not considering the truncated (i.e. discarded) modes of failure can be obtained from the probability of the nodes that are unbranched at the first level [Melchers and Tang, 1985b]:

$$P_\varepsilon = \sum_{i_\varepsilon} P(Z_i \leqslant 0) \tag{5.60}$$

Murotsu *et al.* (1983) proposed that the failure probability of all truncated modes, calculated at the truncation level, be included in (5.59). This will overestimate p_f for the structure since the mode failure probabilities for truncated modes will be overestimated.

In evaluating (5.59), it is to be noted that the events F_i are not represented by a limit state function. Each event is defined by (5.58); this yields a probability estimate, but not a corresponding limit state function. This means that the use of second-moment probability integration (or the FOR method) is not possible unless a limit state function equivalent to each event can be obtained. One operationally satisfactory approach is to use an equivalent limit state function obtained by moving the limit state function for the most dominant mode (i.e. that with the lowest β value) parallel to itself until the probability content that it encloses is identical with that expressed by p_{f_k}. When this is done for each mode of failure, the series system bounds (5.25) and (5.31) or the multinormal result (4.44) together with the identity

$$P(F_1 \cup F_2 \cup \ldots \cup F_m) = 1 - P(F_1 \cap F_2 \cap \ldots \cap F_m) \tag{5.61}$$
$$= 1 - \Phi_n(-\beta, \mathbf{R})$$

may be used to evaluate (5.59).

More generally, an equivalent (linear) limit state function can be obtained in the space of the variables involved in (5.58) by obtaining the direction cosines $\alpha_{E,i}$ (see section 4.3.2) by differentiating the functional expressing $\beta_E = -\Phi^{-1}(p_{f_k})$, since $\beta_E = \beta_E(\mathbf{U})$, where $\mathbf{U}$ is the vector of (independent, standardized) normal variables. Then (see (4.5)): $c_{E,i} = \partial\beta_E/\partial U_i$, $\ell_{E,i}^2 = \sum_i c_{E,i}^2$ and $\alpha_{E,i} = c_{E,i}/\ell_{E,i}$. The main difficulty is to express β_E as a function of $\mathbf{U}$; generally a numerical solution procedure must be adopted [Gollwitzer and Rackwitz, 1983].

Example 5.11
Consider the intersection $E_1 \cap E_2$ in standardized normal space and equations L_1 and L_2 shown in Fig. 5.20(a). The dominant one is L_1; $\beta_1 < \beta_2$. The direction cosines for each are shown.

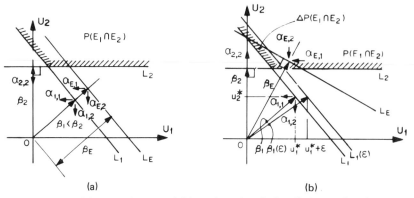

Fig. 5.20 — (a) Approximate and (b) reoriented equivalent limit state functions.

Using the approximate technique, the equivalent limit state L_E is then defined by β_E and the direction cosines $\alpha_{E,i} = \alpha_{1,i}$ are as shown. β_E is given by $\beta_E = -\Phi^{-1}[P(E_1 \cap E_2)]$, which can be evaluated if $P(E_1 \cap E_2)$ can be evaluated.

The more accurate procedure finds a limit state L_E as defined in Fig. 5.20(b), with β_E as before, but with the orientation changed to that defined by the direction cosines $\alpha_{E,i}$. From (4.5) these are defined as $c_{E,i}/\ell$ where $c_{E,i} = \partial\beta_E/\partial u_i$. The change in β_E due to a small change ε in u_1 for β_1, say, can be visualized as shown in Fig. 5.20(b); the checking points u_i^* for both limit states are changed to $(u_1^* + \varepsilon, u_2^*)$. The concomitant change in each β_i is as shown; these changes also change the intersection probability by $\Delta P(E_1 \cap E_2)$. Clearly β_E changes also. It follows that

$$c_{E,i} = \frac{\partial\beta_E}{\partial u_1} = \lim_{\varepsilon \to 0}\left(\frac{\Delta\beta_E}{\varepsilon}\right)$$

$$= \lim_{\varepsilon \to 0}\left\{\frac{(-)\Phi^{-1}P[\beta_1(\varepsilon), \beta_2(\varepsilon)] - (-)\Phi^{-1}P(\beta_1,\beta_2)}{\varepsilon}\right\}$$

where $\beta_1^2(\varepsilon) = (u_i + \varepsilon)^2 + u_2^2$ and $\beta_2(\varepsilon) = \beta_2$ in this case, since the change in u_1 does not affect β_2 for ε small.

5.5.5 Rigid–plastic systems

The truncated enumeration method is a very general procedure to determine failure modes. However, it is expensive in computer time. Fortunately, for specific material behaviour properties, in particular for ideal rigid–plastic behaviour, a number of alternative methods to determine failure modes exist. The traditional approach for the analysis of rigid–plastic- framed structures is that of combination of elementary mechanisms.

These can be systematically obtained from an algorithm given by Watwood (1979). Another algorithm which determines all the collapse mechanisms (elementary and combinations) has also been proposed [Gorman, 1981].

These algorithms have been used by Sigurdsson *et al.* (1985) in a systems reliability calculation procedure which also uses the node replacement (artificial load) approach (but not truncation) for rigid–plastic structures. Procedures essentially similar to truncated enumeration had been described (but not systematically derived) for rigid–plastic node behaviour only, by Murotsu *et al.*, (1977, 1983), Grimmelt *et al.* (1983) and Thoft-Christensen and Sørensen (1982). It is also possible to consider member behaviour in truss structures under the action of both axial load and moment [Murotsu *et al.*, 1985].

The truncated enumeration procedure, and various other similar procedures, are based on the use of an elastic analysis to develop sequences of node failure. As noted, the precise values assigned to the elastic constants for a particular problem should not matter as the procedure is merely an artifice for generating nodal sequences. Gorman (1979) in studying rigid–plastic fameworks found that in certain circumstances, the truncated enumeration type of procedures produced modes of failure based on elastic–plastic calculations which were kinematically inadmissible in terms of rigid–plastic theory. The reason for this arises from the fact that in rigid–plastic structures there can be no deformation whatsoever prior to the formation of a kinematically admissible collapse node. This requires the existence of a particular form of internal stress distribution. In elastic–plastic structures, however, elastic deformation can occur prior to the formation of a plastic collapse mechanism. The associated internal stress distribution will in general be different to that which will exist in a rigid–plastic structure. In particular, there is no guarantee that stress reversal from an elastic to an opposite plastic stress will not occur at the point of plastic collapse. It is therefore necessary, for rigid–plastic structures, to verify the validity of collapse modes generated using elastic–plastic analysis [Melchers and Tang, 1985a].

5.5.6 Other methods for complex structures

Other procedures which may be appropriate for the assessment of structural system reliability include the survival mode approach applied to rigid–plastic structures [Ditlevsen and Bjerager, 1983; Melchers, 1983b]. This appears to be useful in obtaining dominant modes of failure, but less useful in calculating p_f accurately for a structural system [Melchers and Tang, 1985a].

Another possibility is to pose the problem in the space of the applied loads, and to treat the structural response to a given load as the functional to be evaluated before integrating over all possible load situations to obtain the failure probability. Such an approach essentially employs equation (1.18), with $F_R(\ )$ now representing the resistance of the structure rather than an element. The integral of (1.18) must be performed for each loading system. Various aspects of this procedure were explored by Augusti and Baratta

(1973), Schwarz (1980), Melchers (1981), Moses (1982), Gorman (1984) and Lin and Corotis (1985). Conceptually, the procedure is attractive because it should be possible to include all types of structural behaviour, including non-linear effects, in the analysis. However, a major drawback is the need to assume that the failure mode is invariant for a given ratio of applied loads.

Finally, still using the failure mode approach, an approximate method which groups failure modes for rigid–plastic structures according to their correlation has been proposed by Ma and Ang (1981). The most dominant mode is found by maximizing p_f as a function of the parameters which govern the (linear) limit state functions. A linear programming problem results. At the present state of development it appears that the method is better suited to structures with relatively few members.

5.6   CONCLUSION

The analysis of structural systems reliability can proceed directly through the use of the "crude" Monte Carlo method of Chapter 3. Alternatively, the concepts of section 5.2 can be applied to analyse the form of the system as one with series, parallel or conditional subsystems. The more efficient importance-sampling procedure can then be applied.

The bounding theorems of section 5.4 are of particular value when used in conjunction with the FOSM method. Both in this approach and in the importance-sampling technique, it is assumed that the failure modes for the structure are known. As discussed in section 5.3.3, these can be investigated to some extent using Monte Carlo simulation.

For structures with many members, the enumeration of all possible failure modes is a major task. Of particular interest are those modes which make the greatest contribution to the system failure probability. A systematic procedure to select these is described in section 5.5, together with various procedures for limit state definition and the calculation of the corresponding probabilities. A short overview of some other approaches to the problem is also given.

In the present chapter, only trusses and frames were considered since the reliability analysis of two- and three-dimensional systems is still very much at the research stage. However, even for framed structures there are a number of matters which need further investigation. These include the need to extend the range of member properties which can be treated, including the effect of combined internal actions, the need to include connection behaviour, better modelling of applied loading systems such as wave and wind loads, the influence of dynamic effects and vibration, and other time-dependent effects such as fatigue and deterioration of member properties. Many of these matters are under investigation; some involve consideration of time as a parameter. It is to the influence of time that attention is turned in the next chapter.

6

Time dependent reliability

6.1 INTRODUCTION

Consideration will be given in this chapter to so-called "stochastic" or time-dependent random variables and the related failure probability calculations. Initially attention will be confined to situations of the type considered in Chapter 1, i.e. resistance will be measured in terms of material yield strength or some type of permissible stress. Later in this chapter some comments will be made about fatigue reliability and about problems involving dynamic structural behaviour.

As was already noted in Chapter 1, in general the basic variables **X** will be functions of time. This comes about, for example, because loading changes with time (even if it is quasi-static, such as most floor loading) and because material properties change with time, either as a direct result of previously applied loading or because of some deterioration mechanism. Fatigue and corrosion are typical examples of strength deterioration.

The elementary reliability problem (1.15) in "stochastic" (i.e. time-variant) terms with a resistance $R(t)$ and a load effect $S(t)$, at time t becomes

$$p_f(t) = P[R(t) \leqslant S(t)] \tag{6.1}$$

If the instantaneous probability density functions $f_R(t)$ and $f_S(t)$ of R and S respectively are known, the instantanous failure probability $p_f(t)$ can be obtained from the convolution integral (1.18). Schematically the changes in $f_R(t), f_S(t)$ and $p_f(t)$ with time can be depicted as in Fig. 1.7.

Strictly, equation (6.1) only has meaning if the load effect S increases in value at time t if the load (effect) is reapplied precisely at this time. Failure could not occur precisely at the very instant of time t (assuming, of course, that at time less than t the member was safe). Thus, in general, a change in load or load effect is required; this is assured if there are:

(1) discrete load changes (as will be discussed in section 6.2 below);
(2) if for continuous time-varying-loads an arbitrary small increment δt, in time, is considered instead of instantaneous time t.

With this interpretation, it follows that

$$p_f(t) = \int_{G[\mathbf{X}(t)] \leq 0} f_{\mathbf{X}(t)}[(\mathbf{x}(t)] \, d\mathbf{x}(t) \tag{6.2}$$

which, in two-dimensional space, may be represented as in Fig. 1.8 for any given time t. As before $\mathbf{X}(t)$ is a vector of basic variables.

In principle, the instantanous failure probability expressed by (6.1) or (6.2) can be integrated over an interval of time 0–t to obtain the failure probability over that period. In practice, however, the instantaneous value of $p_f(t)$ is usually correlated to the value $p_f(t + \delta t)$, $\delta t \rightarrow 0$, since the processes $\mathbf{X}(t)$ are themselves usually correlated in time. This may be seen in a typical "sample function" or "realization" of a random process load effect Fig. 6.1.

The classical approach to this problem for $R(t)$ and $S(t)$ is to consider the integration transferred to the load or load effect process, which is then assumed to be representable, over the total time period, by an extreme value distribution. The resistance is assumed essentially time invariant. This approach (also called "classical" reliability) formed the basis of discussion in the earlier chapters. The theory for this approach will be outlined in section 6.2.

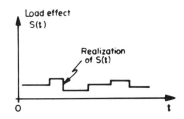

Fig. 6.1 — Typical realization of random process load effect.

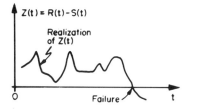

Fig. 6.2 — Realization of safety margin process $Z(t)$ and time to failure.

A refinement is to consider shorter periods of time, such as the duration of a storm, or a year, and to apply extreme value theory within that period.

Simple ideas akin to the return period concept can then be used to determine the failure probability over the lifetime of the structure. This approach, which is described in section 6.3, is quite popular for the practical reliability analysis of major structures such as offshore platforms, towers, etc., subject to definable and discrete loading events.

A somewhat different way of looking at the problem is to consider the safety margin associated with (6.1):

$$Z(t) = R(t) - S(t) \tag{6.3}$$

and to establish the probability that $Z(t)$ becomes zero or less in the lifetime t_L of the structure. This constitutes a so-called "crossing" problem. The time at which $Z(t)$ becomes less than zero for the first time is called the "time to failure" (Fig. 6.2) and is a random variable. The probability that $Z(t) \leq 0$ occurs during t_L is called the "first-passage" probability. The corresponding situation in two-variable space is shown in Fig. 6.3. The probability that the

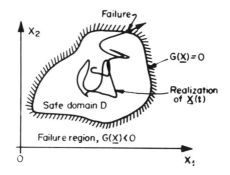

Fig. 6.3 — Outcrossing of vector process $\mathbf{X}(t)$.

process $\mathbf{X}(t)$ will leave the "safe" region $G(\mathbf{X}) > 0$ (i.e. the probability that an "outcrossing" will occur) during the structural lifetime t_L, is again the so-called "first-passage" probability. It is, of course, equal to the probability of failure of the structure since failure is defined by $G(\mathbf{X}) \leq 0$.

The first-passage concept is more general than the classical approaches. In particular, there is no restriction on the form of $G(\mathbf{X})$. However, the determination of the first-passage probability and a proper understanding of the concept requires some knowledge of stochastic processes. An introduction to this topic will be given in section 6.4. A discussion of the time-dependent approach to reliability calculation will then be given in section 6.5.

If the elementary reliability problem (6.1) is to be made to cope with more than one load or load effect, as is required for example in code-writing applications, it is necessary to combine two or more load effects into one equivalent load effect. This may be achieved using the theory and methods of sections 6.4 and 6.5 and is described in section 6.6.

An important strength requirement for structures is fatigue strength. This has not so far been considered, as it is essentially a time-dependent problem (although simplistic "time-independent" reliability analyses have been given). The essential probabilistic considerations for the analysis of fatigue reliability are outlined in section 6.8, following some comments about problems involving dynamic structural behaviour given in section 6.7.

6.2   TIME-INTEGRATED APPROACH

In the time-integrated approach, the whole lifetime $[0, t_L]$ of the structure is considered as a unit and all statistical properties of all random variables must relate to this lifetime. Thus the probability distribution of interest for loads is that for lifetime maximum loads. Similarly, the probability distribution which should be of interest for resistance is the lifetime minimum. However, it is highly unlikely that the occurrence of the lifetime maximum load will coincide with the lifetime minimum resistance, as shown in Fig. 6.4 for typical realizations of R and S. For most problems, however, R may be assumed to be time independent, so that a typical realization of R is now a horizontal line (see Fig. 6.5). The actual location of the realization of R is, of

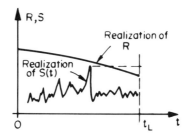

Fig. 6.4 — Typical realizations of load effect $S(t)$ and resistance $R(t)$: non-stationary case.

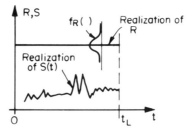

Fig. 6.5 — Typical realizations of load effect $S(t)$ and resistance R, with R a time-independent random variable.

course, governed by the probability density function $f_R()$ if R is a random variable. With this restriction (6.1) becomes

$$p_f(T_L) = P[R \leqslant S_{max}(t_L)] \tag{6.4}$$

where $S_{max}(t_L)$ denotes the maximum load effect in the period $[0, t_L]$. It is of course possible, but unlikely in practice, to interchange the assumptions placed on R and S.

The probability distribution for $S_{max}()$ may be obtained directly by fitting an appropriate probability distribution function to observed extreme value data of past observations and assuming it will be applicable also to the future. Usually, however, data are not available for a sufficiently long period of time, and records for shorter periods must generally be used to synthetize the extreme value distributions.

The time-integrated approach is based on the concept of applying a (one-parameter) loading system Q to the structure at regular intervals in time. In this case the probability of failure of the structure may be considered simply a function of the number N of statistically independent loading applications to cause failure [Freudenthal *et al.*, 1966]:

$$P(N \leqslant n) \equiv F_N(n) = 1 - L_N(n) \tag{6.5}$$

where n is some given number of load applications, $F_N(n)$ is the cumulative distribution function of N, and $L_N(n)$ is defined as the "reliability function". Since N is discrete, it follows that the probability mass function $f_N()$ is given by (cf. (A.9))

$$f_N(n) \equiv P(N = n) = F_N(n) - F_N(n - 1) \tag{6.6}$$

It is also possible to define the so-called "hazard" function:

$$h_N(n) = P(N = n | N > n - 1) = \frac{f_N(n)}{1 - F_N(n - 1)} \tag{6.7}$$

It expresses the probability of failure under the nth loading ($N = n$) given that failure has not occurred under a previous loading ($N > n - 1$). It is more commonly expressed in terms of time (see section 6.3.3 below).

It will be assumed that load applications are independent in time, i.e. there is no dependence between the ith and jth load application ($i \neq j$). It will also be assumed, for convenience, that the resistance R and the load effect S (associated with one application of Q) each have known probability density functions f_R, f_S and cumulative distribution functions F_R, F_S respectively. Both R and S (and therefore Q) will be assumed to be stationary random processes, i.e. f_R and f_S (and so F_R and F_S) do not change with time.

Then, for a series of n loads applied to the structure, the probability that the load effect is less than some value x, say, is given by

$$P(S_1 < x)P(S_2 < x) \ . \ . \ .$$

or

$$P(S^* < x) = F_{S^*}(x) = [F_S(x)]^n \tag{6.8}$$

where S^* is defined as the maximum load effect for n load applications.

If n is very large, (6.8) asymptotically approaches one of the extreme value distributions (see sections A.5.11–A.5.13), which may then be used to describe S^*.

The justification for using extreme value distributions in the time-integrated approach can now be given directly. Using the same argument as in section 1.4.2, the probability of structural failure, i.e. the probability that fewer than n load applications can be supported, is

$$P(N < n) = \int_0^\infty \{1 - F_S(y)]^n\}f_R(y)\ dy \tag{6.9a}$$

$$= \int_0^\infty [1 - F_{S^*}(y)]f_R(y)\ dy \tag{6.9b}$$

and integrating by parts produces (see also (1.19))

$$p_f = F_N(n) = \int_0^\infty F_R(y)f_{S^*}(y)\ dy \tag{6.10}$$

which is identical with (1.18) but with an extreme value distribution for the maximum applied load effect S^*. Note that (6.10) does not specifically contain the number of load applications; this has been absorbed into the derivation of the distribution for S^* (and is therefore assumed to be "large").

In practical situations more than one load system may act. The load systems are usually assumed independent (e.g. live load and wind load) but may not be (e.g. wave load and wind load). In addition, the peaks of different load processes do not usually coincide, so that it would be rather conservative to use an extreme value distribution for each loading system separately. The usual procedure is to derive one process to represent the combined effect of several loading processes. How this might be done is discussed in section 6.6.

6.3   DISCRETIZED APPROACH

If the time period is considered to be 1 year, say, or is taken as the occurrence of a discrete event such as a storm, the problem of interest is the calculation of the probability of failure given that n_L years or events occur during the lifetime t_L of the structure. For a given unit time period n_L is fixed, once t_L is decided. However, for storm events, n_L is not known *a priori*, although the average rate of occurrence of storms may be known.

The discrete time unit may be a day, a month, a year, etc., although a year is commonly adopted. The failure probability calculated using the method of the previous section is then the probability of failure per year. The corresponding resistance and load effect variables of interest are then the extreme values per year, with appropriate probability density functions. Such distributions are obtained from observations (e.g. wind, waves) and are related to the particular time period used as the reference (e.g. 1 year). Thus the maximum wind force per year will have a different probability density function to (say) that of the maximum wind force per day. Only under particular assumptions is it possible to relate these distributions easily, as will be seen below.

The discretized approach can be justified along similar lines to that used in section 6.2 for the time-integrated approach. Two cases may be distinguished:

(1) n_L known deterministically;

(2) n_L a random variable.

6.3.1   Known number of discrete events

Consider again a set of n_L loads applied to the structure. If a given time period is now considered (i.e. 1 year) within which n_L may be considered "large" and for which each load occurrence may be considered to be an independent event, then (6.8) may be used to obtain the extreme value distribution S_1^* for S for 1 year. The probability p_{f_1} of failure per year, is then given by (6.10). However, the probability of failure for a life of n_L years is given by (6.9a), with S_1^* substituted for S and assumed independent between years [Freudenthal, *et al.*, 1966]:

$$p_f(n_L) = F_{N_L}(n_L) = \int_0^\infty \{1 - [F_S^*(y)]^{n_L}\} f_R(y) \, \mathrm{d}y \qquad (6.11)$$

Noting that

$$[F(y)]^n = [1 - \bar{F}(y)]^n \approx 1 - n\bar{F}(\ ) + n(n-1)\frac{\bar{F}^2}{2} - \ldots$$

where $\bar{F}(\ ) = 1 - F(\ )$ and neglecting second-order terms, it follows easily that

$$p_f(n_L) \approx \int_0^\infty \{1 - [1 - n_L \overline{F}_{S_1 \cdot}(y)]\} f_R(y) \, dy \tag{6.12}$$

or

$$p_f(n_L) \approx n_L \int_0^\infty \overline{F}_{S_1 \cdot}(y) f_R(y) dy \tag{6.13}$$

$$\approx n_L p_{f_1} \tag{6.14}$$

where the alternate expression (1.19) has been used for p_{f_1}. The neglect of second-order terms in (6.12) is valid only for those values of y for which $n\overline{F} \ll 1$. It is easily shown that this implies high values of y and that as a consequence the approximation (6.12) and hence (6.13) and (6.14) are most accurate for situations in which $\sigma_S \gg \sigma_R$. Clearly these approximations are also restricted to small values of p_{f_1} and $p_f(n_L)$. Result (6.14) shows that the lifetime failure probability $p_f(n_L)$ can be determined (approximately) from the annual failure probability p_{f_1} simply by multiplying the latter by the number n_L of years in the designated lifetime t_L.

Rather than expressing the life of a structure in terms of the number of load applications, a more natural parameter is time T. The probability that a structure will fail during a time period $[0, t]$ may then be stated as

$$P(T \ll t) = F_T(t) = 1 - L_T(t) \tag{6.15}$$

where $F_T(\ )$ is the cumulative distribution function of T and $L_T(\ )$ is the "reliability function" expresses in terms of time.

If now p_i is the failure probability for the ith time unit (cf. p_{f_1} and (6.10)), then, using the same arguments as for (6.8),

$$P(T \ll t) = 1 - \prod_{i=1}^t (1 - p_i) = 1 - (1 - p)^t \tag{6.16}$$

if $p_i = p$ for all i (as is the case for time-invariant R) and assuming independence of p between time units. If pt is sufficiently small, this may be approximated by

$$P(T \ll t) \approx 1 - \exp(-tp) \approx tp \tag{6.17}$$

which corresponds to (6.14) (and has the same limitations) provided that t and p are appropriately interpreted. This also shows clearly that the choice of time period is arbitrary, provided that it is consistent between t and p.

Nevertheless the choice of 1 year is very common. The probability of failure for a lifetime $[0, t_L]$ is thus

$$p_f(t_L) = 1 - \exp(-t_L p) \approx t_L p \qquad (6.18)$$

6.3.2 Random number of discrete events

The failure probability p_i for the ith time unit depends on the number of load applications which occurs during that time unit. If a time unit is considered to be an "event" such as a storm, the actual number of load applications during the event might be ignored provided that an appropriate extreme value probability distribution is used to describe the maximum load effect S_e^* occurring during the event. The discretization is now over "events" (of indeterminate duration) rather than periods of given duration as in section 6.3.1.

The number of occurrences of the events, however, needs to be known to obtain the lifetime failure probability $p_f(t_L)$. Let this be given by $p_k(t)$, the probability of k events in time $[0, t]$. Then the probability of failure for a period $[0, t]$ may be expressed in time terms as (c.f. (6.11))

$$p_f(t) = F_T(t) = \int_0^\infty \sum_{k=0}^\infty p_k(t)\{1 - [F_{S_e^*}(y)]^k\} f_R(y) \, dy \qquad (6.19)$$

where all possible values of the number of events are considered.

A common assumption is to take $p_k(t)$ as Poisson distributed (see section A.5.4):

$$p_k(t) = \frac{(vt)^k \, e^{-vt}}{k!} \qquad (6.20)$$

where v is the mean rate of occurrence of events. In view of the derivation of the Poisson distribution, this means that the events must be independent and non-overlapping. This is likely to be closely true if v is very low. If (6.20) is substituted into (6.19) and a series approximation similar to that used in deriving (6.12) is applied, it follows readily that

$$p_f(t) \approx 1 - \exp(-vt p_{f_e}) \qquad (6.21)$$

where p_{f_e}, the probability of failure given the occurrence of a single event, is given by (6.10) with S_e^* substituted for S^*. Expression (6.21) corresponds to expression (6.17) when it is recognized that $v p_{f_e}$ is the average failure probability per unit time. Again the same limitations regarding accuracy apply to (6.21) as apply to (6.13) and (6.17).

It should be noted that the calculation of the value of p_{f_e}, the probability of failure given that the event occurs, may well need to take account of some conditional information [Schuëller and Choi, 1977]. Thus, if the event is the occurrence of a storm, say, then S_e^*, the maximum load effect, may depend on the occurrence of a "characteristic" wave height H_k during that storm. Let $S_e^*|H_k$ (having probability density function $f_{S_e^*}|H_k$) represent the maximum load effect given the occurrence of a characteristic wave height H_k. The probability density function $f_{H_k}(h)$ expresses the probability of occurrence of a characteristic wave height between h and $h + \delta h$ as $\delta h \rightarrow 0$. The conditional failure probability $p_{f_e}|H_k$ may then be calculated from (1.18):

$$p_{f_e}|(H_k = h) = \int_0^\infty F_R(y)f_{S_e^*}|H_k(y) \, dy \tag{6.22}$$

It follows from (A11) that the unconditional failure probability is

$$p_{f_e} = \int_0^\infty p_{f_e}|(H_k = h) \, f_{H_k}(h) \, dh \tag{6.23}$$

in which it has been assumed that $H_k > 0$. The probability density functions $f_{H_k}(\ )$ may be obtained directly from field observations. However, determination of $f_{S_e^*}|H_k(\ )$ will require both loading data for a given H_k and a structural analysis using the applied loading Q_e^* corresponding to S_e^*

Example 6.2
An offshore platform is subject to an average of 2.5 storms/year. It has a planned design life of $t_L = 15$ years. Analyses of failure probability given a particular characteristic wave height during a storm are set out in Table 6.1 below. Also shown are the estimated occurrence probabilities $f_{H_k}(\ )$. Δh for each characteristic wave height.

Expression (6.23) may be approximated as $\Sigma \, p_{f_e}|H_k \, f_{H_k} \, \Delta h$ to obtain $p_{f_e} = 2.65 \times 10^{-9}$. Using (6.21), there is obtained

$$p_f(t_L) \approx 1 - \exp[(-2.5)(15)(2.65 \times 10^{-9})]$$

$$\approx (-2.5)(15)(2.65 \times 10^{-9}) \approx 10^{-7}$$

6.3.3 Return period and hazard function
The return period concept was introduced in section 1.3. There it was defined as the mean time between defined probabilistic events, usually taken as the exceedance of some given level or load. This concept can be generalized by defining the events in terms of limit state violation, calculated

Table 6.1 — Analyses of failure probability

Characteristic wave height H_k (m)	$f_{H_k}(\ )\,\Delta h$	$p_{f_c}|H_k$	$p_{f_c}|H_k\,f_{H_k}\,\Delta h$
24	0.100	0.2×10^{-9}	0.020×10^{-9}
26	0.640	0.9×10^{-9}	0.576×10^{-9}
28	0.180	4.0×10^{-9}	0.720×10^{-9}
30	0.060	12.0×10^{-9}	0.720×10^{-9}
32	0.015	24.0×10^{-9}	0.360×10^{-9}
34	0.005	50.0×10^{-9}	0.250×10^{-9}
$\Delta h = 2$ m	$\Sigma = 1.00$		$\Sigma = 2.65 \times 10^{-9}$

using any of the methods of section 1.4, and Chapters 3, 4 and 5. Thus the generalized return period may be defined as

$$\overline{T}_G = \frac{1}{p_{f_1}} \tag{6.24}$$

where p_{f_1} is the probability of limit state violation per unit time (usually 1 year) calculated from (1.18) or (1.31), and $\overline{T}_G$ is measured in similar time units (i.e. years). If there are n_L time units in the design life t_L, and assuming, as before, independence of events between time units, (6.14) may be used to obtain the lifetime failure probability:

$$p_f(t_L) \approx \frac{n_L}{T_G} \tag{6.25}$$

Further, if there are m independent phenomena which contribute to the failure probability, then (see section A.2)

$$p_{f_T} = \sum_i^m p_{f_i}, \tag{6.26}$$

or

$$\frac{1}{T_T} = \sum_i^m \frac{1}{T_i} \tag{6.27}$$

This shows that the reciprocals of return periods may be added, but only under the assumption of independence of events between time units.

Another measure sometimes used in classical reliability theory is the "hazard function". It may be expressed either in the number of load applications (see also section 6.2) or in terms of time. The latter will be used here.

From (6.15), the life of a structure expressed in terms of time as the parameter is given by $P(T < t) = F_T(t)$ while the probability density of the design life is

$$f_T(t) = \frac{\mathrm{d}}{\mathrm{d}t}[F_T(t)] \tag{6.28}$$

This is also called the unconditional failure rate, as it reflects the probability of failure in the time interval t to $t + \mathrm{d}t$ as $\mathrm{d}t \to 0$.

The hazard function (also called "age specific failure rate" or "conditional failure rate") expresses the likelihood of failure in the time interval t to $t + \mathrm{d}t$ as $\mathrm{d}t \to 0$, but now given that failure has not occurred prior to time t, i.e.

$$P \text{ (failure } t \leqslant T \leqslant t + \mathrm{d}t | \text{no failure } t \leqslant T) = \frac{P(t \leqslant T \leqslant t + \mathrm{d}t)}{1 - P(T \leqslant t)}$$

or

$$h_T(t) = \frac{f_T(t)}{1 - F_T(t)} \tag{6.29}$$

It is immediately evident that, for $F_T(t)$ small, $h_T(t)$ is closely similar to $f_T(t)$. Some typical hazard functions are shown in Fig. 6.6.

It is not difficult to show that (6.29) can be recast to

$$F_T(t) = 1 - \exp\left[-\int_0^t h_T(\tau)\,\mathrm{d}\tau \right] \tag{6.30}$$

and

$$f_T(t) = h_T(t) \exp\left[-\int_0^t h_T(\tau)\,\mathrm{d}\tau \right] \tag{6.31}$$

so that knowledge of one of f_T, F_T or h_T is sufficient to obtain the other two.

For a typical, realistic structure, the hazard function usually has the form of a "bath-tub" curve (see Fig. 6.7). This shows the initial high-risk phase

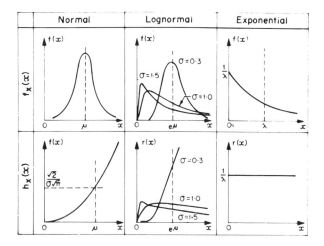

Fig. 6.6 — Typical hazard functions $h_T(t)$.

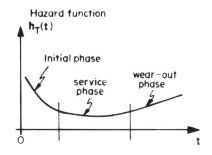

Fig. 6.7 — Typical variation of hazard function with age of structure.

during which the hazard rate reduces sharply as experience is gained with loading the structure. This is essentially a proof-loading stage. As the design life is approached, the hazard rate increases owing to deterioration and wearing-out.

6.4 STOCHASTIC PROCESS THEORY

Before proceeding to discuss the instantaneous approach for dealing with time-dependent problems, it is necessary to introduce some basic ideas about stochastic processes. It is also necessary to describe some elementary processes which will be used in later sections of this chapter. The discussion to follow is based largely on Crandall and Mark (1963), Newland (1984), Papoulis (1965) and Parzen (1962).

6.4.1 Stochastic process

A stochastic process $X(t)$ is a random function of time t such that for any point in time the value of X is a random variable. The outcome x of $X(t)$ is governed by the probability density function $f_X(x, t)$. Of course, the variable t may be replaced by any kind of finite or countable infinite set of values, such as the number of load applications (see section 6.3). However, t is convenient.

For each value of t the observed outcome of $X(t)$ may be plotted; the complete set of such values over a given time interval is called a "realization" or "sample function", such as shown in Fig. 6.8. Since X is a random

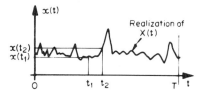

Fig. 6.8 — Realization of basic variable $X(t)$.

variable, the precise form of the realization cannot be predicted. In direct correspondence to the definition (A.10) for a random variable, the mean of all possible realizations at any point in time is simply

$$\mu_X(t) = \int_{-\infty}^{\infty} x f_X(x, t) \, dx \tag{6.32}$$

where $f_X(x, t)$ is the probability density function at time t. The correlation between the realizations at two points in time t_1 and t_2 is termed the "autocorrelation function" since it relates to one realization (see Fig. 6.8):

$$R_{XX}(t_1, t_2) = E[X(t_1)X(t_2)] =$$

$$= \int_{-\infty}^{\infty} \int_{-\infty}^{\infty} x_1 x_2 f_{XX}(x_1, x_2; t_1, t_2) \, dx_1 \, dx_2 \tag{6.33}$$

where $f_{XX}(\ ) = \partial^2 F_{XX}(\ )/\partial x_1 \, \partial x_2$ is the joint probability density function. Here $F_{XX}(\ ) = P\{[X(t_1) \le x_1] \cap [X(t_2) \le x_2]\}$. It is also possible to define an (auto)covariance function analogous to (A.123):

$$C_{XX}(t_1, t_2) = E\{[X(t_1) - \mu_X(t_1)][X(t_2) - \mu_X(t_2)]\}$$

$$= R_{XX}(t_1, t_2) - \mu_X(t_1)\mu_X(t_2) \tag{6.34}$$

As might be expected, if $t_2 = t_1 = t$, the (auto)covariance function becomes the variance function $\sigma_X^2(t)$:

$$\sigma_X^2(t) = C_{XX}(t, t) = R_{XX}(t, t) - \mu_X^2(t) \tag{6.35}$$

analogous to (A.11) and with $\sigma_X(t) = D[X(t)] = $ standard deviation function.

Just as higher "moments" (see section A.4) can be used to provide a description of an ordinary random variable, higher-order correlation functions can also be established. However, these are seldom of much practical interest.

The above concepts for the scalar process $X(t)$ may be generalized to the vector process $\mathbf{X}(t) = [X_1(t), X_2(t), \ldots, X_n(t)]$ having covariance functions $C_{X_i X_j} = \text{cov}[X_i(t_1), X_j(t_2)]$. When $i = j$, these are autocovariance functions for X_i; when $i \neq j$, $C_{X_i X_j}$ is termed a "cross-covariance" function. The collection of functions $C_{X_i X_j}$ is expressed in a covariance function matrix $\mathbf{C}_X$, analogous to the covariance matrix.

Finally, the correlation function (matrix) may be defined analogously to (A.124) as

$$\rho[X_i(t_1), X_j(t_2)] = \frac{\text{cov}[X_i(t_1), X_j(t_2)]}{D[X_i(t_1)]D[X_j(t_2)]}$$

where $D[\ ] = \mathbf{C}_X^{1/2}$ for $t_1 = t_2$, $X_i = X_j$.

Typically, $\rho[X_i(t_1), X_j(t_2)]$ might have the form

$$\rho = e^{-k(t_1 - t_2)(x_1 - x_2)}$$

where k is some defined constant. In this case ρ reduces with greater separation of time points t_1 and t_2, or component processes X_1 and X_2, as might be expected. Also, if $t_1 = t_2$, $X_1 = X_2$, then $\rho = 1$, as would be expected for correlation at a point in time of the same process component.

6.4.2 Stationary processes

When the random nature of a stochastic process does not change with time, it is said to be a "(strictly) stationary" process. This means that all its moments are also independent of time. When only the mean $\mu_X(t)$ and the autocorrelation $R_{XX}(t_1, t_2)$ are independent of absolute time, the process is said to be "weakly stationary" or "covariance stationary". Since the normal distribution is uniquely described by its first two moments, a weakly stationary normal process is also strictly stationary.

A direct consequence of stationarity is that for $C_{XX}(\ )$ and $R_{XX}(\ )$ only the relative difference $(t_1 - t_2) = \tau$, say, is of importance. Thus (6.33), for example, becomes

$$R_{XX}(\tau) = E[X(t)X(t + \tau)] \tag{6.33a}$$

In particular, if $\tau = 0$, $R_{XX}(0) = E(X^2)$, which represents the "mean-square" value of $X(t)$. Also note that $R_{XX}(\tau)$ is an even function in τ, i.e. $R_{XX}(\tau) = R_{XX}(-\tau)$, as is easily verified.

If a process is stationary, by definition it cannot start or stop. Each realization must, theoretically, extend over $-\infty \leqslant t \leqslant +\infty$. This is usually ignored in practical applications. Stationarity can often be assumed to hold a sufficient time after the start of the process. Even if the process is slowly changing with time, it may be acceptable to consider several shorter subprocesses each reasonably stationary over their respective durations.

6.4.3 Derivative process

For the discussion to follow an important property of a process $X(t)$ is its derivative process $\dot{X}(t)$, defined for any realization $x(t)$ as

$$\dot{x}(t) = \frac{d}{dt}[x(t)]$$

The existence of such a derivative implies certain regularity properties of $X(t)$; in particular $X(t)$ is differentiable only (a) if its autocorrelation function $R_{XX}(t)$ has a continuous second-order derivative $R''_{XX}(\tau) = \partial^2 R_{XX}(\tau)/\partial\tau^2$ and (b) if $R''_{XX}(0)$ exists. This may be examined by considering the limit, as $\tau \to 0$, of the incremental form $[x(t + \tau) - x(t)]/\tau$ expressed in terms of R_{XX} and its derivatives [Ditlevsen, 1981b; Papoulis, 1965]. Details will not be given here.

If the second derivative exists, the first derivative $R'_{XX}(\ )$ must also exist and, since R_{XX} is an even function in τ (see previous section), it follows that at $\tau = 0$

$$\frac{\partial R_{XX}(0)}{d\tau} = 0 \tag{6.36}$$

It follows from the previous section that the derivative process $\dot{X}(t)$ will be a stationary process if its autocorrelation function $R_{\dot{X}\dot{X}}(t_1, t_2)$ can be expressed as $R_{\dot{X}\dot{X}}(\tau)$. Consider the cross-correlation between $X(t)$ and $\dot{X}(t)$ [Papoulis, 1965, p. 316]:

$$R_{X\dot{X}}(t_1, t_2) = E[X(t_1)\dot{X}(t_2)] \tag{6.37}$$

which may be written in limit form as

$$R_{X\dot{X}}(t_1, t_2) = \lim_{dt_2 \to 0} \left\{ E \left[X_1(t) \frac{X(t_2 + dt_2) - X(t_2)}{dt_2} \right] \right\}$$

$$= \lim_{dt_2 \to 0} \left[\frac{R_{XX}(t_1, t_2 + dt_2) - R_{XX}(t_1, t_2)}{dt_2} \right]$$

or

$$R_{X\dot{X}}(t_1, t_2) = \frac{\partial R_{XX}(t_1, t_2)}{\partial t_2} \tag{6.38}$$

Using a similar procedure, it follows that

$$R_{\dot{X}\dot{X}}(t_1, t_2) = \frac{\partial^2 R_{XX}(t_1, t_2)}{\partial t_1 \, \partial t_2} \tag{6.39}$$

Further, if $X(t)$ is stationary, with $\tau = t_1 - t_2$, it follows immediately that

$$R_{X\dot{X}}(\tau) = -\frac{dR_{XX}(\tau)}{d\tau} \tag{6.40}$$

and

$$R_{\dot{X}\dot{X}}(\tau) = -\frac{d^2 R_{XX}(\tau)}{d\tau^2} \tag{6.41}$$

which shows that the process $\dot{X}(t)$ is covariance stationary if $X(t)$ is covariance stationary (see section 6.4.2).

Also using (6.36) in (6.40), it follows that for a stationary process $R_{X\dot{X}}(\tau) = E[X(t)\dot{X}(t + \tau)] = 0$ if $\tau = 0$, which means that there is no correlation between a stationary process and its derivative process at any point in time t. (This does not mean, however, that there is no correlation between $X(t_1)$ and $\dot{X}(t_2)$, $t_1 \neq t_2$.)

6.4.4 Ergodic processes
The mean (6.32) and the correlation (6.33) were defined for a stationary process as averages over all realizations. If they can be defined equally well by the time average over a single realization of a stationary process, the process is "weakly ergodic". If the equality holds for all moments of a strictly stationary process, the process is "strictly ergodic". Ergodicity in the mean is defined as

$$\mu_X = \lim_{T \to \infty} \left[\frac{1}{T} \int_0^T x(t) \, dt \right] \tag{6.42}$$

and ergodicity in correlation as

$$R_{XX}(\tau) = \lim_{T \to \infty} \left[\frac{1}{T} \int_0^T x(t+\tau)x(t) \, dt \right] \tag{6.43}$$

This property clearly can hold only for stationary processes. It is of considerable practical value in estimating statistical parameters from one or a few sufficiently long records of the process. The accuracy obtained will depend on the duration T of available records. Stationarity and ergodicity are often assumed in the analysis of stochastic process records unless (and until) there is evidence to the contrary.

6.4.5 First-passage probability

As noted in section 6.1, for time-dependent reliability interest lies mainly in the time that is expected to elapse before the first occurrence of an excursion of the random vector $\mathbf{X}(t)$ out of the safe domain D, defined by $G(\mathbf{X}) > 0$, (see Fig. 6.3). The probability of the first occurrence of such an excursion may be considered to be equivalent to the probability $p_f(t)$ of structure failure during a given period $[0, t]$:

$$p_f(t) = 1 - P[N(t) = 0 \mid \mathbf{X}(0) \in D]P[\mathbf{X}(0) \in D] \tag{6.44}$$

where $\mathbf{X}(0) \in D$ signifies that the process $\mathbf{X}(t)$ starts in the safe domain D at zero time and $N(t)$ is the number of outcrossings in the time interval $[0, t]$.

The general solution of (6.44) is rather difficult to obtain owing to the need to account for the complete history of the process $\mathbf{X}(t)$ in the interval $[0, t]$. The solution will usually depend on the nature of the process $\mathbf{X}(t)$. Series or other approximations have been suggested [e.g. Rice, 1944; Vanmarcke, 1975]. Fortunately, for reliability problems, outcrossings usually occur so rarely that it is often satisfactory for the individual outcrossings to be assumed independent events, and therefore independent of the probability of an outcrossing at $t = 0$. The probability of no outcrossings in $[0, t]$ may then be approximated using the Poisson distribution (A.30) with zero events [Cramer and Leadbetter, 1967]:

$$P[N(t) = 0] = \frac{(vT)^0}{0!} e^{-vt} = e^{-vt} \tag{6.45}$$

where v is the mean outcrossing rate from domain D. (Note that a similar approach was used in section 6.3.2 when dealing with the random occurrence of events such as storms).

In expression (6.44), the term $P[\mathbf{X}(0) \in D]$ is clearly equivalent to $1 - p_f(0)$, where $p_f(0)$ is the probability of failure at $t = 0$. Using also (6.45), expression (6.44) becomes [c.f. Veneziano, *et al.* 1977]

$$p_f(t) = 1 - [1 - p_f(0)] \, e^{-vt} \tag{6.46}$$

$$= p_f(0) + [1 - p_f(0)](1 - e^{-vt}) \tag{6.47}$$

but, since $vt > 1 - e^{-vt}$,

$$p_f(t) \leqslant p_f(0) + [1 - p_f(0)]vt \tag{6.48}$$

In many practical reliability problems vt is very close to zero and $p_f(0) \ll vt$, so that the "first passage" or failure probability becomes

$$p_f(t) \approx 1 - e^{-vt} \approx vt \tag{6.49}$$

If the vector process $\mathbf{X}(t)$ is smooth non-stationary, vt may be replaced by the average outcrossing rate $\displaystyle\int_0^t v(\tau) \, d\tau$.

Other, perhaps more aesthetic, derivations are available. Simulation results to show that (6.48) is indeed an upper bound, mainly because excursions tend to occur in "clumps" and are not strictly independent; these aspects will be ignored herein [Lin, 1970; Yang, 1975].

6.4.6 Distribution of local maxima

A local maximum of the scalar stochastic process $X(t)$ may be defined as the value of $X(t)$ at which $\dot{X}(t) = 0$ and $\ddot{X}(t) < 0$, i.e. when $\dot{X}(t)$ has a zero crossing. Local maxima are therefore all the peaks in a typical realization of $X(t)$ (see Fig. 6.8).

The full distribution function for the local maxima can be obtained, in principle, from an extension of Rice's formula (6.68) to be considered in section 6.4.8.1 below; this is really only tractable for normal processes.

When the mean outcrossing rate is low, as is usual in reliability problems, the distribution of the local maxima can be approximated directly from the first-passage probability given by expression (6.46), using the assumption that the outcrossings follow a Poisson distribution

$$1 - F_m(a) = p_f(t) = 1 - [1 - p_f(0)] \, e^{-vt} \tag{6.50}$$

where $F_m(a)$ is the cumulative distribution function $P(X_m < a)$ for the maximum X_m in the time interval $[0, t]$. Simplifying, it follows that

$$F_m(a) \approx e^{-vt} = 1 - vt \tag{6.51}$$

where v is the mean outcrossing rate in $[0, t]$ out of the safe domain D, defined, for a scalar process $X(t)$, as $x \leq a$. (This is a so-called "level" crossing problem; see section 6.4.8.1.) Evidently, the Poisson assumption becomes less accurate at lower levels $x = a$, and hence $F_m(\ )$ is only well approximated by (6.51) for high values of $x = a$. Result (6.51) is of considerable importance for the discussions to follow.

6.4.7 Discrete processes
A number of discrete processes of increasing complexity will be introduced here. These will be of relevance in the discussion of the time-dependent approach (see section 6.5) or in the discussion of load combinations (see section 6.6).

6.4.7.1 *Borges processes* (w_b)
One of the simplest discrete processes is generated by a sequence Y_k of independently and identically distributed random variables each acting over a given (deterministic) length of time t_b (the so-called "holding time"). A typical realization of this (stationary) process is shown in Fig. 6.9 [Ferry-

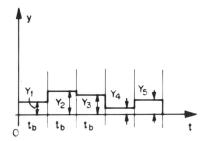

Fig. 6.9 — Realization of Borges process.

Borges and Castenheta, 1971]). In a given time interval $[0, t_L]$, the number of sequences of Y will be $r = t_L/t_b$ so that, owing to independence, the distribution of the extreme value of the sequence is given by

$$P\left\{ \max_{0 \leq t \leq t_L} [w_b(t; t_b, Y)] \leq a \right\} = P\left[\bigcap_{i=1}^{r} (Y_i \leq a) \right] \tag{6.52}$$

$$= [F_Y(a)]^r$$

The first occurrence of the level crossing $Y > a$ must be associated with a maximum value of Y in $[0, t_L]$. It follows therefore that the number of sequences before which $Y > a$ occurs is equal to r and that the probability that this occurs is given by $1 - P\left[\max_{0 \leq t \leq t_L} (w_b \leq a)\right]$ or

$$p_f(t_L) = 1 - [F_Y(a)]^r = 1 - [F_Y(a)]^{t_L/t_b} \tag{6.53}$$

6.4.7.2 *Poisson counting process*

If the time of occurrence of the event, rather than its magnitude, is a random variable, a different but still elementary type of process is obtained; the so-called "Poisson counting process". If $N(t)$ denotes the number of discrete events (or "states") $n = 0, 1, 2, 3, \ldots$ in a given time interval $[0, t]$, then the number of events in $[0, t]$ is given by the Poisson distribution (A.30)

$$P(n, t) = \frac{(vt)^n e^{-vt}}{n!} \tag{6.54}$$

where v is the "intensity" of the process, or the mean rate of occurrence of events per unit time.

The application of the Poisson distribution in the time domain assumes that the probability that an event occurs during any time increment t to $t + \Delta t$ is asymptotically proportional to Δt and that the probability of more than one event in any time interval is negligible as $\Delta t \rightarrow 0$. There is thus no overlap between events. Also $N(0) = 0$.

An important assumption is that the occurrence of events in time increments Δt_i, Δt_j, $i \neq j$, is independent. This means that, for any $t_0 < t_1 < t_2 < \ldots < t_m$, the m random variables

$$N(t_1) - N(t_0), \ldots, N(t_m) - N(t_{m-1}) \tag{6.55}$$

are independent. Clearly if (6.55) is independent also of t_i (but not of $t_{i+1} - t_i$) the process has stationary independent increments [Parzen, 1962].

A typical realization of a Poisson counting process is shown in Fig. 6.10. If v is constant, the process is termed "homogeneous". If $v = v(t)$, it is "non-homogeneous", in which case vt in (6.54) must be replaced by $m(t) = \int_0^t v(\tau) \, d\tau$. In this case the time increments are no longer stationary, since $N(t)$ depends directly on t, rather than on some time increment.

The Poisson counting process is a case of so-called Markov (or memory-less) processes with discrete states $(0, 1, 2, \ldots)$ in continuous time t. It is "memory-less" since each state is independent of previous states.

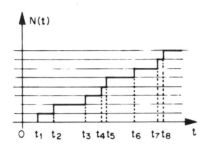

Fig. 6.10 — Realization of Poisson counting process.

The cumulative distribution function $F_W(t)$ for the time W_n which must elapse before the nth event in a Poisson process occurs is obtained by considering all the possible number of events $N(t) < n$, or

$$F_W(t) = 1 - P(W_n > t) = 1 - P[N(t) < n]$$

$$= 1 - \sum_{k=0}^{n-1} \frac{(vt)^k e^{-vt}}{k!} \qquad (6.56)$$

This may be recognized as a gamma distribution (A.40) with mean n/v and variance n/v^2.

Of particular interest is the waiting time before the first event $(n = 1)$ occurs during the time period $[0, t]$, so that, from (6.56) with $n = 1$,

$$p_f(t) = F_W(t) = 1 - e^{-vt} \qquad (6.57)$$

This is equivalent to the first-passage probability (see (6.49)).

6.4.7.3 Filtered Poisson processes
If stochastic properties are now also attributed to the events (states) in a Poisson process, a so-called "filtered Poisson process" is obtained. Typically it is defined as [Parzen, 1962]

$$X(t) = \sum_{k=1}^{N(t)} w(t, t_k, Y_k) \qquad (6.58)$$

where $N(t)$ is a Poisson process of intensity v generating the time points t_k at which the events have a random magnitude Y_k. *The* Y_k are assumed independent and identically distributed. Also $w(t, t_k, Y_k)$ is a "response function", which represents the response contributing (linearly) to $X(t)$ at time t, owing to the magnitude Y_k acting at time t_k. Generally $w(\ ) = 0$ for $t < t_k$.

A filtered Poisson process requires the specification of the intensity ν governing $N(t)$, the distributional properties of Y_k and the form of the response function $w(t, t_k, Y_k)$.

Two forms of filtered Poisson processes have particular relevance to reliability studies. In addition they allow explicit evaluation of their stochastic properties. These processes are:

(1) Poisson "spike" process;
(2) Poisson square wave process.

Both processes find application mainly in modelling loading processes (see Chapter 7). However, many other Poisson processes can be postulated, depending on the choices F_{Y_k} and $w(\ )$ and by allowing the holding time to also be a random variable (so that pulse overlap can occur) [Grigoriu, 1975]. These more complex processes, however, are beyond the scope of this book.

6.4.7.4.   *Poisson spike process*

A process having constant intensity ν and rectangular pulses of height Y_k and length b may be described by a response function

$$w_s(t, t_k, Y_k) = Y_k, \qquad 0 < t - t_k < b$$
$$= 0, \qquad \text{otherwise} \tag{6.59}$$

A typical realization is shown in Fig. 6.11. Here Y_k is a random variable,

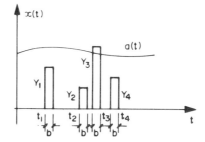

Fig. 6.11 — Realization of Poisson spike process.

independent between pulses and defined by the cumulative distribution function $F_Y(\ )$.

The level crossing rate and the first-passage probability may be obtained directly from the limit as $b \rightarrow 0$. The probability that the process $Y(t)$ has a height greater than $a(t)$ is $1 - F_Y(a)$ and the probability that $Y(t)$ has a height less than $a(t)$ is $F_Y(a)$, so that the probability of an upcrossing v^+ of level $a(t)$ in time $\Delta t \rightarrow 0$ is

$$v_a^+(t) = \lim_{\Delta t \to 0} \left\{ \frac{1}{\Delta t} [P(\text{upcrossing in } \Delta t]v \right\}$$

$$= \lim \left[\frac{1}{\Delta t} (P\{[Y(t) \leqslant a(t)] \cap [Y(t + \Delta t) > a(t)]\}v) \right]$$

$$= F_Y[a(t)]\{1 - F_Y[a(t)]\}v \qquad (6.60)$$

where $v = v(t)$ is the rate of arrival of the pulses. If a is large and time invariant, (6.60) reduces to $v_a^+ \approx [1 - F_Y(a)]v$.

It should be evident that v_a^+ in (6.60) expresses the intensity of a Poisson process. It is then possible to use (6.57) to obtain the first-passage probability, since "failure" is the occurrence of the first event (a spike greater than a); thus

$$p_{f_1}(t_L) = 1 - \exp\{ - [1 - F_Y(a)]vt_L\} \qquad (6.61)$$

where t_L is the nominal lifetime of the structure. As before a, v_a^+ and v may each considered to be functions of time with appropriate substitution of

$$\int_0^t v(\tau) \, d\tau \text{ for } v \text{ and } a(t) \text{ for } a \text{ in (6.61).}$$

The cumulative distribution function for the maximum value of the process $X(t)$ may be obtained by considering $a(t)$ as time invariant. Then the first-passage probability (6.61) represents also the probability that the maximum value of $X(t)$ is greater than a since the first-passage point must be the same as the maximum value of $X(t)$. Hence

$$F_{X_{\max}}(a) = P[X_{\max}(t) < a] = \exp\{ - [1 - F_Y(a)]vt_L\} \qquad (6.62)$$

6.4.7.5   *Poisson square wave process*

When the length b of each rectangular pulse in Fig. 6.11 is given by the random interval $t_n - t_{n-1}$, the square wave process of Fig. 6.12 is obtained. The pulse height Y_k now stays constant until the next event t_{k+1} generates a new pulse height Y_{k+1}. As before, the Y_k are independent, identically distributed random variables.

Proceeding exactly as for a Poisson spike process, and with a constant level a, the level crossing rate v_a^+ is given by (6.60) so that the first-passage probability for a time period $[0, t_L]$ is given by

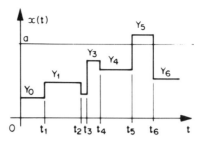

Fig. 6.12 — Realization of Poisson square wave process.

$$p_{f_1}(t_L) = F_{Y_0}(a)(1 - \exp\{ - F_Y(a)[1 - F_Y(a)]\}vt_L) \qquad (6.63)$$

where $F_{Y_0}(a)$ is the probability that $Y_0 < a$, an event independent of subsequent events because of the assumption on Y_k. If a is large $F_{Y_0}(a) \approx F_Y(a) \approx 1$.

It follows readily that the cumulative distribution function for the maximum value of $X(t)$ is derived in a manner similar to that for the Poisson spike process, as

$$F_{X_{max}}(a) = P(X_{max} < a) = F_{Y_0}(a) \exp\{ - [1 - F_Y(a)]vt_L\} \qquad (6.64)$$

6.4.7.6 *Renewal processes*

The Poisson processes above are particular forms of a more general type of process in which events occur along a time axis. Such a process can be taken to define the potential beginning and end times of load events or pulses. It is easily shown that for a Poisson process the time between events is exponentially distributed; more generally any process with events along the time axis governed by an appropriate probability law is termed a "renewal process".

It is possible also to generalize to more complex pulse shapes. However, no general results are available for up-crossing rates in most of these cases, although the results (6.60) for the Poisson processes above may be taken as an approximation.

Further, it is also possible to define the cumulative distribution function of Y_k, the pulse height, such that there is a finite probability p that $Y_k = 0$. This might occur, for example, in some types of floor loading. A typical distribution of Y_k for such a case is shown in Fig. 6.13 [Bosshard, 1975].

In this case (6.60) will be modified to

$$v_a^+ = [p + qF(a)]\{q[1 - F(a)]\}v \qquad (6.65)$$

where $F(\ )$ is the improper cumulative distribution function defined in Fig.

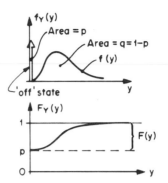

Fig. 6.13 — Probability density function and cumulative distribution function for "mixed" renewal process.

6.13. Note that $F_Y(\ )$ is now $p + qF(\ )$ and that, by differentiation, the probability density function is given by $f_Y(y) = p(\delta y) + qf(y)$, where $f(\ )$ is defined in Fig. 6.13 and $\delta(\ )$ is the Dirac delta function.

In expression (6.65) the first term represents the probability that $Y(t) \leqslant a(t)$ and the second term the probability that $Y(t) > a(t)$. If each pulse automatically returns to zero after it has been applied, the first term in (6.65) is obviously unity [Larrabee and Cornell, 1981]. Processes of this type have been called "mixed processes". Note that the mean arrival rate of pulses for the mixed process is qv, where v is the mean arrival rate of pulses for the unmodified process $Y(t)$.

6.4.8 Continuous processes
Most natural phenomena do not change their characteristics at specific points in time but do so continuously. Thus, for example, wind velocity and wave height are not discrete processes in time but continuous ones.

There are many types of possible continuous processes $X(t)$ which can be postulated. Because of their tractable properties, by far the largest amount of information is available for normal (or Gaussian) processes in which, at any time t, $X(t)$ is normal distributed. This means that for any set of times, t_1, $t_2, \ldots, t_n$, $X(t_i)$, $i = 1, 2, \ldots, n$ is jointly normal distributed, with a given correlation structure.

6.4.8.1 Barrier (or level) upcrossing
As described in section 6.1, an important property in reliability studies is the rate at which a random process $X(t)$ "upcrosses" a "barrier" or "level" $x = a(t)$ (Fig. 6.14). If $X(t)$ represents a loading process, then $x = a(t)$ might represent a (time-dependent) resistance or, if $X(t)$ represents a safety margin, then $x = a(t) = 0$ might represent the limit state. For the present, a scalar process $X(t)$ only will be considered; the more general case for the vector process $\mathbf{X}(t)$ is described in section 6.4.8.2 below.

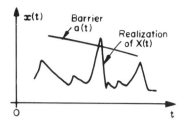

Fig. 6.14 — Typical realization and barrier crossing of process $X(t)$.

Consider the segment of sample function $x(t)$, shown in Fig. 6.15, between the times t_1 and $t_1 + dt$, where $dt \to 0$. Without loss of generality, the time $t_1 + dt$ may be considered to be the time at which the barrier crossing occurs. Also for dt sufficiently small the curves between t_1 and $t_1 + dt$ can be taken as straight lines.

The sample functions which cross $a(t)$ during dt must start below $a(t)$ at t_1 and have sufficient slope $\dot{x}(t)$ at t_1 to pass through $a(t)$ during dt. The limits on the slope $\dot{X}(t)$ are evidently

$$\dot{x}\, dt - \dot{a}\, dt \geqslant a(t) - x(t) \text{ as } dt \to 0 \text{ and } x(t_1) \leqslant a(t_1) \tag{6.66}$$

These are plotted in Fig. 6.16 on orthogonal $(x, \dot{x})$ axes. From section 6.4.3, $X(t)$ and $\dot{X}(t)$ are uncorrelated, so that this representation is straightforward. Now the probability that $X(t)$ is between x and $x + dx$ and that $\dot{X}(t)$ is between $\dot{x}$ and $\dot{x} + d\dot{x}$ is given by the joint probability density function $f_{X\dot{X}}(\ )$. The total number of barrier crossings in dt is given by the fraction of all possible realizations which satisfy the conditions (6.66); this is identical with the probability content contained above the shaded region in Fig. 6.16 or

$$N = \int_{\dot{a}}^{\infty} \int_{a-(\dot{x}-\dot{a})\,dt}^{a} f_{X\dot{X}}(x, \dot{x})\, dx\, d\dot{x} \tag{6.67}$$

As $dt \to 0$, the shaded region in Fig. 6.16 tends to reduce to a narrow strip, or, equivalently, $a - \dot{x}dt \to a$ as integration limit in (6.67), so that upon dividing through by dt and integrating over x the barrier upcrossing rate v_a^+ is

$$v_a^+ = \int_{\dot{a}}^{\infty} (\dot{x} - \dot{a}) f_{X\dot{X}}(a, \dot{x})\, d\dot{x} \tag{6.68}$$

which is a result due to Rice (1944). If a is not time dependent, $\dot{a} = 0$.

The barrier upcrossing rate v_a^+ is an ensemble average, i.e. it is the

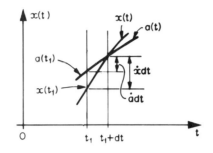

Fig. 6.15 — Segment of sample function $x(t)$ crossing barrier $a(t)$ in dt.

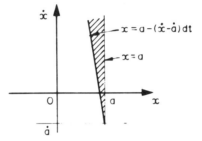

Fig. 6.16 — Integration limits in $(x, \dot{x})$ plane.

average over all realizations of $X(t)$ at time t. Only if the process is ergodic will v_a^+ also be the time average frequency of upcrossings of level $x = a$.

In the special case when $X(t)$ is a stationary normal process, $f_{X\dot{X}}(\ )$ is given by (cf. (A.125) with $\rho = 0$)

$$f_{X\dot{X}}(\ ) = \frac{1}{2\pi\sigma_X\sigma_{\dot{X}}}\exp\left\{-\frac{1}{2}\left[\left(\frac{a - \mu_X}{\sigma_X}\right)^2 + \frac{\dot{x}^2}{\sigma_{\dot{X}^2}}\right]\right\} \qquad (6.69)$$

in which $X(t)$ is normal distributed $N(\mu_X, \sigma_X^2)$ and $\dot{X}(t)$ is $N(0, \sigma_{\dot{X}}^2)$. The mean of $\dot{X}(t)$ is zero for a stationary process. Noting that

$$\int_0^\infty \dot{x}\, \exp\left(-\frac{\dot{x}^2}{2\sigma_{\dot{X}}^2}\right)\, \mathrm{d}\dot{x} = \sigma_{\dot{X}}^2$$

substituting (6.69) into (6.68) and integrating produces

$$v_a^+ = \frac{1}{2\pi}\frac{\sigma_{\dot{X}}}{\sigma_X}\exp\left[-\frac{(a-\mu_X)^2}{2\sigma_X^2}\right] = \frac{\sigma_{\dot{X}}}{(2\pi)^{1/2}}f_X(\ ) \tag{6.70}$$

with $f_X(\ ) = (1/\sigma_X)\phi[(a-\mu_X)/\sigma_X]$, where $\phi(\ )$ is the standardized normal density function (see section A.5.7). Note also that σ_X is obtained from (6.35) as

$$\sigma_X^2(t) = R_{XX}(t,\ t) - \mu_X^2(t) \tag{6.71}$$

which for a stationary process becomes

$$\sigma_X^2 = R_{XX}(\tau = 0) - \mu_X^2 \tag{6.72}$$

By analogy, $\sigma_{\dot{X}}$ is similarly obtained from $\sigma_{\dot{X}}^2(t) = R_{\dot{X}\dot{X}}(t,\ t) - \mu_{\dot{X}}(t)$. Since $R_{\dot{X}\dot{X}}(t,\ t) = R_{\dot{X}\dot{X}}(\tau = 0)$ for a stationary process, using (6.41) and $\mu_{\dot{X}} = 0$ leaves the variance of $\dot{x}$ as

$$\sigma_{\dot{X}}^2 = -\frac{\partial^2 R_{XX}(0)}{\partial\tau^2} \tag{6.73}$$

For non-normal processes, the joint probability density function $f_{X\dot{X}}(\ )$ will usually be much less amenable to definition and integration. Such processes arise, for example, in river flows, mean hourly wind speeds and when normal processes are transformed non-linearly [Grigoriu, 1984]. It is sometimes suggested that for such processes the upcrossing rate may be approximated by (6.70). However, this approximation can be seriously in error.

The Gaussian process with $\mu_X = 0$ has special significance for dynamic problems (see section 6.7) or those involving fatigue (see section 6.8). The upcrossing rate v_a^+ for $a = 0$ then counts the number of loading cycles. For a standardized Gaussian process this is given by $v_0^+ = \sigma_{\dot{X}}/2\pi\sigma_X$.

Finally, all the above results, and in particular expression (6.68), may be extended to smooth non-stationary processes by interpreting v_a^+ and $f_{X\dot{X}}$ as time dependent.

6.4.8.2 Outcrossing rate

The immediate generalization from barrier crossing, involving the scalar process $X(t)$, is the outcrossing concept involving the vector process $\mathbf{X}(t)$. For the two-component process $[X_1(t), X_2(t)]$ a realization of $\mathbf{X}$ might be as shown in Fig. 6.17. The barriers B_i indicated are essentially limit state equations in $\mathbf{X}$ defining the safe domain D [Veneziano et al., 1977].

If the barriers to the domain D† are such that the safe region can be described by $D: Z(t) = G[X(t)] \leqslant 0$ then the scalar process $Z(t)$ can be used in (6.68), provided that $f_{Z\dot{Z}}$ can be determined since, whenever $\mathbf{X}(t)$ outcrosses barriers B_i, $Z(t)$ upcrosses the level zero. Closed-form solutions

---

† In chapters 1–5 $D:G \geqslant 0$ referred to the "failure" region. In the present section only, this notation will refer to the "safe" region to conform to convention for crossing theory.

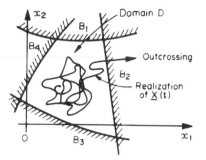

Fig. 6.17 — Two-dimensional outcrossing from domain D.

are available only for a very limited range of problems. Almost all these deal with special types of two-dimensional normal processes, and with open or closed square or circular domains [Hasofer, 1974].

More generally, the domain D is bounded by a series of q discontinuous hypersurfaces B_i defined by†

$$D: G_i[\mathbf{X}(t)] \leqslant 0 \qquad i = 1, 2, \ldots, q \qquad (6.74)$$

This represents the type of problem of interest in structural reliability studies (see section 6.1), but the problem is even less amenable to closed-form solution. It is necessary to obtain approximate solutions even in the case of normal processes.

The first step is to generalize the Rice formula (6.68) to deal with a vector process $\mathbf{X}(t)$. For convenience of exposition consider a two-dimensional vector process $\mathbf{X}(t)$ and an arbitrary domain D (Fig. 6.18). Let $\mathbf{x}(t_1)$ be on the

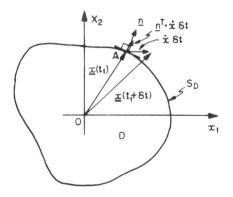

Fig. 6.18 — Vector process realization $\mathbf{x}(t)$, its change in time increment δt and component normal to domain boundary S_D.

† See footnote previous page.

limit state surface at A. For an outcrossing to occur, $\mathbf{x}(t_2)$ with $t_2 = t_1 + \delta t$ where $\delta t \to 0$, must be outside the domain D, as shown. The change in $\mathbf{x}(\ )$ over the time increment δt is also shown as the vector $\dot{\mathbf{x}} \, \delta t$ where $\dot{\mathbf{X}}(t_1)$ is the random vector of time derivatives $[\partial X_1/\partial t, \partial X_2/\partial t]^T$ evaluated at time t_1. If $\mathbf{n}(t)$ represents an outward normal at A, then the component $\dot{\mathbf{x}} \, \delta t$ which contributes to outcrossing (rather than merely moving along the domain boundary) is the scalar product $\mathbf{n}^T \cdot \dot{\mathbf{x}} \, \delta t > 0$ ($= (n_1 \dot{x}_1 + n_2 \dot{x}_2 + \ldots) \delta t > 0$) as shown. For convenience, let this be called $\dot{x}_n$ (a scalar) as $\delta t \to 0$. Comparison with the one-dimensional process described in section 6.4.8.1 and shown in Fig. 6.15 shows that in this case $\dot{x}_n$ corresponds directly to $\dot{x}$. It seems entirely plausible, therefore, to suggest that (6.68) generalized to the m-dimensional vector process is given by

$$v_D^+ = \int_{S_D} d\mathbf{x} \int_0^\infty \dot{x}_n f_{X \dot{X}_n}(\mathbf{x}, \dot{x}_n) \, d\dot{x}_n \qquad (6.75)$$

where the first integral is necessary to ensure that all points of the domain boundary S_D are considered. A formal proof is available [Belyaev and Nosko, 1969]; the result is not restricted to normal processes.

Noting (A.119), expression (6.75) may be rewritten by taking the conditional expectation of $\dot{x}_n$:

$$v_D^+ = \int_{S_D} E(\dot{X}_n | \mathbf{X} = \mathbf{x})^+ f_{\mathbf{X}}(\mathbf{x}) \, d\mathbf{x} \qquad (6.76)$$

where, in (6.75) and (6.76), $\mathbf{X} = \mathbf{X}(t)$ and $\dot{x}_n = \mathbf{n}(t) \cdot \dot{\mathbf{x}}(t) > 0$. The latter condition is denoted by $(\ )^+$; if $\dot{x}_n \leq 0$, $E(\ ) = 0$.

In the form (6.76), the outcrossing rate v_D^+ may be interpreted as follows. For any elemental point $\mathbf{x}(t)$ on S_D, $E(\ )$ in (6.76) represents the expectation that an outcrossing will occur; this is then weighted by the "probability" $f_{\mathbf{X}}(\mathbf{x})$ that $\mathbf{x}$ will occur. This is then summed over the domain boundary S_D. Comparison with (6.68) shows that the integrand in (6.76) might be seen as a one-dimensional solution, appropriately weighted.

The solution of (6.75) or (6.76) is still not straightforward in general. A few theoretically exact solutions for simple time-invariant domains D under the assumption that $\dot{\mathbf{X}}(t)$ and $\mathbf{X}(t)$ at any time t are mutually independent are available [Veneziano et al., 1977], but these are too restrictive for general use in reliability analysis. An approximate bounding technique has been proposed; this will be discussed in section 6.5.2.

In all the above it has been assumed that $\mathbf{X}(t)$ is stationary, although, as usual, the latter restriction may be removed for smooth non-stationary processes by appropriate substitution of $\int_0^t v_D^+(s) ds$ for v_D^+ and by modifying (6.76) appropriately in $f_{\mathbf{X}}(\ )$.

6.5 Time-dependent approach

In the time-dependent approach the reliability of the structure is obtained directly from the first-passage probability, i.e. from the probability that the vector process $\mathbf{X}(t)$ leaves the safe domain D during the nominal life $[0, t_L]$ of the structure (see Fig. 6.17).

As in section 1.4 for the time-independent case, the vector process $\mathbf{X}(t)$ contains all the random variables describing the problem being considered, such as loading, resistances, dimensions, etc. The safe domain D is described by one or more limit state functions $G_i[\mathbf{X}(t)] \geqslant 0$ *for* $i = 1, \ldots, n$.

As noted in section 6.4.8, solution of the time-dependent problem is generally not straightforward and the number of solutions available is limited. Some results are available for certain types of discrete processes and for normal vector processes. These will be considered below.

6.5.1 Discrete processes

Discrete processes, such as Poisson square wave processes (see section 6.4.7.5) are commonly used to describe loads such as floor loading in offices, hospitals, car parks, etc. The outcrossing rate and hence the first-passage probability for the special case in which each of the n independent components $X_i(t)$ of the vector process $\mathbf{X}(t)$ is Poisson square wave will now be considered.

Let the magnitude Y_i of the process be normal distributed, and let the arrival times t be Poisson distributed with a mean arrival rate v_i. A typical realization is shown in Fig. 6.19. The total outcrossing rate v_D^+ will be the

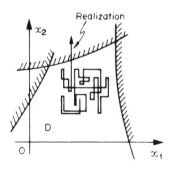

Fig. 6.19 — Typical realization of Poisson square wave vector process $\mathbf{X}(t)$ in two dimensions.

sum over all components of the probability that an outcrossing occurs for each component, weighted by the occurrence probability of that combination of components [Breitung and Rackwitz, 1982]. This may be expressed as

$$v_D^+ = \sum_{i=1}^{n} v_i \int_{-\infty}^{\infty} P(\text{outcrossing due to } Y_i) f_{\mathbf{X}*}(\mathbf{x}) \, d\mathbf{x} \qquad (6.77)$$

where $P(\ ) = P[(Y_i, \mathbf{x}*) \in D] P[(Y_i, \mathbf{x}*) \notin D]$.

X* is **X** without the ith component, evaluated at the time that the ith component has a renewal given by $X_i = Y_i$. Further, $f_{\mathbf{X}*}(\ )$ is the probability density function for **X*** given by

$$f_{\mathbf{X}*}(\ ) = \delta(x_i) \prod_{j=1}^{n} f_{X_j}(x_j), \qquad j \neq i$$

and $\delta(x_i)$ is the Dirac delta function, introduced to ensure that $f_{\mathbf{X}*}(\ ) = 0$ unless the ith component is considered. The multiple integral extends over all the components of **X** and for any one component extends over all its values $-\infty < X_i < \infty$. Particular solutions depend on the shape of the domain. If this is defined as the hypercube $a_j \leq x_j \leq b_j$, and **X** is a vector with standard normal components, it follows that

$$P[(Y_i, \mathbf{x}*) \in D] = P(a_i \leq Y_i \leq b_i) = \Phi(b_i) - \Phi(a_i), \qquad \text{for } a_j \leq x_j \leq b_j; \, j \neq i$$

$$= 0, \qquad\qquad\qquad\qquad \text{otherwise}$$

and

$$P[(Y_i, \mathbf{x}*) \notin D] = P[(Y_i < a) \cup (Y_i > b_i)] = 1 - \Phi(b_i) + \Phi(a_i) \quad (6.78)$$

so that

$$v_D^+ = \sum_{i=1}^{n} \left\{ v_i \int_{a_1}^{b} \cdots \int_{a_n}^{b} [\Phi(b_i) - \Phi(a_i)][1 - \Phi(b_i) + \Phi(a_i)] \right.$$

$$\left. \times \prod_{\substack{j=1 \\ j \neq i}}^{n} \phi(x_j) \, dx_1 \ldots dx_{i-1} \, dx_{i+1} \ldots dx_n \right\} \qquad (6.79)$$

$$= \sum_{i=1}^{n} \left\{ v_i [\Phi(b_i) - \Phi(a_i)][1 - \Phi(b_i) + \Phi(a_i)] \prod_{\substack{j=1 \\ j \neq i}}^{n} [\Phi(b_j) - \Phi(a_j)] \right\}$$

$$\qquad\qquad\qquad\qquad\qquad\qquad\qquad\qquad\qquad\qquad (6.80)$$

This expression has a direct infirst term [ ] is the probability that the renewal Y_i of X_i is in the safe region D, the second term [ ] the probability that Y_i is outside D and the π[ ] term is the probability that all the remaining components are in D, i.e. that X_i is actually the only component outcrossing at this time. Evidently (6.80) may be contracted to

$$v_D^+ = \prod_{\substack{j=1 \\ j \neq i}}^{n} [\Phi(b_j) - \Phi(a_j)] \sum_{i=1}^{n} \{v_i[1 - \Phi(b_i) + \Phi(a_i)]\} \qquad (6.81)$$

in which the π[ ] term denotes the probability that all components are in the safe region D and the summation term represents the sum of each component that is independently out of D.

By a parallel argument it may be shown that for a hyperplane defined by $\beta + \boldsymbol{\alpha} \cdot \mathbf{X} = 0$ where $\boldsymbol{\alpha}$ is the vector of direction cosines of the normal to the hyperplane and β the distance from the origin perpendicular to the plane (cf. section 4.3.2), the outcrossing rate is given by

$$v_D^+ = \sum_{i=1}^{n} \{v_i[\Phi(-\beta) - \Phi_2(-\beta, -\beta, \rho)]\} \qquad (6.82)$$

where $\rho_i = 1 - \alpha_i^2$ and $\Phi_2(\ )$ is the bivariate normal integral [Breitung and Rackwitz, 1982].

Extension to other forms of limit state functions $G_i[\mathbf{x}(t)] = 0$ is not necessarily easy since the formulation of the term $P(\ )$ in (6.77) is generally more difficult than that for the above examples.

When just two (independent) load processes $X_1(t)$ and $X_2(t)$ and a linear limit state function given by $x_1(t) + x_2(t) - a = 0$ are considered, the range of solutions for the mean outcrossing rate and hence the first-passage probability are somewhat greater. Exact solutions for the first-passage problem when X_1 and X_2 are Poisson square wave processes have been given by Bosshard (1975), Hasofer (1974) and Gaver and Jacobs (1981). Larrabee and Cornell (1981) have given an exact solution for the mean level upcrossing rate of rectangular renewal processes. Approximate solutions or bounds for the mean level crossing rate of the process $X_1(t) + X_2(t)$ are also available for renewal processes having more general pulse shapes [Larrabee and Cornell, 1981; Madsen et al., 1979; Waugh, 1977]. Because of their limited range of application, discussion of these results will be deferred until section 6.6 on load combinations.

6.5.2 Continuous processes
Except for normal processes, the determination of outcrossing rates and first-passage probabilities for continuous processes and general limit state

functions is generally not yet feasible. For normal processes some progress can be made, particularly when the safe region is described by convex (hyper)polyhedrals (as limit state functions). In this case the problem of outcrossing rate determination has a close affinity with the first-order second-moment (FOSM) method. The first-passage probability may then be bounded from above using (6.46)–(6.49). The possibility of extension of the solutions to non-normal processes, in the manner of the first-order reliability (FOR) method of Chapter 4 will be considered in section 6.5.3 below. First, however, normal processes with convex polyhedral limit state functions will be considered.

Recall that for the safe domain D, bounded by S_D, the outcrossing rate v_D^+ of a vector process $\mathbf{X}(t)$, is given by (6.76):

$$v_D^+ = \int_{S_D} E(\dot{X}_n | \mathbf{X} = \mathbf{x})^+ f_{\mathbf{X}}(\mathbf{x}) \, d\mathbf{x} \tag{6.83}$$

where $\dot{x}_n = \mathbf{n}(t) \cdot \dot{\mathbf{x}}(t) > 0$ is the unit rate vector normal to S_D at $\mathbf{x}$ and $(\ )^+ \geq 0$ for $\dot{x}_n > 0$, and otherwise $(\ )^+ = 0$. The term $f_{\mathbf{X}}(\mathbf{x})$ denotes the probability density function for $\mathbf{X}(t)$.

Let it now be assumed that S_D consists of a set of time-invariant piecewise continuous convex (hyper)planes. It then follows that on any one (hyper)plane, the expectation term $E(\ )^+$ will be independent of the precise value of $\mathbf{X}$, since the unit normal $\dot{X}_n$ will have the same direction anywhere on the (hyper)plane. Let it further be assumed that $X_i(t)$ and $\dot{X}_j(t)$ are independent at any given time t (cf. section 6.4.3, which shows that X_i and $\dot{X}_i$ are independent). Then (6.83) may be rewritten as

$$v_D^+ = E(\dot{X}_n)^+ \int_{S_D} f_{\mathbf{X}}(\mathbf{x}) \, d\mathbf{x} \tag{6.84}$$

where the integral over the domain surface S_D represents the probability that $\mathbf{X}$ lies on S_D and the $E(\ )$ the probability that an outcrossing will occur for $\mathbf{X}$ on S_D. For any one (the lth) (hyper)plane H_l, containing the partial domain surface ΔS_l, the partial outcrossing rate is then

$$v_{\Delta D_l}^+ = E(\dot{X}_{nl})^+ \int_{\Delta S_l} f_{\mathbf{X}}(\mathbf{x}) \, d\mathbf{x} \tag{6.85}$$

The integral in (6.85) represents the probability that $\mathbf{X}$ lies on ΔS_l. This can be considered also in two parts: firstly, the probability that $\mathbf{X}$ lies on the (hyper)plane H_l and, secondly, that $\mathbf{X}$ lies within ΔS_l given that $\mathbf{X}$ is on H_l.

It should be noted that, for linear limit state functions, ΔS_l as projected on H_l is a (hyper)polygon of dimension $n - 1$; the probability content within ΔS_l may be obtained directly by integrating $f_{\mathbf{X}}^*(\ )$ over ΔS_l, where $f_{\mathbf{X}}^*(\ )$ is the

$n-1$ probability density function obtained from $f_{\mathbf{X}}(\ )$ by integrating in the direction normal to H_l. Rather than determining $f_{\mathbf{X}}^{*}(\ )$ first, $f_{\mathbf{X}}(\ )$ may be integrated over the $n-1$ dimensions of ΔS_l, provided that the result is weighted by the probability that X lies on H_l. Hence the integral in (6.58) may be replaced by two terms:

$$v_{\Delta D_l}^{+} = E(\dot{X}_{nl})^{+} \int_{H_l} f_{\mathbf{X}}(\mathbf{x}) \, d\mathbf{x} \int_{\substack{\Delta S_l \\ n-1}} f_{\mathbf{X}}(\mathbf{x}) \, d\mathbf{x} \qquad (6.86)$$

where it is now understood that integration of $f_{\mathbf{X}}(\ )$ over ΔS_l is in $n-1$ dimensions and that the integral over H_l is in one dimension, perpendicular to H_l [*Veneziano et al.* 1977].

The second integral in (6.86) may be evaluated by using the linear limit state system bounds of Chapter 5 but now used to determine the probability content p_l contained within ΔS_l rather than that outside ΔS_l. Clearly the probability content required is the complement of that obtained directly from the system bounds. Example 6.3 below will illustrate this.

The first integral in (6.86), which represents the probability that **X** lies on H_l, may be evaluated by considering what happens when the hyperplane H_l is displaced laterally to itself.

If the lth linear safety margin has the expression $Z_l = a_0 + a_1 X_1 + a_2 X_2 + \ldots + a_n X_n$, with $a_0 > 0$, a_i all constants, it follows from the discussion in section 4.3 that the limit state $Z_l = 0$ has a normal on H_l defined by $\mathbf{a}_l$:

$$\mathbf{a}_l = (\alpha_{1l}, \alpha_{2l}, \ldots)^{\mathrm{T}} \qquad (6.87)$$

where

$$\alpha_{il} = \frac{a_i}{l}$$

with   (6.88)

$$l = \left(\sum_{i=1}^{n} a_i^2 \right)^{1/2}$$

If the (hyper)plane is displaced a distance δ in the direction of the normal $\mathbf{a}_l$, the change in the value of Z_l is (see Fig. 6.20)

$$Z_l(>0) - Z_l(=0) = [a_0 + a_1(X_1 + \delta\alpha_1) + a_2(X_1 + \delta\alpha_1) + \ldots] - (a_0 + a_1 X_1 + \ldots)$$

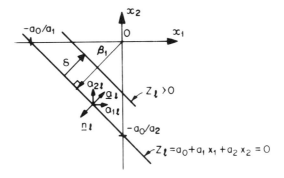

Fig. 6.20 — Parallel shift of hyperplane H_l.

$$= \delta \sum_{i=1}^{n} a_i \alpha_{il}$$

$$= \delta \left(\sum_{i=1}^{n} a_i^2 \right)^{\frac{1}{2}} \tag{6.89}$$

using (6.88). The change in the corresponding reliability index $\beta = \mu_Z / \sigma_Z = E(Z_l)/D(Z_l)$ is given by

$$\Delta \beta_l = \frac{E(Z_l > 0) - E(Z_l = 0)}{D(Z_l)} \tag{6.90}$$

The probability content between the two hyperplanes is clearly

$\Delta \Phi(-\beta_l) = \Phi(-\beta_l) - \Phi[-(\beta_l - \Delta \beta_l)]$. The probability density $\int_{H_l} f_{\mathbf{x}}(\ ) \, d\mathbf{x}$ is then the probability change as δ, the distance between the planes, reduces to zero:

$$\int_{H_l} f_{\mathbf{x}}(\mathbf{x}) \, d\mathbf{x} = \lim_{\delta \to 0} \left[\frac{\Delta \Phi(-\beta_l)}{\delta} \right] = \lim_{\delta \to 0} \left[\phi(\beta_l) \frac{\Delta \beta_l}{\delta} \right] \tag{6.91}$$

or, using (6.89) and (6.90),

$$\int_{H_l} f_{\mathbf{x}}(\mathbf{x}) \, d\mathbf{x} = \frac{\phi(\beta_l)}{D(Z_l)} \left(\sum_{i=1}^{n} a_i^2 \right)^{1/2} \tag{6.92}$$

When all X_i are standard normal random variables, it follows readily that (6.92) reduces to $\phi(\beta_l)$ [Veneziano *et al.* 1977].

The result (6.92) together with the expectation term $E(\ )^+$ in (6.86) may be seen as a correction term defined by

$$K_l = E(\dot{X}_{nl})^+ \int_{H_l} f_{\mathbf{X}}(\mathbf{x}) \, d\mathbf{x} \tag{6.93}$$

This term may be simplified by noting that the (hyper)polyhedron ΔS_l may be defined on H_l by a series of linear limit state functions

$$Z_j = a_{0j} + \sum_{i=1}^{n} a_{ij} X_i, \qquad i = 1, \ldots, n; \, i \neq l$$

defined such that $Z_j > 0$ inside ΔS_l, and with $Z_l = 0$.

Noting that $\dot{Z}_l = \sum_{i=1}^{n} a_i \dot{X}_i$ and $\dot{X}_{nl} \equiv \mathbf{n}_l^{\mathrm{T}} \cdot \dot{\mathbf{X}}$, where $\mathbf{n}_l$ is the outward normal vector to the plane H_l and therefore equal to $-\boldsymbol{\alpha}_l$ (see Figs. 6.18 and 6.20), it follows that

$$\dot{X}_{nl} = -\boldsymbol{\alpha}_l^{\mathrm{T}} \cdot \dot{\mathbf{X}} = -\sum_{i=1}^{n} \alpha_{il} \dot{X}_i \tag{6.94}$$

$$= -\frac{\sum\limits_{i=1}^{n} a_i \dot{X}_i}{\left(\sum\limits_{i=1}^{n} a_i^2 \right)^{1/2}} \tag{6.95}$$

$$= -\frac{\dot{Z}_l}{\left(\sum\limits_{i=1}^{n} a_i^2 \right)^{1/2}} \tag{6.96}$$

using (6.88). Collecting terms (6.96) and (6.92) into (6.93) produces

$$K_l = E(-\dot{Z}_l)^+ \frac{\phi(\beta_l)}{D(Z_l)} \tag{6.97}$$

This expression may be further simplified by noting that Z_l and therefore $\dot{Z}_l$ is normally distrubuted (since **X** is normal). Let the mean and standard deviation of $\dot{Z}_l$ be μ and σ respectively. Then the probability density function $f_{\dot{Z}_l}(\ )$ is also given by $(1/\sigma)\phi[(\dot{Z}-\mu)/\sigma]$ while the expectation conditional on having $-\dot{Z}_l > 0$ is given by (A.10):

$$E(-\dot{Z}_l)^+ = -\int_{-\infty}^{0} v \frac{1}{\sigma} \phi\left(\frac{v-\mu}{s}\right) dv \tag{6.98}$$

where the integration limits have been chosen to achieve the positive part of expectation. Integrating by parts and substituting into (6.97),

$$K_l = \phi(\beta_l)\frac{\sigma}{D(Z_l)}\left[\phi\left(\frac{\mu}{\sigma}\right) - \frac{\mu}{\sigma}\Phi\left(-\frac{\mu}{\sigma}\right)\right] \tag{6.99}$$

where $\mu = E(\dot{Z}_l)$, $\sigma = D(\dot{Z}_l)$ and $\phi(\ )$ and $\Phi(\ )$ are the usual probability density function and cumulative distribution function for the standardized normal distribution respectively. In the important case when the limit states are time invariant, $\mu = E(\dot{Z}_l) = 0$ and (6.99) reduces to

$$K_l = \frac{\phi(\beta_l)}{(2\pi)^{\frac{1}{2}}} \frac{D(\dot{Z}_l)}{D(Z_l)} \tag{6.100}$$

The outcrossing rate v_D^+ for the whole domain is obtained as the sum of the outcrossing rate for each hyperplane H_l, $l = 1, \ldots, k$; thus,

$$v_D^+ = \sum_{l=1}^{k} v_{D_l}^+ = \sum_{l=1}^{k} K_l p_l \tag{6.101}$$

where $p_l = \int_{\Delta S_l} f_{\mathbf{X}}(\mathbf{x}) \, d\mathbf{x}$ as described for equation (6.86). If p_l is bounded, then (6.101) also results in bounds, but now on v_D^+. The upper bound to the first-passage probability then follows directly from (6.46)–(6.49).

The above has been obtained assuming differentiability of each safety margin Z_l for all t. In particular, this implies that for the existence of the

mean $E[\dot{Z}_i(t)]$ the mean value function $E[Z_i(t)]$ must be differentiable for all

t and that for the existence of the standard deviation $D[\dot{Z}_i(t)]$ the covariance function $\text{cov}[Z_i(t_1), Z_j(t_2)]$ must be differentiable with $i = j$ and for $t_1 = t_2$; thus,

$$E[\dot{Z}_i(t)] = \frac{d}{dt}\{E[Z_i(t)]\} \tag{6.102}$$

$$\text{var}[\dot{Z}_i(t)] = \text{cov}[\dot{Z}_j(t), \dot{Z}_i(t)] = \left.\frac{\partial^2}{\partial t_1\,\partial t_2}\text{cov}[Z_i(t_1), Z_j(t_2)]\right|_{\substack{i=j \\ t_1=t_2}} \tag{6.103}$$

For purely normal vector processes, a lower bound to the first-passage probability may be based on the second-order system bounds (5.25) and (5.31) of Chapter 5. Examples have been given by Ditlevsen (1983b) and Wickham (1985).

Example 6.3   [*Adapted from Ditlevsen, 1983b*]
The rigid–plastic frame shown in Fig. 6.21 is loaded by two stationary

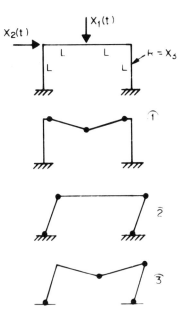

Fig. 6.21 — Rigid–plastic frame and collapse modes.

Gaussian stochastic load processes $X_1(t)$ and $X_2(t)$. It has a random variable resistance $R = X_3$. Disregarding possible dynamic effects, the most significant collapse modes are as shown. The expected values are $E[X_1(t)] = \mu$, $E[X_2(t)] = 0.5\mu$, $E(X_3) = 2\mu$ and the covariance functions are, because of stationarity, functions of the time difference $t_1 - t_2 = \tau$ only: $c_{11}(t_1, t_2) = c_{22}(t_1, t_2) = \sigma^2 \rho(\tau)$, $c_{12}(t_1, t_2) = c_{21}(t_1, t_2) = 0.5\sigma^2 \rho(\tau)$, $c_{33}(t_1, t_2) = 0.2\sigma^2 \rho(2\tau)$ and all others are zero. If $\tau = 0$ then $\rho = 1$. Further, let $\sigma = 0.25\mu$ and $L = 4$ units.

(i) *Preliminaries*

From collapse mode analysis, the safety margin expressions are

$$
\begin{aligned}
Z_1 &= -4x_1 &&+ 4x_3 \\
Z_1 &= &&-4x_2 + 4x_3 \\
Z_1 &= -4x_1 &&- 4x_2 + 6x_3
\end{aligned}
\tag{6.104}
$$

rom which the mean values of Z_i are 4μ, 6μ and 6μ. To determine the standard deviations of Z_i, expression (A.162) might be applied, and (A.163) may be used to find the covariances between the Z_i. All the required results are more easily obtained by matrix manipulation as given by (B.12)

$$
C_Z = A C_X A^T
\tag{6.105}
$$

where A is the matrix $Z = AX$ given by (6.104) and C_X is the matrix of covariance functions c_{ij}, between X_i and X_j, given by

$$
C_X = \sigma^2 \begin{bmatrix} \rho(\tau) & 0.5\rho(\tau) & 0 \\ 0.5\rho(\tau) & \rho(\tau) & 0 \\ 0 & 0 & 0.2\rho(2\tau) \end{bmatrix}
\tag{6.106}
$$

Now (6.105) becomes

$$
C_Z = 16\sigma^2 \begin{bmatrix} -1 & 0 & 1 \\ 0 & -1 & 1 \\ -1 & -1 & 1.5 \end{bmatrix} \begin{bmatrix} a & b & 0 \\ b & 1 & 0 \\ o & o & c \end{bmatrix} \begin{bmatrix} -1 & 0 & -1 \\ 0 & -1 & -1 \\ 1 & 1 & 1.5 \end{bmatrix}
\tag{6.107}
$$

or

$$
C_Z = 16\sigma^2 \begin{bmatrix} a+c & b+c & a+b+1.5c \\ b+c & a+c & a+b+1.5c \\ a+b+1.5c & a+c+1.5c & 2(a+b)+(1.5)^2 c \end{bmatrix}
\tag{6.108}
$$

where $a = \rho(\tau)$, $b = 0.5\rho(\tau)$ and $c = 0.2\rho(\tau)$.

If $\tau = t_1 - t_2 = 0$ as is required to obtain the variances of Z_1, Z_2 and Z_3 according to (6.35) and (6.33a), $\rho = 1$ and (6.108) becomes

$$C_Z = 16\sigma^2 \begin{bmatrix} 1.2 & 0.7 & 1.8 \\ 0.7 & 1.2 & 1.8 \\ 1.8 & 1.8 & 3.45 \end{bmatrix} \qquad (6.109)$$

The diagonal terms represent the variances, so that the standard deviations $D(Z_i)$ are $4\sqrt{1.2}\sigma$, $4\sqrt{1.2}\sigma$ and $4\sqrt{3.45}\sigma$ respectively. The collapse mode reliability indices $\beta_i = E(Z_i)/D(Z_i)$ are $4/\sqrt{1.2} = 3.65$, $6\sqrt{1.2} = 5.48$ and $6/\sqrt{3.45} = 3.23$ since $\sigma = 0.25\mu$.

(ii) *Initial failure probability*   The probability of failure at time $t = 0$, given by $p_f(0)$ can now be calculated using the methods of Chapter 5. Here the bounds (5.23) and (5.29) together with the approximation (3.8) for the calculation of $P(F_i \cap F_j)$ will be applied. To use the latter, the correlations $\rho_{ij} = \rho(Z_i, Z_j)$ are required. These are obtained from (6.109) as

$$\rho_{12} = \frac{\text{cov}(Z_1, Z_2)}{\sigma_{Z_1}\sigma_{Z_2}} = \frac{0.7}{(\sqrt{1.2})\,(\sqrt{1.2})} = 0.583$$

$$\rho_{13} = \rho_{23} = \frac{1.8}{(\sqrt{1.2})\,(\sqrt{3.45})} = 0.885$$

$$\rho_{11} = \rho_{22} = \rho_{33} = 1$$

To use (3.8), the terms $(\beta_i - \beta_j\rho_{ij})/(1 - \rho_{ij}^2)^{\frac{1}{2}}$ need to be evaluated. It may be shown that this term represents the "conditional" value $\beta_{i|j}$ (see Example 6.4). The results are

$$\beta_{i|j} = \begin{bmatrix} - & 0.560 & 1.700 \\ 4.126 & - & 5.625 \\ 0.0 & -3.476 & - \end{bmatrix} \qquad (6.110)$$

Using the worst appropriate bound (i.e. additive bound) from (3.8) and ordering the reliability indices according to increasing values produces the following lower bound on $p_f(0)$:

$$p_f(0) > \Phi(-3.23) + \{\Phi(-3.65) - [\Phi(-3.65)\Phi(0) + F(-3.23)\Phi(-1.700)]\}^+$$

$$+ \{\Phi(-5.48) - [\Phi(-5.48)\Phi(-5.560) + \Phi(-3.65)\Phi(-4.126)]$$

$$- [\Phi(-5.48)\Phi(-3.476) + \Phi(-3.23)\Phi(-5.625)]\}^+$$

or

$$p_f(0) > 6.57 \times 10^{-4} \qquad (6.111)$$

Using (5.29), the upper bound (with the appropriate "max" form of (3.8)) is

$$p_f(0) < \Phi(-3.23) + \Phi(-3.65) + \Phi(-5.48) - \Phi(-3.65)\Phi(0)$$

$$- \max[\Phi(-5.48)\Phi(-3.476), \ \Phi(-3.23)\Phi(-5.625)]$$

or

$$p_f(0) < 6.77 \times 10^{-4} \qquad (6.112)$$

(iii) Conditional probability on H_3, H_2 *and* H_1 The condition probability $\int_{\Delta S_l} f_{\mathbf{x}}(\mathbf{x}) \, d\mathbf{x}$ of equation (6.86) for $l = 3$ requires the projection of the limit state functions 1 and 2 onto the plane H_3 representing the third limit state function. The projections of the reliability indices β_1 and β_2 are also required (see Fig. 6.22(a)). It may be shown that the β projections are the conditional reliability indices $\beta_{i|j}$ already calculated above (see Example 6.4). Also required are the conditional correlations $\rho_{ij|k}$, representing, typically, the angle $v_{12|3}$ shown in Fig. 6.22(b), since $\rho = \cos v$ as in (5.42). It may be shown, either using a geometrical argument (see Example 6.4) or from a regression consideration [Ditlevsen, 1983b], that

$$\rho_{ij|k} = \frac{\rho_{ij} - \rho_{ik}\rho_{jk}}{[(1 - \rho_{ik}^2)(1 - \rho_{jk}^2)]^{1/2}} \qquad (6.113)$$

which results in

$$\rho_{12|3} = -0.924, \quad \rho_{13|2} = \rho_{23|1} = 0.976 \qquad (6.114)$$

The conditional probability content within ΔS_3 on the plane H_3 can now be bounded using (6.110) and (6.114) and using the most conservative combinations in (3.8). Noting that the bounds produce the probability content outside the region ΔS_3, it follows that the probability content within ΔS_3 is bounded by

$$1.0 - p_3 > \Phi(-1.700) + \{\Phi(-5.625) - [\Phi(-5.625)\Phi(-A) +$$

$$\Phi(-1.700)\Phi(-B)]\}$$

$$1.0 - p_3 < \Phi(-1.700) + \Phi(-5.625) - \max[\Phi(-5.625)\Phi(-A),$$

$$\Phi(-1.700)\Phi(-B)]$$

where

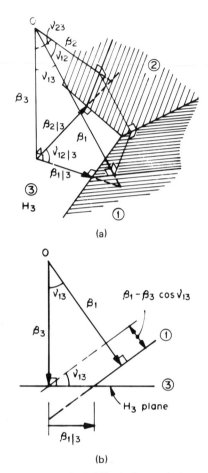

Fig. 6.22 — Projections of reliability indices and planes 1, 2 and 3.

$$A = \frac{1.700 - (-0.924)(5.625)}{[1 - (-0.924)^2]^{1/2}} = 18.04$$

$$B = \frac{5.625 - (-0.924)(1.700)}{[1 - (-0.924)^2]^{1/2}} = 18.82$$

or

$$0.955\ 435 < p_3 < 0.955\ 435 \tag{6.115}$$

Similarly, it may be shown that for H_2 and H_1

$$0.999\ 746 < p_2 < 0.999\ 746 \tag{6.116}$$

and

$$0.5 < p_1 < 0.5 \tag{6.117}$$

(iv) *Outcrossing rate and first-passage probability*   In this example the limit state expressions are time invariant: thus the correction factor K_l is given by (6.100). The terms $D(\dot{Z}_l)$ are obtained from the square root of the diagonals of $\text{var}(\dot{Z}_l)$ given by (6.103).

To carry out the required differentiation, the general expression for C_Z given by (6.107) is used and τ is replaced by $\tau = t_1 - t_2$. Finally, putting $\tau = t_1 - t_2 = 0$, produces the variance of $\dot{Z}_i$ on the diagonals $i = j$, from which $D(\dot{Z}_l) = \sigma(\sqrt{1.8}\gamma, \sqrt{1.8}\gamma, \sqrt{4.8}\gamma)$ for $l = 1, 2, 3$, with $\gamma = [-\rho''(0)]^{\frac{1}{2}}$

The factors K_l, $l = 1, 2, 3$, become

$$K_l = \left[\frac{\phi(3.65)\sqrt{1.8}\gamma}{(2\pi)^{1/2}(4\sqrt{1.2})}, \frac{\phi(5.48)\sqrt{1.8}\gamma}{(2\pi)^{1/2}(4\sqrt{1.2})}, \frac{\phi(3.23)\sqrt{4.8}\gamma}{(2\pi)^{1/2}(4\sqrt{3.45})} \right] \tag{6.118}$$

and, according to equation (6.101), the mean outcrossing rate v_D^+ is then given by

$$v_D^+ = [0.1222\gamma\phi(3.65)](0.5) + [0.1222\gamma\phi(5.48)][(0.999\ 746)] + $$
$$+ [0.117\ 64gf(3.23)](0.955\ 435) \tag{6.119}$$

where the bounds have been ignored owing to their closeness. Evaluation results in $v_D^+ = 2.7 \times 10^{-4}\gamma$ so that, using (6.48), $p_f(t_L) \leqslant p_f(0) + vt_L$, the first-passage probability $P_f(t_L)$ is bounded:

$$(6.57 + 2.7\gamma)10^{-4} < p_f(0) + v_D^+ t_L < (6.77 + 2.7\gamma)10^{-4} \tag{6.120}$$

where $[0, t_L]$ is the period of interest. In practice $\gamma = [-\rho''(0)]^{1/2}$ can be evaluated if an analytic expression is available for $\rho(\tau)$, e.g. $\rho = A \exp[-B(t_1 - t_2)]$.

Example 6.4
The reliability indices β_i have projections $\beta_{i|j}$ on H_j as shown in Fig. 6.22(b). These may be interpreted as conditional reliability indices in the sense of Example 6.3. The angles v_{ij} are also shown; by (5.44) these are related to the correlation coefficients through $\rho_{ij} = \cos v_{ij}$. Expressions for $\beta_{i|j}$ and $\rho_{ij|k}$ may be derived directly from the three-dimensional geometry shown in Figs. 6.23 and 6.22(b); thus for $\beta_{1|3}$

$$\frac{E(Z_1|Z_3 = 0)}{D(Z_1|Z_3 = 0)} = \beta_{1|3} = \frac{\beta_1 - \beta_3 \cos v_{13}}{\sin v_{13}} = \frac{\beta_1 - \beta_3\rho_{13}}{(1 - \rho_{13}^2)^{\frac{1}{2}}} \tag{6.121}$$

as also given in expression (3.9).

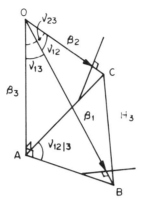

Fig. 6.23 — Geometry to determine conditional reliability indices and correlation coefficients.

To obtain $\rho_{12|3}$, the angle $v_{12|3}$ in Fig. 6.23 must be calculated. This can be found by twice applying the cosine rule of trigonometry ($a^2 = b^2 + c^2 - 2bc \cos A$, with A opposite a):

$$BC^2 = OB^2 + OC^2 - 2(OB)(OC) \cos v_{12}$$

and

$$BC^2 = AB^2 + AC^2 - 2(AB)(AC) \cos v_{12|3}$$

With $AB = \beta_\beta \tan v_{13}$, $AC = \beta_3 \tan v_{23}$, $OB = \beta_3/\cos v_{13}$, $OC = \beta_3/\cos v_{23}$ and with $\rho_{ij} = \cos v_{ij}$, it follows readily that

$$\rho_{12|3} = \frac{\rho_{12} - \rho_{13}\rho_{23}}{[(1 - \rho_{13}^2)(1 - \rho_{23}^2)]^{\frac{1}{2}}} \qquad (6.122)$$

These results also apply, or course, to any three-dimensional subset of n-dimensional space. As will be evident from Example 6.3 such a subset is sufficiently detailed for consideration of (hyper)polyhedral safe regions.

6.5.3 General processes

When the safe region is not necessarily convex, the determination of the outcrossing rate becomes more difficult. If the unsafe region can be visualized or expressed mathematically in terms of unions and intersections of component planar (half-space) regions, some conceptual results for unions and intersections of unsafe regions may be useful [Rackwitz, 1984]. Alternatively, a simplified convex (or spherical) region boundary may be employed to produce bounds on the outcrossing rate [e.g. Veneziano *et al.*, 1977]. However, such bounds may be extremely conservative.

A further complication arises with non-linear limit state functions. As in FOSM theory, these may be linearized. Intuitively, an appropriate expansion point is now the point of greatest local outcrossing density, in a way similar to that in which the point having $f_\mathbf{X}(\ )$ a maximum is used as expansion point in time-invariant theory (see Chapter 4) [Breitung and Rackwitz, 1982]. These points are not usually the same, unless the local outcrossing rate at any point in $\mathbf{X}$ has components which are indifferent with respect to direction, in the same sense that $f_\mathbf{X}(\ )$ is unbiased with respect to direction (i.e. isotropic). This means that, for all i, $v_i = v$ is required. However, preliminary results suggest that the precise choice of linearization point is not usually critical. Once linearization points have been chosen, the calculation of the mean outcrossing rate may follow the procedure outlined in the previous section.

Another possibility is that the vector process $\mathbf{X}(t)$ is not normal nor completely continuous in all its components. In principle it is possible to separate the calculation of the outcrossing rate into that concerned with continuous processes, that concerned with (various) discontinuous processes and that concerned with time-invariant random variables [Rackwitz, 1984]. The corresponding outcrossing rates from the time-variant components may be added provided that it is (reasonably) assumed that the different groups of component processes are independent.

Non-normal component processes may be transformed to equivalent normal processes in the manner of section 4.4, using the point of maximum local outcrossing rate as expansion point. In the context of vector processes, such a transformation has also been termed a "translation" of the normal process, in order to recognize that each component is independently subject to a univariate non-linear transformation which ignores dependence between components [Grigoriu, 1984].

It should be evident that the theory for time-dependent reliability assessment is not as fully developed as might be desired, although the basis for its complete development in FOSM and in FOR terms has been clearly indicated above. For many problems of practical significance a fully time-dependent approach is only required when the resistance basic variables are time dependent or when more than one loading case must be considered. This last has been a matter of considerable research activity in order to find an equivalent single loading to use with time-independent reliability calculcations. How the theory so far introduced can be used to address this problem is the subject of the next section.

6.6 LOAD COMBINATIONS

6.6.1 Introduction

A special case of the outcrossing problem arises when the probability distribution for the combined effect of two or more stochastic loads is sought such that one equivalent loading system is obtained. The need for this result arises in the application of the time-integrated approach (see section 6.2) in which the time-dependent reliability problem has been converted to a time-independent one through the use of extreme value distributions. This is the approach currently used for code calibration (see Chapter 9) and relatively simple rules are therefore required. These will be considered later in this section.

If $X_1(t)$ and $X_2(t)$ represent stationary, mutually independent continuous load processes, the probability distribution in a time interval $[0, t_L]$ for the linear sum $X = X_1 + X_2$ can be obtained from a consideration of the upcrossing rate of $X(t)$ as a function of the barrier level $x = a$ through (6.51). The main problem is thus the calculation of the upcrossing rate for X, for the barrier a or, equivalently, the outcrossing rate for (X_1, X_2) out of the domain bounded by the plane $X_1(t) + X_2(t) = a$.

If X_1 and X_2 are each normal stationary processes, the sum $X = X_1 + X_2$ is also normal and stationary, with mean and variance given by (A.160) and (A.162). The upcrossing rate for a stationary normal process then follows directly from the result (6.70) for a single process $X(t)$.

Unfortunately not all load processes can be adequately described, even under instantaneous conditions, by a normal process and, as noted in section 6.4.8.1, the use of (6.70) for non-normal processes is seldom sufficiently accurate.

6.6.2 General formulation

In principle, the expected upcrossing rate may be determined using Rice's formula (6.68) for the sum process $X(t)$. To use it, the joint probability density function $f_{X\dot{X}}(\ )$ is required. This can be expressed in terms of $f_{X_1\dot{X}_1}(\ )$ and $f_{X_2\dot{X}_2}(\ )$ by means of the convolution integral

$$f_{X\dot{X}}(a, \dot{x}) = \int_{-\infty}^{\infty} \int_{-\infty}^{\infty} f_{X_1\dot{X}_1}(x_1\dot{x}_1) f_{X_2\dot{X}_2}(a - x_1, \dot{x} - \dot{x}_1) dx_1 \, d\dot{x}_1$$

(6.123)

Noting that $\dot{x} = \dot{x}_1 + \dot{x}_2$, $x_1 = x - x_2$ and changing the integration variables, and order, the resulting triple integral form of (6.68) is

$$v_X^+(a) = \int_{-\infty}^{\infty} \int_{-\infty}^{\infty} \int_{\dot{x}=-\dot{x}_1}^{\infty} (\dot{x}_1 + \dot{x}_2) f_{X_1\dot{X}_1}(x, \dot{x}_1) f_{X_2\dot{X}_2}(a - x, \dot{x}_2) \, d\dot{x}_2 \, d\dot{x}_1 \, dx$$

(6.124)

which is not usually analytic. However, bounds can be established by changing the integration region for $\dot{x}_1$ and $\dot{x}_2$. If the region of integration is increased to $0 \leqslant \dot{x}_1 \leqslant \infty$, $-\infty \leqslant \dot{x}_2 \leqslant \infty$ for the $\dot{x}_1$ component of (6.124), and to $-\infty \leqslant \dot{x}_1 \leqslant \infty$ and $0 \leqslant \dot{x}_2 \leqslant \infty$ for the $\dot{x}_2$ component, an upper bound is obtained. Integrating over $\dot{x}_1$ and $\dot{x}_2$ leaves

$$v_X^+(a) \leqslant \int_{u=-\infty}^{\infty} v_1(u) f_{X_2}(a-u) \, du + \int_{u=-\infty}^{\infty} v_2(u) f_{X_1}(a-u) \, du$$

$$(6.125)$$

where $v_i(u)$ is the upcrossing rate for the process $X_i(t)$. The $v_i(u)$ are readily evaluated using the results for a single variable. For some common processes they are given in sections 6.4.7 and 6.4.8. The $f_{X_i}(\ )$ are the probability density functions of the $X_i(t)$, for arbitrary time t. They are also known as the "arbitrary-point-in-time" probability density functions. Expression (6.125) is sometimes referred to as the "point-crossing" formula.

A lower bound can also be obtained, and the results may be extended to more than two loads acting simultaneously for non-linear combinations and to non-stationary load processes [Ditlevsen and Madsen, 1983].

For an important class of process combinations, (6.125) represents an exact solution. This class is largely that for which one of the two processes X_1 or X_2 has a discrete distribution, such as the square wave (section 6.4.7.5) or impulse (section 6.4.7.4) type. More generally (6.125) is the exact result whenever the processes in the sum satisfy [Larrabee and Cornell, 1981]

$$P[\dot{X}_i(t) > 0 \quad \text{and} \quad \dot{X}_j(t) < 0] = 0 \tag{6.126}$$

for all processes i, j and all times t. This condition ensures that one process does not cancel out the act of upcrossing of the other process by decreasing in value. Typical processes which satisfy this condition are shown in Fig. 6.24.

If X_1 and X_2 are each stationary normal processes, with means μ_{X_1} and μ_{X_2} and standard deviations σ_{X_1} and σ_{X_2} respectively, then, as usual

$$f_{X_i}(x_i) = \frac{1}{\sigma_{X_i}} \phi\left(\frac{x_i - \mu_{X_i}}{\sigma_{X_i}}\right)$$

and, from (6.70), the upcrossing rate for X_i is

$$v_i^+(a) = \frac{1}{(2\pi)^{1/2}} \frac{\sigma_{\dot{X}_i}}{\sigma_{X_i}} \phi\left(\frac{a - \mu_{X_i}}{\sigma_{X_i}}\right)$$

When these are substituted into (6.125) the upcrossing rate for $X = X_1 + X_2$ becomes

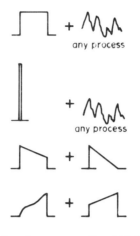

Fig. 6.24 — Combinations of random processes for which the point-crossing formula is exact.

$$v_X^+(a) \leqslant \frac{1}{(2\pi)^{1/2}} \frac{\sigma_{X_1} + \sigma_{X_2}}{\sigma_X} \phi\left(\frac{a - \mu_X}{\sigma_X}\right)$$

with $\mu_X = \mu_{X_1} + \mu_{X_2}$ and $\sigma_X^2 = \sigma_{X_1}^2 + \sigma_{X_2}^2$. The exact result is obtained from (6.70) with $\sigma_X^2 = \sigma_{X_1}^2 + \sigma_{X_2}^2$. The error is thus indicated by the ratio $(\sigma_{X_1} + \sigma_{X_2})/(\sigma_{X_1}^2 + \sigma_{X_2}^2)^{1/2}$. This has a maximum of $\sqrt{2}$ when $\sigma_{X_1} = \sigma_{X_2}$. For most other combinations of load processes, the error is less [Larrabee and Cornell, 1981].

6.6.3 Discrete processes
It will be instructive to consider the sum of two non-negative rectangular renewal processes of the "mixed" type (see section 6.4.7.6). Typical traces are shown in Fig. 6.25. From (6.65) the upcrossing rate for each process is

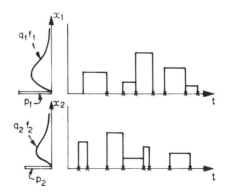

Fig. 6.25 — Typical realizations of mixed rectangular renewal processes with given probability density functions f_{X_1} and f_{X_2}.

$$v_i^+(a) = v_i[p_i + q_iF_i(a)][q_i(1 - F_i(a)]$$ (6.127)

where the mean rate of arrival of pulses in the mixed process is $v_iq_i = v_{mi}$, say. Also the arbitrary-point-in-time distribution for a mixed process is $f_{X_i}(x_i) = p_i\delta(x_i) + q_if_i(x_i)$, where $\delta(\ )$ is the Dirac delta function (see Fig. 6.25). Writing $G_i(\ ) = 1 - F_i(\ )$, and substituting for $v_i^+(\ )$ and $f_{X_i}(\ )$, with $i = 1, 2$ into (6.125) and integrating will lead to

$$v_X^+(a) = v_{m1}p_2[p_1 + q_1F_1(a)]G_1(a) + v_{m2}p_1[p_2 + q_2F_2(a)]G_2(a)$$

$$+ v_{m1}p_1q_2\left[\int_0^a G_1(a - x)f_2(x)\ dx\right]$$

$$+ v_{m2}p_2q_1\left[\int_0^a G_2(a - x)f_1(x)\ dx\right]$$

$$+ v_{m1}q_1q_2\left[\int_0^a F_1(a - x)G_1(a - x)f_2(x)\ dx\right]$$

$$+ v_{m2}q_1q_2\left[\int_0^a F_2(a - x)G_2(a - x)f_1(x)\ dx\right]$$ (6.128)

Fortunately, this rather fearsome result has a simple explanation, as may be shown by looking at the problem from first principles.

For one process, the probability at each renewal that the process is active, i.e. that $X(t)$ has a non-zero value, is simply q (see Fig. 6.25). Similarly, the probability that each renewal results in a zero value (inactive) of $X(t)$ is p (see Fig. 6.25), again not influenced by the value of X in the previous time increment. The process is therefore memory-less, or Markovian.

For two processes, the possible combinations of active and inactive states which allows the sum of the two processes to cross from below the barrier level a to above it are shown in Table 6.2 [Waugh, 1977; Larrabee and Cornell, 1979].

For each state change (i.e. the change of $X_1 + X_2 < a$ to the state $X_1 + X_2 > a$), the contribution to the upcrossing rate $v_X^+(a)$ is given by

$$v_{(j)}^+ = \lim_{\Delta t \to 0}\left[P\left(\begin{array}{c|c}\text{upcrossing} & \text{change of}\\ \text{of level } a & \text{state in } \Delta t\end{array}\right) P\left(\begin{array}{c}\text{change of}\\ \text{state in } \Delta t\end{array}\right)\right]$$ (6.129)

Table 6.2 — Initial conditions and state changes for upcrossings

State change	Process 1	Process 2	Upcrossing due to
(a)	Inactive	Active or inactive	Process 2
(b)	Active	Inactive	Process 2
(c)	Active	Active	Process 2
(d)	Active or inactive	Inactive	Process 1
(e)	Inactive	Active	Process 1
(f)	Active	Active	Process 1

For state change (a) of Table 6.2, this becomes

$$v_{(a)}^+ = \lim_{\Delta t \to 0} \left(P\left\{ [X_1(t) = 0] \cap [0 \leqslant X_2(t) < a] \cap [X_2(t + \Delta t) > \right.\right.$$

$$\left.\left. > a] \; \left| \; {\text{change} \atop \text{of state}} \right\} P\left({\text{change} \atop \text{of state}} \right) \right)\right.$$

$$= \{ p_1[p_2 + q_2 F_2(a)] \; G_2(a) \} v_2 q_2 \tag{6.130}$$

which corresponds to the second term in (6.128) when it is remembered that therein $v_2 q_2 = v_{m2}$

For state change (b)

$$v_{(b)}^+ = \lim_{\substack{\Delta t \to 0 \\ \text{all } x}} \left(P\left\{ [0 < X_1(t)] \cap [X_2(t) = 0] \cap [X_2(t + \Delta t) > (a - x) | X_1 = \right.\right.$$

$$\left.\left. = x] \; \left| \; {\text{change} \atop \text{of state}} \right\} P\left({\text{change} \atop \text{of state}} \right) \right)\right.$$

$$= \left\{ q_1 p_2 \left[\int_0^a G_2(a - x) f_1(x) \; dx \right] \right\} v_2 q_2 \tag{6.131}$$

which corresponds to the fourth term in (6.128) with $v_2 q_2 = v_{m2}$. In a similar fashion it follows that the state change (c) is

$$v_{(c)}^+ = \lim_{\substack{\Delta t \to 0 \\ \text{all } x}} \left(P \left\{ [0 < X_1(t)] \cap [0 < X_2(t) < a - x] \cap [X_2(t + \Delta t) > \right. \right.$$

$$\left. \left. > (a - x)|X_1 = x] \begin{vmatrix} \text{change} \\ \text{of state} \end{vmatrix} P \begin{pmatrix} \text{change} \\ \text{of state} \end{pmatrix} \right) \right) \quad (6.132)$$

which is identical with the sixth term in (6.128). The other terms in (6.128) arise from state changes (d)–(f) of Table 6.2. If each pulse returns to zero prior to the commencement of the next pulse, the term $p_2 + q_2 F_2(a)$ in (6.130) is unity by definition, as is the term p_2 in (6.131). The fifth and sixth terms in (6.128) do not exist in this case.

With the upcrossing rate determined, the cumulative distribution function $F_X(\ )$ for the total load $X = X + X_2$ may be estimated using (6.51). The error in using (6.51) together with (6.128) has been investigated by comparison with the few known exact results for square wave processes [Hasofer, 1974; Bosshard, 1975; Gaver and Jacobs, 1981; Larrabee and Cornell, 1979] and found typically to give about a 20% overestimate for high barrier levels a, and about a 60% overestimate for lower levels of a.

6.6.4 Simplifications
6.6.4.1 Load coincidence method
For impulse type loading with the pulses returning to zero, (6.128) may be considerably simplified (see above). Further, noting, for example that

$$\int_0^a G_1(a - x) f_2(x) dx = G_{12}(a) - G_2(a), \quad \text{with} \quad G_{12}(\ ) = 1 - F_{12}(\ ) \quad \text{where}$$

$F_{12}(\ )$ is the cumulative distribution function for the total height of the two pulses allows (6.128) to be simplified to

$$v_X^+(a) = v_{m1}(p_2 - v_{m2}\mu_1)G_1(a) + v_{m2}(p_1 - v_{m1}\mu_2)G_2(a) + v_{m1}v_{m2}(\mu_1 + \mu_2)G_{12}(a) \quad (6.133)$$

Here $v_{mi}\mu_i$ has been substituted for q_i. As before, v_{mi} is the mean pulse arrival rate of the process $X_i(t)$ (see (6.127)). The terms μ_i are the mean durations of the pulses of $X_i(t)$. If the pulses are of short duration, $\mu_i \to 0$ while, if they are of relatively infrequent occurrence, $p_i \to 1$. Hence, a reasonable approximation for the upcrossing rate of the total process $X = X_1 + X_2$ is

$$v_X^+(a) \approx v_{m1}G_1(a) + v_{m2}G_2(a) + v_{m1}v_{m2}(\mu_1 + \mu_2)G_{12}(a) \quad (6.134)$$

a result first given by Wen (1977a) using rather different reasoning. The first and second terms of (6.134) represent upcrossings of each process acting alone, while the third term represents upcrossings when both processes are active, i.e. when the pulses overlap. When the pulses are of very short duration, the third term may be neglected.

It has been found from comparisons with simulation results that expression (6.134) yields surprisingly good estimates of the upcrossing rate $v_X^+(a)$ even with the active fraction for each process as high as 0.2 and for reasonably high levels of barrier a [Larrabee and Cornell, 1979]. Pulse shapes other than rectangular and dependence between pulse renewals have also been considered [Wen, 1977b, 1981].

6.6.4.2 Borges processes

The combination of Borges processes is of particular interest in relation to code calibration work because in this context it provides a sufficiently good estimate of the maximum combined load probability distribution [Turkstra and Madsen, 1980].

When each process $X_i(t)$ in the sum $X = X_1 + X_2 + X_3 + \ldots$ is represented as a Borges process (see section 6.4.7.1) such that the duration of pulses is $\tau_1 < \tau_2 < \tau_3 < \ldots$ respectively for each process, with $\tau_i = m_i \tau_{i+1}$, m_i integer, as shown in Fig. 6.26, the theory used above can still be applied. The

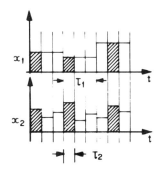

Fig. 6.26 — Realizations of two Borges processes, with $\tau_1 = m\tau_2$, where m is an integer.

occurrence rates are now $v_i = n_i/t_L$ where n_i is the integer number of pulses of process X_i in the period $[0, t_L]$. With the upcrossing rate determined according to one of the formulae above, the cumulative distribution function of the maximum value of X may be obtained through the application of (6.51). However, another approach is possible and will be outlined below.

It may be shown that cumulative distribution function $F_{\max X}(\ )$ for the maximum value of X is given by (the convolution integral) [Grigoriu, 1975; Turkstra and Madsen, 1980]

$$F_{\max X}(x) = \left\{ \int_{-\infty}^{x} f_{X_1}(v) [F_{X_2}(x-v)]^m \, dv \right\}^n \qquad (6.135)$$

for a two-component process with $\tau_2 = \tau_1/m$, m integer, and where there are n pulses of τ_1 in $[0, t_L]$. This integral becomes more complex for more than two loads, but it may be evaluated for such cases using an FOSM algorithm (Chapter 4). The essential concept can be demonstrated using a linear combination of three loads.

The maximum in the period $[0, t_L]$ can be rewritten as

$$\max_{t_L}[X_1(t) + X_2(t) + X_3(t)] = \max_{t_L}[X_1(t) + Z_2(t)] \qquad (6.136)$$

where $Z_2(t)$ is another Borges process given by

$$Z_2(t) = X_2(t) + X_3^9(t, \tau_2) \qquad (6.137)$$

in which $X_3^9(\ )$ represents the maximum pulse of X_3 over the period τ_2 (there are τ_2/τ_3 such pulses; cf. Fig. 6.26). $X_3^9(t, \tau_i)$ is also known as the τ_i duration envelope of $X(t)$.

$F_{Z_2}(\ )$, the cumulative distribution function of Z_2, is given by the convolution integral (6.135), now with $m = \tau_3/\tau_2$ and $n = 1$ since only the interval τ_2 is considered. The maximization can then be rewritten

$$\max_{t_L}[X_1(t) + X_2(t) + X_3(t)] = \max_{t_L}[Z_1(t)] \qquad (6.138)$$

with now

$$Z_1(t) = X_1(t) + Z_2^9(t, \tau_i) \qquad (6.139)$$

where $Z_2^9(\ )$ represents the total pulse of Z_2 over the time interval τ_1. $F_{Z_1}(z)$ is again obtained from (6.135). It is clear that the above reformulation process can be extended to any number of loads in the sum provided that the $F_{Z_i}(\ )$ can be obtained and the additions of the processes in $Z_i(t)$ carried out. This can be done readily using FOSM theory [Rackwitz and Fiessler, 1978].

The procedure may be summarized as follows. The cumulative distribution functions $F_{X_i}(\ )$ of each process pulse may be approximated by a normal distribution (if X_i is not normal) on the basis of a set of trial "checking points" x_i^*. (Recall from section 4.4 that in so doing the tail area ($= F_{X_i}(x_i^*)$ is preserved, and the same $f_{X_i}(\ )$ is ensured.) This transformation is carried out first for X_2, and also separately, but using the same x_i^*, for $F_{X_3^\circ}(x_3^\circ) = [F_{X_3}(x_3)]^m$. Recall that $m = \tau_2/\tau_3$ is the number of pulses of X_3 in an X_2 pulse.

As a result of the transformation, the mean $\mu_{X^*_2}$ and variance $\sigma^2_{X^*_2}$ for the equivalent normal distribution for process X_2 are obtained, as well as $\mu_{X^\circ_3}$ and $\sigma^2_{X^*_3}$ for $X_3^\circ(\ )$. Hence the addition of expression (6.137) can be performed to obtain the mean μ_{Z_2} and variance $\sigma^2_{Z_2}$ for $Z_2(t)$ as $\mu_{Z_2} = \mu_{X^*_2} + \mu_{X^*_2}$, $\sigma^2_{Z_2} = \sigma^2_{X^*_2} + \sigma^2_{X^*_3}$. It is now possible to determine $Z_2^\circ(t, \tau_i)$ and to find the equivalent normal distribution for $X_1(t)$ and $Z_2^\circ(\ )$ in expression (6.139) and hence the equivalent normal distribution for $Z_1(t)$. This, of course, is the result sought, but it is dependent on the initial choice of the "checking point" $\mathbf{x}^*$.

The complete distribution function $F_{\max X}(x)$ can be obtained by systematically choosing different values of $x = x^*$ and repeating the above process. In direct correspondence to the concepts of chapter 4 the individual checking point values x_i^* for the variables $X_i(t)$ are selected so as to maximize the local probability density $f_{X_1}(x_i^*)f_{X_2}(x_2^*)f_{X_3}(x_3^*) \ldots .$. The equivalent normal probability density functions $f_{X^*_i}(\ ) = f_{X_i}(\ )$ may be used for this purpose. Of course, the values x_i^* must sum to $x^* = \sum_{i=1} x_i^*.$

The use of this algorithm for checking-code-based load combination rules has been described by Turkstra and Madsen (1980).

6.6.4.3 *Deterministic load combination: Turkstra's rule*

The procedures discussed in the previous sections are, despite various simplifications, still too complex for routine design and for use in design codes. These currently have simple additive forms of load combination rules. The most primitive form consists simply of adding stresses due to different loads without regard to different load uncertainties; others make some allowance for this with an appropriate set of multipliers. The question of interest here is the justification of such multipliers.

A deterministic load combination rule may be derived from a consideration of Borges processes [Turkstra and Madsen, 1980] or, as will be seen, directly from the "point-crossing" formula (6.125) [Larrabee and Cornell, 1981].

If the approximation $G_{\max X}(a) = 1 - F_{\max}(a) \approx v_X^+(a)t$ given by (6.51) is substituted in (6.125) there is obtained

$$G_{\max X}(a) \approx \int_{-\infty}^{\infty} G_{\max X_1}(x)f_2(a-x)\ dx + \int_{-\infty}^{\infty} G_{\max X_2}(x)f_1(a-x)\ dx$$

$$(6.140)$$

where, as before, the maxima are taken over the time interval of interest, usually the life of the structure $[0, t_L]$.

Noting that, if $Z = W + V$, with W, V independent, the complementary cumulative distributive function $G_Z(\ )$ is given by the convolution integral

$$G_Z(z) = \int_{-\infty}^{\infty} G_W(z-v)f_V(v)\ dv = \int_{-\infty}^{\infty} G_W(w)f_V(z-w)\ dw$$

$$(6.141)$$

(see also Chapter 1), so that the two integrals in (6.140) represent, respectively, max $X_1 + \tilde{X}_2$ and $\tilde{X}_1 + $ max X_2, with $\tilde{X}_i$ representing the "arbitrary-point-in-time value of X_i. Similarly, if $Z = $ max(W, V), then it may be shown that

$$G_Z(a) = G_W(a) + G_V(a) - G_W(a)G_V(a) \qquad (6.142)$$

where, for high barrier levels a, the last term may be ignored. Thus using (6.141) to represent each integral in (6.140) and then applying (6.142) it follows, loosely, that

$$\text{max } X \approx \text{max}[(\text{max } X_1 + \tilde{X}_2); (\tilde{X}_1 + \text{max } X_2)] \qquad (6.143)$$

This result is known as "Turkstra's rule" [Turkstra, 1970]: "design for the largest of the lifetime maximum of load 1 plus the value of load 2 when the maximum of 1 occurs or the lifetime maximum of load 2 plus the value of load 1 when the maximum of 2 occurs".

The rule is readily extended to more than two loads. It can also be applied to load effects. For n loads,

$$\text{max } X \approx \text{max}\left(\text{max } X_i + \sum_{j=1}^{n}\tilde{X}_j\right), \qquad j \neq i; i = 1, \ldots, n \qquad (6.144)$$

In this form the rule is seen as being of form similar to that in many existing code formats; it is also evident that it has some affinity with the probabilistic requirements for load combination. However, Turkstra's rule does not fix the load levels to be used with the rule; these have to be separately selected. As will be seen in Chapter 9, max X_i is commonly selected as the 95 percentile value of the load while $\tilde{X}_i$, the arbitrary-point-in-time value, may be taken as the mean value, provided that the process X_i is stationary.

Although a simple and convenient rule for code calibration work and sometimes other situations, Turkstra's rule is not really suitable for accurate computation of probabilities. This is because expression (6.140) is not an upper bound, as was (6.125), and because the ( ) terms in (6.144) are not independent, as implicit in (6.142) [Larrabee and Cornell, 1981].

6.7 DYNAMIC ANALYSIS OF STRUCTURES

A dynamic structural analysis is necessary when the structure interacts with the time-dependent loading acting on it in such a way as to affect the structural response to the loading. Thus both deformations and stresses are affected. The traditional approach to the dynamic analysis of a structure is to work in the so-called "time domain", which means that the equations of motion for the structure are integrated with respect to time. The loading must, of course, be specified in terms of its variation in time. The response calculated for the structure is in terms of stresses and deformations as a function of time. This procedure is very accurate and can be applied to quite complex structures. Material and structural properties may be non-linear, although in such cases the analysis is usually iterative and hence time consuming. Details are outside the scope of this book but excellent treatments exist [e.g. Clough and Penzien, 1975].

If the loading (or the structure, or both) has random properties, the "time domain" solution scheme is of limited usefulness since a unique description of the loading as a function of time is of course not available. Naturally a realization (see section 6.4.1) for the loading can be generated and the structural response analysed for this load system; such a procedure might be repeated many times and the statistics of the response determined. In essence, therefore, the limit state function $G(\mathbf{X})$ can be generated if a design criterion such as maximum allowable stress or deformation is also specified. In principle, it is therefore possible to use the Monte Carlo technique to determine structural reliability. Unfortunately, in practice, such an approach is often impractical, since a large series of time domain analyses are required, each of which is already time consuming. However, in certain situations, such as when non-linear structural behaviour must be accurately considered, it may be the only viable procedure.

When the structural behaviour is linear, i.e. for elastic structures under small-deflection assumptions, or more generally when the "transfer function" between input (e.g. loads) and output (e.g. stresses) is linear, an alternative procedure termed the "frequency domain" method may be employed. It is also known as "spectral analysis" and is extensively used in the analysis of the random vibration of mechanical systems [e.g. Newland, 1984]. While the theory is essentially outside the scope of this book, it is useful to review some aspects of the method briefly since it has application in structural reliability analysis for systems such as dynamically sensitive large towers and offshore structures, particularly in relation to their fatigue life (see also section 6.8).

A stationary stochastic process $X(t)$ may be analysed by decomposing it into an infinite series of sine and cosine waves, each occurring with random magnitude but with their own constant frequency ω of oscillation. Such a representation is known as a Fourier integral representation. The coefficients for the cosine terms are obtained from integration of the product of the stochastic process $X(t)$ and $\cos \omega t$ and similarly for the sine terms. Each such term is a Fourier transform of $X(t)$. Since the random properties of $X(t)$ depend on its (auto)correlation function $R_{XX}(\tau)$ (see section 6.4.1) it might be imagined that the Fourier transform of $X(t)$ can be expressed as a function of $R_{XX}(\tau)$. Indeed the infinite number of cosine coefficients can be represented by a continuous function of ω:

$$S_X(\omega) = \frac{1}{2\pi} \int_{-\infty}^{\infty} R_{XX}(\tau) \cos \omega\tau \; d\tau \qquad (6.145)$$

Since $R_{XX}(\tau)$ is a symmetric function, the equivalent to (6.145) for the sine terms is zero; also $S_X(\omega)$ is symmetric (Fig. 6.27(a)). $S_X(\omega)$ is commonly known as the (mean-square) spectral density. Evidently, if $X(t)$ is completely chaotic, $R_{XX}(\tau)$, given by (6.33a), is zero (except at $\tau = 0$!) and it follows that $S_X(\omega)$ is a constant for all ω. This situation is known also as "white noise", since no particular frequency predominates over any other (see Fig. 6.27(b)). Similarly, if the stochastic process has a dominant frequency, around ω_0, say, the spectral density takes the form shown in Fig. 6.27(c). This is known as a narrow-band process, and it is of major interest in dynamic structural analysis because most structures have only one (or at most very few) dominant mode of vibration. This mode is, of course, that associated with the natural frequency of the structure.

If in (6.145) τ is put to zero and both sides integrated over the range $-\infty$ to $+\infty$, it may be shown that $R_{XX}(0) = \displaystyle\int_{-\infty}^{\infty} S_X(\omega)\, d\omega$, which together with (6.33a) and (6.72) leaves, provided that $u_X = 0$,

$$\sigma_X^2 = E[X(t)^2] = \int_{-\infty}^{\infty} S_X(\omega)\, d\omega \tag{6.146}$$

This shows that the area under the curve $S_X(\omega)$ is the mean-square value of the stationary stochastic process $X(t)$ and is also equal to its variance (see Fig. 6.27(a)). The requirement that $\mu_X = 0$ merely indicates that the stochastic part of the problem is of interest here. A separate structural analysis for the steady state condition $\mu_X \neq 0$ can be superimposed on the results from a stochastic analysis with $\mu_X = 0$.

Apart from the variance of the process, the probability distribution of its peaks is also of interest. Let this be denoted $F_p(a)$, i.e. the probability that the peaks of $X(t)$ lie below the level $x(t) = a$, with $\mu_X = 0$. Immediate use may be made of the results in section 6.4.8.1, for narrow-band processes sufficiently smooth to have maxima only above $x = 0$. Then the proportion of cycles for which $X > a$ is simply v_a^+ / v_0^+, where v_a^+ is the upcrossing rate given by (6.70) and v_0^+ is the rate at which cycles occur (i.e. the rate at which $X(t)$ upcrosses $x = 0$). It follows directly that, for $0 \le a \le \infty$,

$$1 - F_p(a) = \frac{v_a^+}{v_0^+} = \exp\left(-\frac{a^2}{2\sigma_X^2}\right) \tag{6.147}$$

and, by differentiation with respect to the level $x = a$,

$$f_p(a) = \frac{a}{\sigma_X^2} \exp\left(-\frac{a^2}{2\sigma_X^2}\right) \tag{6.148}$$

Expressions (6.147) and (6.148) represent the Rayleigh distribution (Fig.

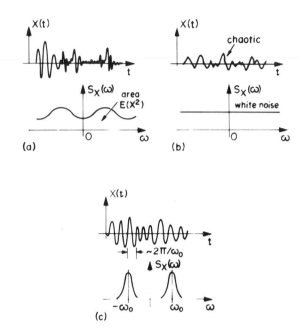

Fig. 6.27 — Realizations of and spectral densities for (a) wide-band, (b) white noise and (c) narrow-band random process.

6.28). Evidently the maximum value of $f_p(a)$ is at $a = \sigma_X$ which is the magnitude of the majority of peaks. No peaks occur below $a = 0$.

If the process is not completely smooth, e.g. if more than one maximum may occur per zero crossing, or if the stochastic process $X(t)$ is not a normal process, more general extreme value distributions such as *EV–III* may be appropriate to describe the probability distribution of the peaks of the process [Newland, 1984].

Finally, it should be noted that the spectral density of the structural deflections or of the stress at some point in the structure may be obtained from consideration of the excitation–response relationship(s) for linear structures when working in the frequency domain. In essence this relationship is given by

$$S_Y(\omega) = |H(\omega)|^2 S_X(\omega) \tag{6.149}$$

for a single input or load process $X(t)$ generating a single output (e.g. stress) $Y(t)$. The function $H(\omega)$ is known as the frequency response function. Expression (6.149) may be generalized for multiple independent inputs simply by (linear) superposition:

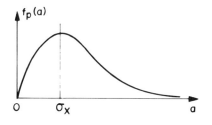

Fig. 6.28 — Probability density function for Rayleigh distribution.

$$S_Y(\omega) = \sum_{i=1}^{n} |H_i(\omega)|^2 S_{Xi}(\omega) \qquad (6.150)$$

A somewhat more complex relationship might be used when the inputs are correlated [Newland, 1984].

Figure 6.29 illustrates the general principle of using frequency response

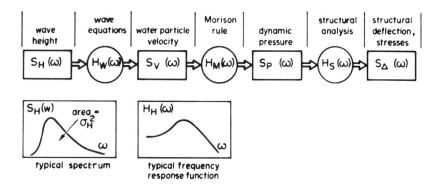

Fig. 6.29 — Relationship between input and output spectral density functions for offshore structural analysis.

functions to obtain an output spectral density from input spectral information. The latter is generally deduced from observations and analysis of the physical process(es). $H(\omega)$ depends on the system being analysed, and may, for example, be obtained by integration of the response of a structure to a given impulse [Newland, 1984; Clough and Penzien, 1975]. These details are outside the scope of the present discussion; for the purposes of reliability calculation for narrow-band processes, expressions (6.146) and (6.148) are the essential tools, given that a spectral analysis has been carried out.

6.8 FATIGUE ANALYSIS

6.8.1   General formulation

A very important case of time-dependent reliability analysis concerns fatigue. It is also typical of limit states which are governed by the number of load applications (and their intensities) rather than by extreme events. A brief review of procedures for dealing with fatigue in a reliability analysis is therefore both of particular and of general interest.

The safety margin or limit state function (1.15) may be expressed more generally as

$$Z = X_a - X_r \qquad (6.151)$$

where X_a represents the actual performance or strength of the structure and where X_r represents the required performance or applied load(s) during the structure's life. Thus, if fatigue life is expressed in terms of the number of cycles of stress, X_a becomes the number of cycles required to cause failure of the material and X_r the required number of cycles to satisfy design requirements for a given lifetime. A corresponding situation exists if fatigue life is measured in terms of crack size or in terms of a damage criterion. In general both X_a and X_r are uncertain quantities. Their precise description depends on the form of fatigue model employed.

6.8.2   The *S–N* model

The traditional model to describe the fatigue life N_i of a component or structure under constant-amplitude repeated loading is given by, for example, the American Society of Civil engineers (ASCE) (1982)

$$N_i = KS_i^{-m} \qquad (6.152)$$

where K and m are conventionally taken as constants, but as random variables here, and N_i is the number of stress cycles at constant stress amplitude S_i. Test results normally allow m and K to be estimated. Typically, conservative values are used such that (6.152) produces a safe estimate of fatigue life N_i. For a reliability analysis, the model (6.152) must be a realistic rather than a conservative predictor, so that typical values for m and K quoted in the literature may not be appropriate.

The safety margin (6.151) may now be written as

$$Z = KS_i^{-m} - N_0 \qquad (6.153)$$

where N_0 is the number of cycles which the structure must be able to sustain for satisfactory performance. N_0 may be subject to uncertainty.

In practice the amplitude of the stress cycles is not constant but is a random variable. If the number of cycles which occur at each amplitude level can be measured or estimated, the empirical Palmgren–Miner hypothesis

$$\sum_{i=1}^{l} \frac{n_i}{N_i} = \Delta \tag{6.154}$$

is often adopted. Here n_i represents the actual number of cycles at the stress amplitude S_i. If the stress amplitudes are all different (6.154) reduces to

$$\sum_{i=1}^{N} \frac{1}{N_i} = \sum_{i=1}^{N} K^{-1} S_i^m = \Delta \tag{6.155}$$

where N is the total number of (random) variable-amplitude cycles. The damage parameter Δ is conventionally taken as unity but may typically lie in the range 0.9–1.5. Hence Δ reflects the (large) uncertainty arising from the empirical nature of (6.154); a lognormal distribution with unit mean and a coefficient of variation of about 0.4–0.7 appears appropriate [Madsen, 1982; ASCE, 1982].

To use (6.155), N must be determined, together with the corresponding probability distribution for S_i. Various cycle-counting techniques to estimate N from broad-band random data exist; for discussion of these see, for example, ASCE (1982).

In principle the safety margin (6.151) can now be formulated directly in terms of the damage parameter Δ as

$$Z = \Delta - X_0 \sum_{i=1}^{N} K^{-1} S_i^m \tag{6.156}$$

where the parameter X_0 has been introduced to allow for model uncertainty in applying (6.152) to a realistic problem, such as might arise from the difficulty in measuring S accurately.

Expression (6.156) presents difficulties if, as might be expected, N is an uncertain quantity. One approach is to put the stress amplitudes into l groups (with l a given number) and to let the number of cycles N_i in each group be an uncertain quantity; thus

$$Z = \Delta - X_0 \sum_{i=1}^{l} K^{-1} N_i S_i^m \tag{6.157}$$

6.8.3 Fracture mechanics models
An alternative but not always accepted approach to fatigue modelling is to consider crack growth under repeated or random load systems [ASCE, 1982; Schijve, 1979]. Based on experimental evidence, the crack growth rate

da/dN may be related to the range of stress intensity factor Δk (at the crack tip) by

$$\frac{da}{dN} = C(\Delta K)^m \tag{6.158}$$

where a is the current crack length, N the number of stress cycles, and C and m are experimental "constants", which usually depend on cycling frequency, the mean stress and the environmental conditions, including, it seems, test laboratories. Both C and m should be treated as uncertainties in a reliability analysis. The range of stress intensity factor ΔK, for situations in which the conventional stress intensity factor $K(a)$ does not change significantly with stress level, may be obtained from

$$\Delta K = K(a)\ \Delta S\ (\pi a)^{1/2}, \quad \Delta K > \Delta K_{th} \tag{6.159}$$

where ΔS represents the range of applied stress and where $K(a)$ is a function of current crack length a, local geometry and the nature of the stress field. ΔK_{th} is the threshold level of ΔK; below it $\Delta K = 0$.

The variation of crack length a with the number N of applied stress cycles can be obtained in principle from integration of (6.158) and using (6.159):

$$a(N) = a[a_0,\ K(a),\ \Delta S,\ C,\ m,\ \Delta K,\ \Delta K_{th},\ N] \tag{6.160}$$

where a_0 is the initial crack length. Expression (6.160) can be used to obtain the mean and variance of $a(N)$ given that statistical parameters are known for the parameters involved. For variable-amplitude loading ΔS will depend on the loading sequence, and it will be a random variable. Methods for dealing with this include application of (6.160) in an incremental manner, or using an effective ΔK approach [ASCE, 1982]. In any case, the limit state function (6.151) may be written as

$$Z = a_a - a(N) \tag{6.161}$$

where a_a is the performance requirement on crack length given a lifetime t_L in which N cycles of loading are expected to occur. An alternative limit state formulation is in terms of the crack-tip-opening displacement which is related to the material "toughness" [ASCE, 1982].

6.9 CONCLUSION

This chapter has been concerned with the incorporation of time as a parameter in reliability analysis. The traditional time-integrated and discrete time approaches were reviewed. An overview of stochastic process theory was given in order to introduce the fully time-independent approach to reliability. An example application of this was given for the case in which all processes are normal processes.

Three further topics were also introduced. The discussion of load combinations was required to obtain simplified procedures currently employed in code calibration (see Chapter 9). This was followed by an outline of spectral methods as used for the analysis of dynamically sensitive structures, and it was shown how results so obtained can be used in reliability calculation procedures. Finally, an overview of the reliability formulation for fatigue problems was given. The integration of these various aspects, dynamic analysis and fatigue, is an important area of practical application, e.g. in offshore structure reliability assessment [Karadeniz *et al.*, 1984; Baker, 1985].

7

Load and load effect modelling

7.1 INTRODUCTION

The loads which may act on a structure can be broadly divided into two groups: those due to natural phenomena, such as wind, wave, snow and earthquake loading, and those due to man-imposed effects, such as dead loads and live loads (e.g. floor loading).

The magnitude of most loads varies with time and with location. This means that loads can be represented as stochastic processes, such as those discussed in Chapter 6. In addition, dynamic effects may occur as a result of load–structure interaction. Because of these possibilities, the modelling of load processes may be quite difficult. Perfect models are not possible owing to insufficient data, imperfect understanding and the necessity to predict future loading. Appropriate models are therefore sought, particularly since loading is usually the most uncertain factor in a reliability analysis. Efforts spent on loading data collection and on load modelling may be more productive than refinement of the reliability estimation techniques.

In the discussions to follow, usually only single load processes will be considered, since loading combinations can be treated, if required, using the methods already described in section 6.6.

The process of constructing a model of a particular load is as follows:

(1) identification and definition of the random variables which can be used to represent the uncertainties in loading description (this can be quite difficult and depends on the understanding of the load process);
(2) selection of appropriate probability distributions for each random variable (this may involve physical reasoning);
(3) selection of the distribution parameters using available data and standard parameter estimation techniques such as;

    (a) method of moments;

(b) method of maximum likelihood;

(c) order statistics.

The main emphasis in the present chapter will be on items (1) and (2). Standard statistics texts may be consulted for item (3).

If observations of a physical phenomenon are available over a period of time, the statistical properties of the load can be estimated directly from data records, and yearly maxima and daily maxima can be extracted to produce extreme value distributions for use in time-integrated analysis. If instantaneous records are available, the instantaneous probability density function may be hypothesized, and a complete time-dependent reliability analysis (see section 6.5) may be possible.

For man-imposed loads such as building live loads, there are usually insufficient long-term data with which to work, and statistical properties of loading must be derived mathematically, with appropriate and with plausible assumptions. The derivation of such relationships, particularly for floor loads, has been a central problem in code calibration work. Largely to reflect these differences in load modelling, only three types of loading will be discussed in this chapter; wind loading in section 7.2, wave loading in section 7.3 and floor loading in section 7.4. Snow and earthquake loadings are also very important in certain geographical locations; the reader should consult the specialist literature for a discussion of these load types and their statistical properties. However, even for wind and wave loading, only limited attention will be given to the physics of these processes; attention will be focused mainly on the statistical properties.

7.2 WIND LOADING

Wind loading can be derived from statistical data for wind speeds. Relatively few data exist for direct wind force (or even localized wind pressure), although it is generally considered that wind is a chaotic phenomenon, at least in the micro-time scale. In principle, the correct probabilistic model for 'instantaneous' wind speed at a point is therefore a normal process [Davenport, 1961]. In practice, departures from the idealized model have been noted [e.g. Melbourne, 1977].

A complete description of wind speed as acting on a structure to generate wind forces will require consideration of the variation of wind speeds from point to point on the structure, and the response of the structure itself. However, these are matters for wind mechanics and will not be considered here [e.g. Simiu and Scanlan, 1978].

To convert instantaneous wind speed $V(t)$ to wind pressure $W(t)$, acting on a particular part of a structure, use is normally made of the standard hydro-dynamic relationship:

$$W(t) = \tfrac{1}{2}\rho C V(t)^2 \qquad\qquad (7.1)$$

where ρ is the density of air (about 1.2 kg/m³) and C is the "wind pressure coefficient", a quasi-static quantity which depends on the size and orientation of the structure.

Expression (7.1) for the instantaneous wind pressure $W(t)$ can be used directly with the fully time-dependent reliability calculation approach of section 6.5 if the dynamic response of the structure is not significant. However, for flexible structures, account will have to be taken of dynamic effects. In this case the conventional procedure is to use a spectral analysis (see section 6.7). Because expression (7.1) is non-linear, it must be linearized before it can be used in a spectral analysis to relate wind speeds, modelled as a normal process, to wind forces. The conventional approach is to consider the wind speed $V(t)$ to be composed of a time-independent mean value $\bar{V}$ and an additive stationary fluctuating component $v(t)$, assumed to be much smaller than $\bar{V}$. Then (7.1) can be linearized to [Davenport, 1961]

$$W(t) \approx \tfrac{1}{2}\rho C[\bar{V}^2 + 2\bar{V}v(t)] \qquad (7.2)$$

where $v(t)^2$ has been ignored. If there is more than one wind velocity component, $\bar{V}$ and $v(t)$ should be replaced by vectors in (7.1) and (7.2); when the structure also responds with a velocity of its own, $v(t)$ should be replaced by the relative velocity. The spectral density for $v(t)$, which represents the frequency components present in the random process $v(t)$, (see section 6.7), and knowledge of which is required for a spectral analysis, has been given in different forms by Davenport (1967), Harris (1971) and Kaimal et al. (1972). Details are outside the scope of this book; readers are urged to consult specialist wind engineering texts [e.g. Simiu and Scanlan, 1978].

The output from the spectral analysis, such as the spectrum of forces, or deflections, is then converted to statistical properties for these quantities as outlined in section 6.7.

For use with the time-integrated and the discrete time approaches of sections 6.2 and 6.3, lifetime maximum and annual maximum wind speeds and their probability density functions are required. These can be derived either directly or indirectly from wind speed records. Wind speed records of reasonable reliability have, for many areas, been obtained as part of meteorological data collection.

Wind speed is measured either in terms of the "3 s gust" speed, i.e. the average speed over a gust of duration about 3 s, or as the "fastest-mile wind speed" (USA) which is the wind speed averaged over the passage of 1 mile of wind as measured by the anemometer propellor tips. Clearly the two are not equivalent, particularly at low wind speeds. These results may be converted to mean hourly wind speeds. Irrespective of the choice of basis of measurement, it is generally agreed that for non-cyclonic regions, the annual maximum wind speed as obtained from wind speed records can be described by an extreme value type I distribution [Simiu et al., 1978; Simiu and

Filliben, 1980]. This choice is consistent with the use of a normal process for instantaneous wind speed at least if readings are assumed to be independent (see section A.5.11). Other distributions also offer a reasonable fit to the data, particularly at higher wind speeds. Thus the Frechet (extreme value type II), Rayleigh and Weibull distributions have also been advocated [Thom, 1968; Davenport, 1983; Melbourne, 1977]. One (theoretical) difficulty is that the *EV-I* distribution permits negative V values; this is not the case with the Frechet distribution, which is also considered to be somewhat more conservative at high V values.

The meteorological mechanism for cyclones and hurricanes is entirely different from that of thunderstorms [Batts *et al*, 1980]. This is evident also from some typical recorded data of the associated wind speeds (Fig. 7.1)

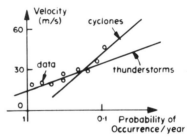

Fig. 7.1 — Typical wind Gust speeds for cyclones and thunderstorms [Gomes and Vickery, 1976].

[Gomes and Vickery, 1976]. There is some doubt whether long-term cyclonic wind speed predictions can be made on the basis of the relatively short-term records available, since recent experience (e.g. El Nino phenomenon) suggests that long-term atmospheric processes may exist to render the usual stationarity assumptions false [Walker, 1984]. This points up the difficulty of achieving "perfect" probabilistic models; despite these difficulties, it has been suggested that the annual maximum wind speeds for cyclonic winds might also best be described by an *EV-I* distribution [Russell and Schuëller, 1974; Gomes and Vickery, 1976; Simpson and Riehl, 1981].

The annual maximum wind speeds for cyclonic and non-cyclonic winds, referred to above, are suitable for use with the discrete time reliability calculation procedure (see section 6.3). The lifetime maximum is also of importance, particularly for the time integrated approach of section 6.2. If $F_V(\ )$ denotes the cumulative distribution function of V, then, for independent annual maxima, $F_{VL} = [F_V(v)]^L$ where L is the lifetime in years. If V has the *EV-I* distribution, then F_{VL} will also be *EV-I* distributed (see Appendix A). Some typical values and parameters are shown in Table 7.1. It

Table 7.1 — Typical average values of mean hourly wind speed data

	Annual maximum			50 – year maximum		
	mean speed (m/s)	(miles/h)	Coefficient of variation	mean speed (m/s)	(miles/h)	Coefficient of variation
USA[a]	15.5	(34.7)	0.12–0.17	24.1	(18.6)	0.11–0.14
Australia[b]	14.9	(33.3)	0.12	24.7	(55.2)	0.12
Cardington, UK[c]	15.5	(34.7)	0.24	23	($\approx$52)	0.12

[a]Converted from "fastest-mile" wind records [Simiu and Filliben, 1980].
[b]Converted from "3-s" wind records [Pham et al., 1983].
[c]Converted from "3-s" wind records [Shellard, 1958].
Mean hourly speeds $\approx 0.77 \times$ fastest-mile speeds.
Mean hourly speeds $\approx (1/1.55–1/1.7) \times$ 3-s gust speeds.

is seen that, for $L = 50$ years, the ratio of annual mean to 50-year mean wind speed is roughly about 0.7.

The daily maximum wind speed is of interest in determining daily maximum wind loads as these may be considered to be good approximations for the "average-point-in-time" loadings. Wind speed data of this type can be obtained from meteorological centres. An *EV-I* distribution again appears to be appropriate [Ellingwood et al, 1980].

The wind pressure loading W may be obtained from wind speed using again the standard hydrodynamic relationship (7.1), which can be rewritten for particular structures or surfaces of structures, as

$$W = kC_pV^2 \tag{7.3}$$

where C_p is the pressure coefficient (which may be a function of V) and $k = cEG$ is a "constant" for a given structure (c is a is a "constant", E an exposure coefficient and G the "gust factor", which depends on wind turbulence and the dynamic interaction of the wind and the structure).

Standard wind engineering texts [e.g. Simiu and Scanlan, 1978] or codes of practice should be consulted for appropriate values for the above coefficients, which depend on structure size, orientation with respect to wind direction, structure geometry, surface roughness, etc.

Expression (7.3) may be used directly in a limit state function without formally deriving the statical parameters and distribution of W. However, if these are nevertheless required, such as in code calibration work, it must be noted that the probability distribution for W cannot be obtained in closed form (since the square of *EV-I* has no simple distribution) and techniques such as Monte Carlo simulation must be used to determine an appropriate distribution. With inclusion of the data for the other parameters, it has been found that the *EV-I* distribution yields a good empirical fit for loads greater than the 90 percentile [Ellingwood et al., 1980] Distributions and data for C_p, G and E must be estimated; however, it is by no means clear that C_p, for example, is a constant as commonly assumed; a lognormal distribution has been suggested. Typically, means for these parameters can be obtained from wind loading codes, and V_{C_p}, V_G and V_E are of the order 0.12–0.15, 0.11 and 0.16 respectively [Ellingwood *et al.*, 1980; Schuëller *et al.*, 1983]. These results are suitable for use in code calibration (see Chapter 9).

For dynamically sensitive structures, annual or lifetime maximum probability density functions for loads are not directly obtainable. Recourse must be made to spectral analysis to convert instantaneous results to equivalent annual or lifetime maxima [Simiu and Scanlan, 1978].

As noted above, directional effects may be important. Wind is a turbulent phenomenon and hence will change direction locally, as well as having an overall direction. Meteological data can be used for overall directionality statistics; it is not uncommon to assume a uniform distribution for wind direction for locations for which records are insignificant.

Local turbulence effects are not always of importance in reliability calculations, although they are important in studying wind loads. Details are beyond the scope of this book [Simiu and Scanlon, 1978].

7.3 WAVE LOADING

As might be imagined, the determination of statistical descriptions for wave loads has many features in parallel with that for wind loads, since both involve structure–fluid interaction. It is not proposed to discuss the physics of the problem in detail here; reference might be made, for example, to Sarpkaya and Isaacson (1981) for an extensive treatment of wave forces. Only a very brief account will be given here, with emphasis on statistical properties [see also Schuëller, 1981].

As a first approximation the instantaneous velocity $U(t)$ and the acceleration $\dot{U}(t)$ of a water particle in a water wave (see Fig. 7.2) can be represented by independent normal processes. The velocity has a mean equal to the current [Borgman, 1967]. If records for $U(t)$ and $\dot{U}(t)$ are

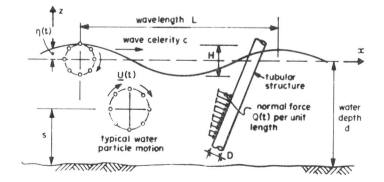

Fig. 7.2 — Schematic representation of water particle motion, wave shape and force exerted on a tubular structural member.

available, e.g. from current measurements, σ_U and $\sigma_{\dot{U}}$ can be estimated directly. Where this is not the case μ_U, σ_U and $\sigma_{\dot{U}}$ can be calculated from wave theory and knowledge of the frequency components of wave height (i.e. the wave spectrum). The latter, of course, must again be obtained from measurements, but now of wave height, wave period and wind speed. It is usually assumed that the sea elevation $\eta(t)$ is a stationary normal process, at least during the duration of a storm, and that it is composed of only a small range of frequencies, i.e. it is assumed to be a "narrow band" process (see section 6.7). These assumptions allow the sea elevations to be related to the wave spectrum $S_\eta(\omega)$, where ω is the wave frequency ($\omega = 2\pi/T$, where T is the period). The spectrum obviously depends on the local conditions, such as fetch, for the development of wind-generated waves and water depth. As a result, different spectra have been proposed, e.g. JONSWAP for the North Sea by Hasselmann *et al.* (1973) and a spectrum for a fully developed sea by Pierson and Moskowitz (1964).

Once the wave spectrum is known, the statistical properties of the probability distribution of wave height can be determined. For sinusoidal waves, the mean is at the mean water depth, so that $\mu_H = 0$; the variance can be estimated from (6.146) as $\sigma_H^2 = \text{var}[\eta(t)] = \int_0^\infty S_\eta(\omega)\, d\omega$. In principle this means that, for a fully developed sea, for which independence between successive wave heights might be assumed, the Rayleigh distribution (see section A.5.13) should apply for wave height [Longuet-Higgins, 1952; Holmes *et al.*, 1983]. However, each of the extreme value distribution types *EV-I–EV-II* (see sections A.5.11–A.5.13) have been proposed as appropriate, depending on the geographical location being considered. All these extreme value probability distributions should be considered conditional, in the sense that they apply provided that a storm occurs.

Water particle velocities $U(t)$ are obtained from a knowledge of wave height H and wave length L (see Fig. 7.2). Although many wave theories have been proposed [Sarpkaya and Isaacson, 1981] it is usually sufficient for wave force calculations to employ the so-called linear theory due to Airy [Lighthill, 1978]. This theory is sufficiently simple to be used in spectral calculations.

According to the Airy theory, the water surface elevation η is given by a sine wave

$$\eta(x, t) = \frac{H}{2} \sin (\omega t - kx) \tag{7.4}$$

where $\omega = 2\pi/T$ is the wave frequency, $T = 2\pi/\omega$ is the wave period, t is the time, H is the wave height, L is the wave length (distance between crests), $k = 2\pi/L$ is the wave number and x is the horizontal distance. The wave celerity c is given by $c = \omega/k = L/T$ (see Fig. 7.2). The horizontal components of velocity U and acceleration $\dot{U}$ are given by [e.g. Lighthill, 1978; Weigel, 1964]

$$U_h = \omega \, \frac{H}{2} \, \frac{\cosh[k(z+d)]}{\sinh(kd)} \, \sin(\omega t - kx) \tag{7.5a}$$

$$\dot{U}_h = \frac{dU}{dt} = \omega^2 \, \frac{H}{2} \, \frac{\cosh[k(z+d)]}{\sinh(kd)} \, \cos(\omega t - kx) \tag{7.5b}$$

where d is the water depth and z is the location of U, $\dot{U}$ below water surface.

The vertical component of velocity has a similar form, with sinh( ) for cosh( ) and cos( ) for sin( ). The acceleration component follows directly.

With known statistical properties for H, those for U and $\dot{U}$ can be determined directly from (7.5a) and (7.5b). As noted earlier, the particle velocity and acceleration are generally considered to be represented by normal processes, with the mean for U_h being the steady current, and the mean for $\dot{U}$ being zero.

Once the water particle velocity and acceleration are known at any depth z (Fig. 7.2), the normal force per unit length exerted at a particular location of a slender cylinder such as typical in steel offshore structures is given by [Morison et al., 1950]

$$Q(t) = k_d U_n |U_n| + k_m \dot{U}_n \tag{7.6}$$

where U_n is the vector of incident water particle velocity normal to the cylinder. This expression was originally proposed for vertical piles but is commonly assumed to be more generally valid. The first term in (7.6) relates to the drag force exerted by the wave on the pile; the second term relates to the influence of the mass of the water displaced by the cylinder on its oscillatory behaviour. For a circular cylinder of diameter D, and water of density ρ, the coefficients k_d and k_m are

$$k_d = \frac{C_d \rho D}{2} \text{ and } k_m = \frac{C_m \rho \pi D^2}{4} \tag{7.7}$$

where the dimensionless drag and mass coefficients C_d and C_m have been extensively experimentally investigated, including for oscillating flows of given frequency [e.g. Morison et al., 1950; Tickell, 1977; Holmes and Tickell, 1979]. The results show a large degree of scatter, particularly for cylinders not oriented in a vertical direction. Some typical values are shown in Table 7.2.

The probability density function for Q at a particular location (x, z) can be found from (7.6) and the known properties for U and $\dot{U}$ using convolution or by Monte Carlo simulation. The mean and variance can be approximated by (A.178) and (A.179) respectively. It is usually found that the probability density function for Q has more extensive upper and lower tails than a normal distribution with the same mean and variance, so that approximation by a normal distribution would be unconservative for higher wave load

Table 7.2 — Indicative values for coefficients C_d and C_m for smooth clean cylinders.

Coefficient	Mean	Typical range	Coefficient of variation
C_d	0.65	0.6–0.75	≈ 0.25
C_m	1.5	1.2–1.8	≈ 0.20–0.35

estimation. Nevertheless, a common assumption [Borgman, 1967] is that the extreme cylinder force is Rayleigh distributed (see section A.5.13), which implies the further assumption of a narrow-band process (see section 6.7). This also tends to underestimate the probability of occurrence of higher wave loads.

The extreme value distribution for $Q(t)$ can also be obtained from calculation of the upcrossing rate through the use of (6.51). The upcrossing rate can be found, in principle, from Rice's formula (6.68). However, this requires the existence of the fourth moment of the velocity spectrum, a rather restrictive requirement [Borgman, 1967].

In most cases "jacket"-type offshore structures are sufficiently flexible for dynamic effects to be important. The usual approach is to employ spectral analysis (see section 6.7). The procedure starts from the wave height spectrum $S_H(\omega)$ and requires linear relationships to convert wave height successively to velocities and accelerations ((7.4a) and (7.4b)), wave forces and structural load effects (see Fig. 6.29). Linearization of Morison's equation (7.6) in the velocity term is required, a considerable approximation since the mean of the velocity is either zero or equal to the (generally small) current, unlike the linearization required for wind velocities (see section 7.1). Details of the spectral analysis and the required simplifications are described in the classic paper by Borgman (1967), where correlation between wave forces at different locations is also considered.

In general, the resulting load effects are assumed to be normal distributed, with variance obtained from the load effect spectrum through (6.146). Mean values must be obtained from a mean value analysis.

7.4 FLOOR LOADING

7.4.1 General

Live loading on floors must be modelled, since long-term records are not available, and since there are many possible parameters which may influence it. Attention will be confined herein mainly to office live loads, but similar considerations apply to other types of floor loading.

Originally data used in design consisted of estimates of dense crowd loading. Thus, in 1883, UK floor loads were set at 140 lbf/ft² (approximately

7 kPa) for domestic dwellings and 170–225 lbf/ft^2 (8.3–11.0 kPa) for public areas. This suggests that a typical man (weight, 170 lbf) occupied about 1 ft^2 in a crowd situation. For multistorey buildings it is unlikely that all floors will be subject to crowd loading at the same time in corresponding locations. This led to the concept of "live load reduction", which allows a reduction in load per unit area as the area being considered is increased. Such a rule was already in evidence in New York in the early 1900s.

Later work, directed mainly at setting design loadings rather than complete probability distributions, identified additional particular features of floor live loading.

(a) The loading consists of sustained effects plus short-term transient effects (see Fig. 7.3).

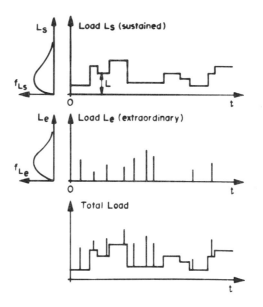

Fig. 7.3 — Time histories of typical live loads.

(b) Occupancy changes produce changes in the sustained loading (Fig. 7.3).
(c) There is variation of loading within rooms (Figs 7.4(a) and 7.4(b)).
(d) Variation of loading between rooms shows some dependence.
(e) There is some dependence between loadings on different floors (Fig. 7.5).

In addition, there is the area dependence effect already mentioned (Fig. 7.6). Ideally, all these factors must be accounted for in a model of floor

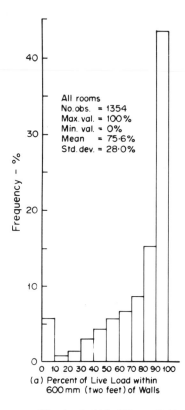

Fig. 7.4a — Percentage of live load within 600 mm (2 ft) of walls [Culver, 1976].

loading. In order to make progress, the sustained loading will be discussed first, followed by the transient load effect. These will then be combined using an appropriate load combination procedure.

7.4.2 Sustained load representation

The approach to be adopted below is to derive an expression for the floor load intensity at some arbitrary location in a building and then to convert this to an equivalent uniformly distributed loading. This fits in with conventional design loads and simplifies application in reliability analyses. Allowance is then made for tenancy changes before the probabilistic model is completely developed.

The load intensity $w_{ij}(x, y)$ on an infinitesimal area ΔA at location (x, y) on the ith floor of the jth building might be modelled in a simplified way as [Pier and Cornell, 1973]

$$w_{ij}(x, y) = m + \gamma_{\text{bld}} + \gamma_{\text{flr}} + \varepsilon(x, y) \tag{7.8}$$

where m is the "grand" mean load intensity, γ_{bld} the deviation of floor load

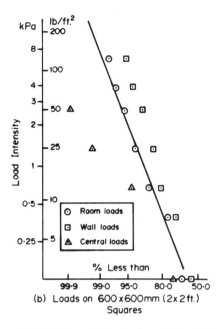

Fig. 7.4b — Observed distributions of floor loads on 600 mm × 600 mm (2 ft × 2 ft)
squares [Lind and Davenport, 1972].

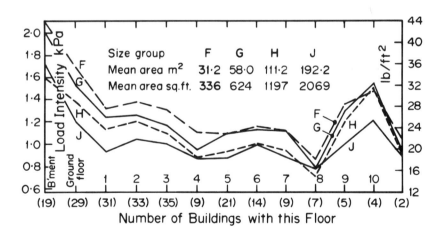

Fig. 7.5 — Variation of 95% level of occupancy loading with floor level number; one
occupation only (i.e. 95% loads < values plotted) [Mitchell and Woodgate, 1971].
Reproduced from Current Paper CP 3/71, Building Research Station, by permission
of the Controller, HMSO. Crown copyright.

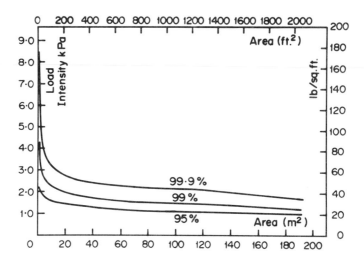

Fig. 7.6 — Variation of loading intensity probabilities with tributary area; floors other than lowest basements and ground floors; after one occupancy only [Mitchell and Woodgate, 1971]. Reproduced from Current Paper CP 3/71, Building Research Station, by permission of the Controller, HMSO. Crown copyright.

intensity from m for building j, γ_{flr} the deviation of floor load intensity from m for floor i for all buildings and ε the spatial uncertainty of floor loading for a given floor (also termed a zero mean "random field").

The parameters γ and ε are random variables with zero means and will be assumed independent, although this is obviously not always the case. The term $m + \gamma_{bld}$ represents the variation of average floor load from building to building; survey data such as those of Mitchell and Woodgate (1971) would allow γ_{bld} to be evaluated if the data were more extensive. A similar argument applies to γ_{flr}. Given the uncertainty in the model, a "second-moment" representation is sufficient, so that the mean of w_{ij} is simply $\mu_w = m$ and its variance is given by

$$\sigma_w^2 = \sigma_{bld}^2 + \sigma_{flr}^2 + \sigma_\varepsilon^2 \tag{7.9}$$

In order to use model (7.8), σ_{bld}, σ_{flr} and σ_ε must be evaluated. This can be done from survey data. Consider a floor of given area A, of size $a \times b$. The total load on this area is given by $L(A)$ or

$$L(A) = \int_0^a \int_0^b w(x, y) \, dx \, dy \tag{7.10}$$

with

$$E[L(A)] = \int_0^a \int_0^b E[w(x, y)] \, dx \, dy = \int\int m \, dx \, dy = mA \qquad (7.11)$$

and

$$var[L(A)] = \int_0^a \int_0^a \int_0^b \int_0^b \text{cov}[w(x_1, y_1), w(x_2, y_2)] \, dx_1 \, dx_2 \, dy_1 \, dy_2 \qquad (7.12)$$

$$= \int_0^a \int_0^b \int_0^b \{\sigma_{\text{bld}}^2 + \sigma_{\text{lr}}^2 + \text{cov}[\varepsilon(x_1 y_1), \varepsilon(x_2 y_2)]\} \, dx_1 \, dx_2 \, dy_1 \, dy_2$$

The latter expression can be evaluated if cov[ ] is known [cf. Vanmarcke, 1983]. One approach is to recognize that the covariance will probably be inversely proportional to the distance between points 1 and 2, e.g. thus

$$\text{cov}[\ ] = \rho_c \sigma_\varepsilon^2 \, e^{-r^2/d} \qquad (7.13)$$

where $r^2 = (x_1 - x_2)^2 + (y_1 - y_2)^2$, d is a constant, ρ_c is a correlation coefficient to allow for the so-called "stacking effect" vertically between floors (the tendency for occupants to load floors in a similar pattern) and σ_ε^2 is the variance in ε.

Clearly r represents a horizontal and ρ_c a vertical spatial parameter. In general ρ_c would vary from location to location, but this will be ignored here for simplicity. The constant d and the parameters ρ_c and σ_ε^2 may be evaluated from observations. For one floor, $\rho_c = 1$, and expression (7.12) may be shown to reduce to

$$var[L(A)] = (\sigma_{\text{bld}}^2 + \sigma_{\text{flr}}^2)A^2 + A\pi d \, \sigma_\varepsilon^2 K(A) \qquad (7.14)$$

where

$$K(A) = [erf \left(\frac{A}{d}\right)^{1/2} - \left(\frac{d}{A\pi}\right)^{1/2} (1 - e^{-A/d})]^2$$

and erf is the error function.

The load $L(A)$ can be converted to a load per unit area, averaged out over the area A, by dividing through by A, so that $L(a)/A \equiv U(A)$, for which the mean is $E[U(A)] = m$ and the variance is given by

$$\text{var}[U(A)] = \sigma^2_{\text{bld}} + \sigma^2_{\text{flr}} + \pi d\sigma^2_\varepsilon \frac{K(A)}{A} \tag{7.15}$$

For n floors of equal area A the variance becomes [Pier and Cornell, 1973]

$$\text{var}[U(A)] = \sigma^2_{\text{bld}} + \frac{\sigma^2_{\text{flr}}}{n} + \pi d\sigma^2_\varepsilon \frac{K(A)}{nA} + \rho_c\sigma^2_\varepsilon \frac{(n-1)\,K(A)}{nA} \tag{7.16}$$

Survey data such as those shown in Figs 7.5–7.7 [Mitchell and Woodgate, 1971] may now be used to estimate the parameters in expressions (7.15) and (7.16). For example $\sigma^2_{\text{bld}} + \sigma^2_{\text{flr}}$, σ^2_ε and d may be obtained from the relationship between the coefficient of variation $\{\text{var}[U(A)]\}^{1/2}/m$ and area A, while ρ_c and σ^2_{bld} may be obtained from the relationship between the coefficient of variation of the column load $\{\text{var}[U(A_n)]\}^{1/2}/m$ and n the number of storeys supported [Pier and Cornell, 1973].

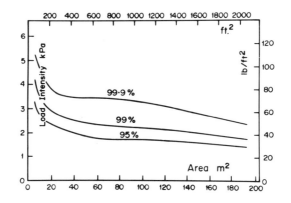

Fig. 7.7 — Variation of loading intensity probabilities with tributary area; floors other than lowest basements and ground floors; after 12 occupancies [Mitchell and Woodgate, 1971]. Reproduced from Current Paper CP 3/71, Building Research Station, by permission of the Controller, HMSO. Crown copyright.

7.4.3 Equivalent uniformly distributed load

With the estimation of the variances in (7.9) the load intensity $w(x, y)$ given by expression (7.8) is completely defined (in second-moment terms). However, it will generally be the case that the equivalent uniformly distributed load ($EUDL$) producing the same particular internal action (or load effect) as w is more useful. This is also the loading specified in structural design codes.

The meaning of ($EUDL$) may be illustrated for the beam shown in Fig. 7.8. A typical "realization" of $w(x)$ is shown in Fig. 7.8(a). What is sought is

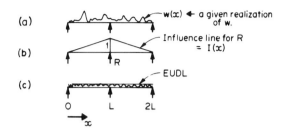

Fig. 7.8 — Equivalent uniformly distributed load (*EUDL*) related to actual load realization.

the load (*EUDL*) of Fig. 7.8(c) such that, for example, the reaction R is properly estimated. Use is made of the influence line $I(x)$ shown in Fig. 7.8(b). The reaction R can then be given both in terms of $w(x)$ and in terms of (*EUDL*) as

$$R = \int_0^{2L} w(x)\, I(x)\, dx = \int_0^{2L} EUDL\, I(x)\, dx \qquad (7.17)$$

but, since (*EUDL*) is not a function of x,

$$(EUDL) = \frac{\displaystyle\int_0^{2L} w(x)\, I(x)\, dx}{\displaystyle\int_0^{2L} I(x)\, dx} \qquad (7.18)$$

By using the appropriate influence line, similar expressions can be derived for other actions (or load effects) while, in two dimensions, influence surfaces take the place of influence lines. By direct analogy

$$(EUDL) = L = \frac{\displaystyle\int_{A_1}\!\!\int w(x, y)\, I(x, y)\, dx\, dy}{\displaystyle\int_{A_1}\!\!\int I(x, y)\, dx\, dy} \qquad (7.19)$$

where A_1 is the so-called "influence area". It corresponds directly to the total beam length, or "influence length" $2L$ in Fig. 7.8, which is seen to be twice that of the tributary length (equal to L) as would be expected for a reaction (half-span length for each span). It might be expected, therefore, that the influence area A_1 is not, in general, identical with the tributary area

A_T commonly used with reduction formulae for floor loadings in structural design codes. Some typical influence lines and surfaces are shown in Fig. 7.9.

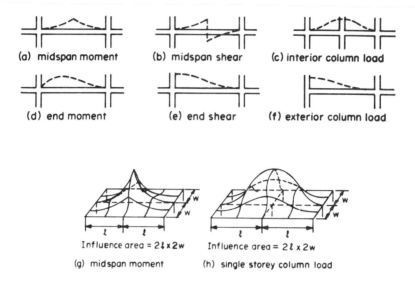

(a) midspan moment   (b) midspan shear   (c) interior column load

(d) end moment   (e) end shear   (f) exterior column load

Influence area = $2l \times 2w$ Influence area = $2l \times 2w$

(g) midspan moment   (h) single storey column load

Fig. 7.9 — Typical influence lines (in plane frames) and typical influence surfaces [from McGuire and Cornell 1974].

It follows readily that the mean for L is $E(L) = m$ and its variance is

$$\text{var}(L)$$

$$= \frac{\int_{A_I}\int\int_{A_I}\int I(x_1, y_1)\, I(x_2, y_2)\text{cov}[w(x_1, y_1), w(x_2, y_2)]\, dx_1\, dy_1\, dx_2\, dy_2}{[\int_{A_I} I(x, y)\, dx\, dy]^2}$$

(7.20)

The latter can be bounded by assuming independent $w(x, y)$, or cov[ ] = 0 and by substituting (7.15):

$$\text{var}(L) \leqslant \sigma_{\text{bld}}^2 + \sigma_{\text{flr}}^2 + \pi d\sigma_{\varepsilon}^2 \frac{K(A)}{A} k$$

(7.22)

where

$$k = \frac{\int_{A_I}\int I^2(x, y) \, dx \, dy}{\left[\int_{A_I}\int I(x, y) \, dx \, dy\right]^2} \tag{7.22}$$

Provided that appropriate influence areas A_I are used, the value of the equivalent uniformly distributed load L is relatively insensitive to the action being considered. This can be seen by the value k in expression (7.21) [McGuire and Cornell, 1974; Ellingwood and Culver, 1977]:

end moments in beams: $k = 2.04$
column loads: $k = 2.2$
midspan beam moments: $k = 2.76$

In general, the values of calculated $(EUDL)$ will be approximately similar if the influence areas are approximately similar and the influence surface shapes are approximately similar. The only exception is the $(EUDL)$ value for midspan shear, which, as might be expected, becomes comparable only if half the influence area is used [McGuire and Cornell, 1974].

7.4.4 Distribution of equivalent uniformly distributed load
By taking live load survey data and converting them directly to $(EUDL)$, it is possible to produce histograms showing the relative occurrence of different levels of equivalent loadings. This is shown in Fig. 7.10 for increasing floor

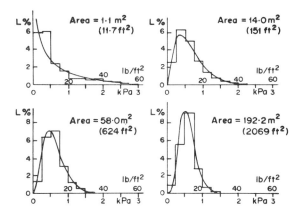

Fig. 7.10 — Histograms of floor load intensity as modelled by distribution functions [Pier and Cornell, 1973].

areas [Pier and Cornell, 1973]. As might be expected (see section A.5.8), the probability density function required to model these results changes from highly skewed to approximately normal as the contributing floor area is increased. An appropriate distribution model is

$$f_L(x) = \frac{\lambda(\lambda x)^k e^{-\lambda x}}{\Gamma(k)}, \; x \geqslant 0 \qquad (7.23)$$

The parameters λ and k may be obtained from the mean and variance of L (see section A.5.6). This distribution appears to fit the observed data better than the normal distribution or the lognormal distribution in the >90 percentile range [Corotis and Doshi, 1977]. The extreme value type I distribution is another possibility and is appropriate on physical grounds since the $(EUDL)$ value reflects the maximum loading for a time interval.

As already noted in section 6.4.7.6, it would be expected that for any structure there will be periods of zero live load and hence the probability density function should strictly have a probability "spike" at the origin (see Fig. 6.13). However, this is often (conservatively) ignored in live load modelling.

The fact that the value of $(EUDL)$ is relatively insensitive to the type of internal action (stress resultant) considered can also be shown by plotting the load values at a given percentile of, say, the gamma distribution against the influence area. A typical plot is given in Fig. 7.11 [McGuire and Cornell, 1974].

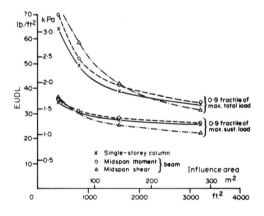

Fig. 7.11 — Single-storey beam and column fractiles as function of influence areas [McGuire and Cornell, 1974].

7.4.5 Maximum (lifetime) sustained load

When the time-integrated approach to structural reliability is used (see section 6.2), it is desired to know the distribution of the lifetime maximum sustained load L_s. This is the maximum load which might act during the life of the structure.

It is clear from Fig. 7.3 that L_s will depend on the changes in the occupancy and/or use of the floor during its lifetime. In formulating a model for L_s, it will be assumed that the loadings relevant to each occupancy are independent. The maximum sustained load L_s which occurs during the lifetime t_L of the structure can be represented as

$$L_s = \max_{t}[L(t)], \ 0 < t < t_L \tag{7.24}$$

with $F_{L_s}(x) = P(L_s \leqslant x)$ denoting the cumulative distribution function of L_s, also given in terms of the level crossing concept (section 6.4.8.1) by

$$F_{L_s}(x) = P[L(0) \leqslant x] \ P[\text{no "upcrossing" by } L(t) \text{ of } x, \text{ in } 0 \leqslant t\text{-}t_L] \tag{7.25}$$

Here $P[L(0) \leqslant x]$, which equals $F_L(x)$, denotes the probability that the initial sustained load is less than or equal to x. It is assumed in expression (7.25) that the maximum sustained load is not caused by the first tenant, i.e. the probability $P[L(0) > x]$ has been ignored. An "upcrossing" in the present context corresponds to a change of occupancy for the floor being considered. Data [Mitchell and Woodgate, 1971; Harris *et al.*, 1981] suggest that this may be approximately (and asymptotically) represented by a Poisson counting process (cf. section 6.4.7.2), so that the cumulative distribution function for L_s becomes:

$$F_{L_s}(x) = F_L(x) \ e^{-v_x t} \tag{7.26}$$

where v_x is the average rate of upcrossings. Clearly, an upcrossing can occur only at a change of occupancy (Fig. 7.3).

It is possible to relate the rate of upcrossings to the rate of occupancy changes. During a small interval Δt of time, the probability that an upcrossing occurs is given by

$$v_x \ \Delta t = P \text{ (one upcrossing in time } t \text{ to } t + \Delta t)$$
$$= P \text{ (upcrossing in this } \Delta t \mid \text{occupancy change in this } \Delta t) \ v_o \tag{7.27}$$

where v_o is the rate of occupancy changes. Expression (7.27) may also be written as

$$v_x \ \Delta t = P\{[L(t) < x] \cap [L(t + \Delta t) \geqslant x] \mid \text{occupany change in this } \Delta t\} \ v_o \ \Delta t \tag{7.27a}$$

Cancelling Δt, and assuming that successive loads L are independent, leaves (cf. section 6.4.8.1)

$$v_x = F_L(x) \, [1 - F_L(x)] \, v_o \qquad (7.28)$$

Substituting into (7.26) produces the cumulative distribution function for L_s in terms of that for the average-point-in-time load L as

$$F_{L_s}(x) = F_L(x) \, \exp\{ - F_L(x) \, [1 - F_L(x)]v_o t\} \qquad (7.29)$$

For extreme loads, $F_L(x) \to 1$ so that (7.29) may be approximated by

$$F_{L_s}(x) \approx \exp\{ - v_o t[1 - F_L(x)]\} \qquad (7.30)$$

Actual survey data [e.g. Mitchell and Woodgate, 1971] suggest that there is approximately 8 years on average between office occupancy changes (see also Fig. 7.12). Assuming ergodicity (see section 6.4.4), it follows that $v_o = 0.125$.

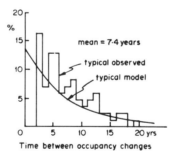

Fig. 7.12 — Frequency distribution for time between changes in occupancy: observed and modelled [Pier and Cornell, 1973].

The magnitude of the maximum lifetime sustained load L_s is well described by either the type I extreme value distribution or the gamma distribution in the region of high load values. Methods for fitting the distributions to data have been reviewed by Harris *et al.* (1981).

With multiple zones (floors, bays, etc.) contributing to the total load effect, such as in column loads, it is unlikely that all zones change tenancy at the same time [Chalk and Corotis, 1980]. However, no data on the necessary correlation appear to be available; ignoring this effect is obviously conservative [Pier and Cornell, 1973].

7.4.6 Extraordinary live loads

Extraordinary live loads are those resulting from groups of people and from crowds. Such loadings are not usually considered in live load surveys. There is considerable uncertainty about extraordinary loading since data has been gathered mainly from questionnaires.

Extraordinary live loads may be modelled as clusters of loads (e.g. persons) acting within a series of load cells randomly distributed over the floor area of interest. If a single occurrence of an extraordinary live load is denoted by E_1, then a plausible model might be [McGuire and Cornell, 1974; Ellingwood and Culver, 1977]

$$E_1 = QN\lambda \qquad (7.31)$$

where Q is the weight of a single person, (typically $\mu_Q = 70\,\text{kgf}\,(150\,\text{lbf})$ and $\sigma_Q \approx 12$ kgf (25 lbf)), N is the number of loads (i.e. persons) per cell (typically $\mu_N \approx 4$ and $\sigma_N \approx 2$) and λ is the mean number of cells per specific area A. The load E_1 may be reasonably assumed to be gamma or *EV-I* distributed. Using second-moment concepts, the equivalent uniformly distributed load denoted L_{el}, has mean and variance given by (A.167) and (A.169):

$$\mu_{el} = \mu_Q\,\mu_N\,\frac{\lambda}{A} \qquad (7.32)$$

and

$$\sigma_{el}^2 = (\mu_Q^2\sigma_N^2 + \mu_N^2\sigma_Q^2 + \sigma_Q^2\sigma_N^2)\,\frac{\lambda k}{A^2} \qquad (7.33)$$

The parameter k is the influence area parameter given by expression (7.22) and has here been assumed to be identical for each load Q; this need not be so in general.

It is reasonable to assume that the arrival of each extraordinary live load is governed by a Poisson counting process (cf. section 6.4.7.2). Then, following a derivation parallel to that for expression (7.30), the cumulative distribution function for the maximum extraordinary live load acting over a period t_L may be approximated by

$$F_{L_e}(x) \approx \exp\{-v_e t_L\,[1 - F_{el}(x)]\} \qquad (7.34)$$

where it has been assumed that the probability distribution for a single occurrence of an extraordinary live load can be used as a surrogate for the probability distribution for a (random) number of occurrences.

Although data are scarce, it seems reasonable to assume that v_e, the average rate of extraordinary live load arrivals, is about one per year, and that λ which is a function of area A, lies in the range 2 cells per 17 m² for "small" areas, to an average of 1 cell per 17 m² for "large" areas, with a "small" area defined as 17 m² (180 ft²). An empirical relationship $\lambda = (1.72A\text{-}24.6)^{1/2}, A > 14.4$ m² (or $\lambda = (0.16A - 24.6)^{1/2}, A > 155$ ft²) has been suggested to relate λ to the total floor area A [McGuire and Cornell, 1974].

7.4.7 Total live load

The proper combination of sustained live load modelled as a Poisson square wave process and extraordinary live load modelled as a Poisson spike process is, as already noted in Chapter 6, not necessarily an easy matter. First-passage solutions have been proposed by Hasofar (1974) and Gaver and Jacobs (1981). However, a very much simplified and time-integrated approach specifically developed and verified for floor live loads will be briefly described here. It is based on earlier work in which the main emphasis was placed on deriving design loadings (i.e. upper fractiles) rather than complete probability distributions [McGuire and Cornell, 1974].

In accordance with Turkstra's rule (6.144), the total live load L_t during the lifetime of the structure can be considered as $L_t = \max_i(L_{ti})$. Intuitive argument suggests that the L_{ti} terms are likely to be one of the following:

$$
\begin{aligned}
\text{case I:} &\quad L_{t1} = L_s + L_{e1} \\
\text{case II:} &\quad L_{t2} = L + L_e \qquad\qquad\qquad (7.35) \\
\text{case III:} &\quad L_{t3} = L_s + L_e
\end{aligned}
$$

Here L and L_s are interpreted as the arbitrary-point-in-time sustained live load and the maximum lifetime live load respectively, L_e as the maximum lifetime extraordinary live load and L_{e1} as the maximum extraordinary live load during one sustained loading period. Each case therefore consists of a maximum load and an arbitrary-point-in-time load (cf. (6.144)). In each case the loads are equivalent uniformly distributed loads. Although Turkstra's rule assumes that each load case is independent, simulation has shown that often case I and case II provide similar L_{ti} values and occur together. Case III was found to have a small probability of occurrence and the probability that a maximum total load occurs when neither component load process is at a lifetime maximum was found to be negligible [McGuire and Cornell, 1974; Chalk and Corotis, 1980].

With this method of load combination, a probability distribution for L_t can only be derived at the expense of some (conservative) simplifications. Thus, case II is simplified by assuming that L is given approximately by the deterministic mean value m of sustained load in equation (7.8) and that, since $L_e \gg L$ in general, $\sigma_{t2} \approx \sigma_e$. Further, if $E(\tau)$ is the average or expected duration of L_s, and t_L is the expected structure lifetime, then the probability that L_s and L_e do not occur at the same time is $p = [t_L - E(\tau)]/t_L$. Assuming now that cases I and II are independent, the cumulative distribution function of the total load is given approximately by

$$F_{L_t}(x) = P(L_t < x) \approx P(L_s + L_{el} < x) \; P(m + L_e < x) \; p$$
$$+ \; P(L_s + L_e < x) \; (1 - p) \tag{7.36}$$

Since both case I and case III represent lifetime maxima, it is reasonable to model them by extreme value type I distributions. For case II, L_e was already represented by the extreme value type I distribution. With these representations [Chalk and Corotis, 1980]

$$F_{L_t}(x) = \exp[-\exp(-w_1)] \; \exp[-\exp(-w_2)] \; p$$
$$+ \; \exp[-\exp(-w_3)] \; (1 - p) \tag{7.37}$$

where $w_i = \alpha_i(x - \beta_i)$ represent the reduced variate forms for the extreme value type I distribution. Note that $i = 1$, $i = 2$, $i = 3$ correspond to case I, case II and case III respectively. The parameters α_i and β_i are distribution parameters (cf. Appendix A). Hence the cumulative distribution function $F_{L_t}(\;)$ can be determined if the moments of each term in (7.36) are known. Table 7.3 shows some basic load parameters, load process statistics based on the models given above and the relative importance of cases I–III for office floor loading; similar results are available also for other occupancy classes [e.g. Chalk and Corotis, 1980; Harris *et al.*, 1981].

7.4.8 "Permanent" loads

Permanent loads are those that do not vary significantly through the life of the structure, even though their actual value may be uncertain. Dead loads are typically of this type. They result from the self-weights of the materials used in construction and from permanent installations. Because these individual permanent loadings are additive, the variability of the total permanent load is less than that of the individual items (as measured by the variance); it also suggests that the central limit theorem applies. Hence dead loads are commonly assumed to be closely approximated by the normal distribution, typically with a mean equal to the nominal load, and a coefficient of variation of 0.05–0.10. However, there is limited evidence that dead loads are underestimated [Ellingwood *et al.*, 1980] and a mean somewhat greater (say 5%) than the nominal may be appropriate. Such systematic errors on the part of designers need further investigation.

The variability in dead load appears to be mainly due to non-structural claddings, services and permanent installations rather than to the variability of the load-bearing materials themselves.

7.5 CONCLUSION

Only three types of loading were discussed in this chapter, mainly to indicate the thinking involved in their modelling. If observations over a suitably long time period are available, load statistics may be deduced directly (perhaps through a hydrodynamic relationship and/or a spectral analysis), such as was

Table 7.3 — Some basic loads, modelling parameters and load combinations
for office floor loadings [Chalk and Corotis, 1980; Harris *et al.*, 1981]

Basic data			
Typical influence area	A_I	18.6 m^2	(200 ft^2)
Typical design life		50 years	
Instantaneous uniformly distributed sustained load L			
Mean	μ_L	0.53 kPa	(10.9 lbf/ft^2)
Standard deviation	σ_L	0.37 kPa	(7.6 lbf/ft^2)
Average number of occupancy changes per year	v_o	0.125 per year	
Expected occupancy duration	$E(\tau) = 1/v_o$	8 years	
Instantaneous uniformly distributed extraordinary live load			
Mean	μ_{el}	0.39 kPa	(8 lbf/ft^2)
Standard deviation	σ_{el}	0.40 pKa	(8.2lbf/ft^2)
Average number of extraordinary live load occurrences per year	v_e	1 year	
Theoretical values (simulated values agree reasonably closely)			
Maximum sustained load L_s			
Mean		1.21 kPa	(24.9 lbf/ft^2)
Standard deviation		0.33 kPa	(6.9 lbf/ft$^{2)}$)
Maximum extraordinary load during one sustained load L_e			
Mean		1.20 pKa	(24.7 lbf/ft^2)
Standard deviation		0.37 pKa	(7.6 lbf/ft^2)
Lifetime maximum extraordinary load L_e			
Mean		1.79 kPa	(36.7 lbf/ft^2)
Standard deviation		0.41 kPa	(8.4 lbf/ft^2)
Simulated values for combinations			
Case I : $L_s + L_{el}$ (occurrence rate, 30%)	Mean	2.50 kPa	(51.2 lbf/ft^2)
Case II : $L + L_e$ (occurrence rate, 41%)	Mean	2.40 kPa	(49.1 lbf/ft^2)
Case III: $L_s + L_e$ (occurrence rate, 17%)	Mean	2.79 kPa	(57.2 lbf/ft^2)
Other (occurrence rate, 12%)	Mean	2.15 kPa	(44.2 lbf/ft^2)

Note that 1 kPa $\equiv$ 1 kN/m^2.

outlined for wind loading. Wave, snow and earthquake loadings are often
considered to be in this category, depending on the geographical location.

Floor loading is of a special nature since it is influenced by many factors
including the possibility of human intervention. This, and the fact that
relatively few observations are available, tends to produce greater uncertainty than might be expected in the modelling of natural loadings.

The methodology used for floor load modelling can also be applied to
other situations, e.g. truck loading on bridges and roads, train loading,
crane loads, etc. As a result of the nature of many of these loads, the study of
load modelling has become rather specialized and reference to the appropriate literature is recommended.

8

Resistance modelling

8.1 INTRODUCTION

The uncertainties associated with strength properties (and some stiffness properties) will be considered in this chapter to complement the discussion of loading given in the previous chapter.

To describe adequately the resistance properties of structural elements, information about the following is required:

(1) statistical properties for material strength and stiffness;
(2) statistical properties for dimensions;
(3) rules for the combination of various properties (as in reinforced concrete members);
(4) influence of time (e.g. size changes, strength changes, deterioration mechanisms such as fatigue, corrosion, erosion, weathering, marine growth effects);
(5) effect of "proof loading", i.e. the increase in confidence resulting from prior successful loading;
(6) influence of fabrication methods on element and structural strength and stiffness (and perhaps other properties);
(7) influence of quality control measures such as construction inspection and in-service inspection;
(8) correlation effects between different properties and between different locations of members and structure.

Only relatively little information is actually available in statistical terms, mostly for items (1)–(3). A useful summary of time-independent statistical properties for reinforced and prestressed concrete members, metal members and components, masonry and heavy timber structures has been given by Ellingwood *et al.* (1980). Because of space limitations, and to illustrate the essential thinking, the present chapter will be confined mainly to a review of the statistical properties of hot-rolled structural steel.

8.2 BASIC PROPERTIES OF HOT-ROLLED STEEL MEMBERS

8.2.1 Steel material properties

Steel material properties data have long been available from tests taken on billets produced at steel mills [e.g. Julian, 1957; Alpsten, 1972] and from more specific testing programs [e.g. Johnston and Opila, 1941]. There are also incidental data from individual research projects [Galambos and Ravindra, 1978].

The applicability of such data in reliability assessment must be evaluated. Thus mill test data are often considered unsuitable since the tests are performed at a loading rate greater than that likely in real structures. Further, the steel tested is not necessarily typical of the "as supplied" steel. It is a practice, in some cases, to supply rejected higher-grade steel with the next lowest grade [Lay, 1979] and this tends to cause a possible second peak in the probability density function for the yield strength. A further difficulty is that mill test samples are commonly taken from the webs of rolled sections, whereas in practice the (usually lower strength) properties of the (thicker) flanges are of more interest. Finally, there is evidence that (probably unintentioned) bias is present in mill test results as a result of the effect of different mills [Lay, 1979].

8.2.2   Yield strength F_y
The strength of steel is dependent on the material properties of the alloy, and hence statistical properties must be related closely to the specified steel type. It is normal practice to sample each billet of steel and only if a specified minimum strength is achieved is the steel accepted for further processing. The data so obtained are extensive but, as already noted, have certain flaws if they are to be used for statistical properties of complete steel members.

Typical mill test data for steel hot-rolled shapes is given in Table 8.1, for

Table 8.1 — US yield stress F_y data

Mean mill F_y/ specified F_y	Estimated mean static F_y/ specified F_y	Coefficient of variation	Number of samples	Reference
1.21	1.09	0.087	3974	[Julian, 1957]
1.21	1.09	0.078	3124	[cf. Tall and Alpsten, 1969]

Data apply to samples taken from webs of hot-rolled sections.
Specified nominal mill strength F_y = 228 MPa (33 ksi).
There is no overlap between the two sources of data [Galambos and Ravindra, 1978].

ASTM A7 steel from US mills. Both sets of data cover a number of steel mills, many shapes of section and a time span of more than 40 years prior to 1957. Table 8.2 summarizes the British mill test data given by Baker (1969) for both plates and structural sections to BS 15 and BS 968 while a summary of Swedish mill test data for different nominal strength is given in Table 8.3.

Table 8.2 — British yield stress F_y data [adapted from Baker, 1969]

Type of steel	Plate thickness (mm)	Mill	Mean mill $F_y/$ specified F_y	Estimated mean static $F_y/$ specified F_y	Coefficient of variation
Structural	10–13	Y	1.15	1.04	0.09
carbon steel	10–13	W	1.14	1.03	0.05
plates	37–50	Y	1.03	0.92	0.12
	37–50	W	1.07	0.96	0.05
High-strength	10–13	M	1.11	1.03	0.08
steel plates	10–13	K	1.11	1.03	0.04
	37–50	M	1.06	0.98	0.06
	37–50	L	1.15	1.17	0.05
Structural	10–13	Q	1.20	1.09	0.05
carbon steel webs of shapes	16–20	L	1.19	1.10	0.12
High strength	6–10	N	1.19	1.11	0.06
steel webs of shapes	37–50	L	1.06	0.98	0.05
Structural	3.7	—	1.27	1.16	0.05
carbon steel tubes	6.4	—	1.32	1.21	0.08
High-strength	5.9	—	1.18	1.10	0.05
steel tubes	6.4	—	1.15	1.07	0.08

Nominal F_y: structural, 250 MPa (36 ksi); high-strength, 360 MPa (50 ksi).
Note that 25.4 mm = 1.0 in.

As noted, mill tests are invariably conducted at a higher strain rate than is normal for conventional "quasi-static" loads on structures. This leads to an overestimate of the yield strength. An empirical correction factor to obtain a static stress F_y from a test producing a higher strain rate stress F_{yh} is given by [Rao et al., 1966]

$$F_{yh} - F_y = 22 + 6900\varepsilon \quad \text{(MPa)}$$
$$= 3.2 + 1000\varepsilon \quad \text{(ksi)}$$

where ε is the strain per second, $(0.0002 \leqslant \varepsilon \leqslant 0.0016)$.

The difference $F_{yh} - F_y$ has been found to be approximately normally distributed with a mean of about 24 MPa (3.5 ksi) and a coefficient of variation of 0.13 [Mirza and MacGregor 1979b]. This allows mill test data to be converted to static stress values.

Using an upper range value for ε, Galambos and Ravindra (1978) suggested that as a first approximation the mill F_y can be reduced by 28 MPa

Table 8.3 — Swedish yield stress F_y data [adapted from Alpsten, 1972]

Nominal F_y (MPa)	Mean mill F_y/ nominal F_y	Estimated mean static F_y/ nominal F_y	Coefficient of variation	Number of samples
220	1.234	1.11	0.103	19 857
260	1.174	1.06	0.099	19 217
360	1.108	1.03	0.057	11 170
400	1.092	1.02	0.054	2 447

Note that 1.0 MPa = 1.0 N/mm^2 = 0.145 ksi.

(4 ksi). The effect on the coefficient of variation might be ignored. This approach has been used to obtain the adjusted ratios $F_y/F_{y\ specified}$ given in the tables. The resulting values might be compared with those given in Table 8.4 obtained by Galambos and Ravindra (1978) from limited numbers of

Table 8.4 — Static yield stress data [Galambos and Ravindra, 1978]

Location	Specified F_y (MPa)	Mean static F_y/ specified F_y	Coefficient of variation	Number of samples
Flange	228	1.00	0.12	34
Flange	345	1.08	0.08	13
Flange	24, 345, 448	1.08	0.09	16
Web and flange	379	1.00	0.05	24
Web	238	1.05	0.13	36
Box	248	1.06	0.07	80

Note that 1.0 MPa = 1.0 N/mm^2 = 0.145 ksi.

experimental projects. It is seen that there is reasonable agreement for corresponding sample types.

From the work of Alpsten (1972) and Baker (1969), it is possible to observe that the extreme value type I distribution, the lognormal distribution, and, to a lesser degree, the truncated normal distribution all fit the experimental data (see Fig. 8.1). These distributions are all positively skewed as would be expected since the minimum value of the yield strength is the specified value, and hence mill test data are truncated in the left tail. As noted, tests which do not meet the strength criteria are either downgraded or re-used in the steel-making process [Alpsten, 1972; Lay, 1979].

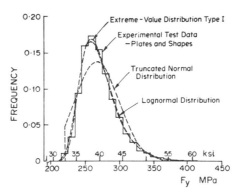

Fig. 8.1 — Typical histogram for yield strength of mild steel plates and shapes together with three fitted probability density functions [adapted from Alpsten 1972].

The distribution types which fit the data do not appear to be affected by mill or testing laboratory, nor by whether the tensile or compressive yield strength is sought [Johnston and Opila, 1941].

8.2.3 Moduli of elasticity
A summary of collated data for the elastic moduli E (elastic modulus in tension or compression), υ (Poisson's ratio) and G (shear modulus) is given in Table 8.5. The data cover a period of more than 20 years. At least two different (US) steel mills were involved and, although all tests were performed at the same laboratory, the data are not comprehensive enough to indicate clearly an appropriate probability distribution.

8.2.4 Strain-hardening properties
A useful discussion of the difficulties of determining strain-hardening properties for steel has been given by Alpsten (1972). No standard test procedures exist. Only the strain-hardening modulus appears to have been investigated. Work by Doane, quoted in Galambos and Ravindra (1978), indicates that E_{ST} = 3900 MPa (570 ksi) in tension and 4600 MPa (670 ksi) in compression. It was suggested that a coefficient of variation of 0.25 might be appropriate.

8.2.5 Size variation
Relatively few data for cross-sectional dimensions of hot-rolled section shapes are available. Typical distributions of a cross-sectional dimension are given in Fig. 8.2 [Alpsten, 1972]. Height and width variation appear to be quite small, typically with a coefficient of variation of 0.002. There is slightly greater variation in thickness.

Of somewhat more importance for strength is the variation in section properties such as cross-sectional area A, second moments of area I, weight W per unit lenth and elastic section modulus Z (S in US terminology). Some typical histograms are shown in Fig. 8.3 [Alpsten, 1972]. Most of the

Table 8.5 — Elastic moduli of structural steel [adapted from Galambos and Ravindra, 1978]

Property	Mean/ specified	Coefficient of variation	Number of tests	Type of test	Reference
E	1.01	0.010	7	Tension coupon	[Lyse and Keyser, 1934]
E	1.02	0.014	56	Tension coupon	[Rao, et al., 1966]
E	1.02	0.01	67	Tension coupon	[Julian, 1957][a]
E	1.02	0.01	67	Compression coupon	[Julian, 1957][a]
E	1.03	0.038	50	Tension and compression coupon	[Johnston and Opila, 1941]
E	1.08	0.060	94	Tension coupon and stub column	[Tall and Alpsten, 1969]
ν	0.99	0.026	57	Tension coupon	[Julian, 1957][a]
ν	0.99	0.021	48	Compression coupon	[Julian, 1957][a]
G	1.08	0.042	5	Torsion coupon	[Lyse and Keyser, 1934]

Specified values: $E = 200000$ MPa (≈ 29000 ksi; $\nu = 0.03$; $G/E = 0.385$.
[a] As attributed by Galambos and Ravindra (1978), but no data are given in report as published.

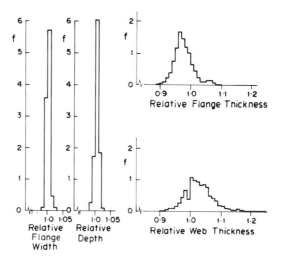

Fig. 8.2 — Typical histograms of cross-sectional dimensions for hot-rolled sections [Alpsten, 1972].

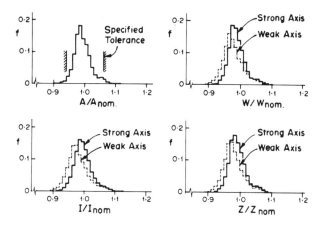

Fig. 8.3 — Typical histograms for section properties for hot-rolled mild steel sections
[Alpsten, 1972].

variation is due to flange thickness variation. A value of unity for the ratio of mean/specified geometric properties, and an average coefficient of variation of 0.05 have been suggested [Ellingwood *et al.*, 1980].

8.2.6   Properties for reliability assessment

Some of the material and dimensional properties described above relate to "as-milled" properties from a number of mills in aggregate (e.g. Table 8.1) while other data appear to cover results from just one mill (e.g. Table 8.2). It would be expected that there is greater variability in results as the data included in the data base expands from:

(1) a billet of steel;
(2) all billets from one grade of steel and one mill;
(3) several mills;
(4) steel delivered to site, without guarantee that it is from one mill;
(5) steel of different sizes and strength grades.

For a reliability assessment, appropriate statistical properties must be used. This implies knowledge of sources of material supplies, and their quality. If such knowledge is not available or is limited to national or regional averages, rather conservative estimates for the coefficient of variation (and perhaps the probability density functions) must be made. Such a situation applies, for example, to structural code calibration activities.

8.3   PROPERTIES OF STEEL REINFORCING BARS

The sources of variability and the physical properties of interest for reinforcing bars are rather similar to those for hot-rolled steel shapes. A useful overview has been given by Mirza and MacGregor (1979b). They suggest that, after adjustment for rate of testing, and after allowing for nominal cross-section areas of bars, the probability density function for steel of 410 MPa (grade 60) yield stress can be represented by a beta distribution:

$$f_{F_y} = 7.587\left(\frac{F_y - a}{c}\right)^{2.02}\left(\frac{b - F_y}{c}\right)^{6.95}$$

for $a \leqslant F_y \leqslant b$ where a, b and c are 372, 703 and 331 MPa or 54, 102 and 48 ksi.

A rather similar distribution type was found to be appropriate for ultimate tensile strength. The modulus of elasticity for steel reinforcing bars was considered to have a mean value of $E = 2.01 \times 10^5$ MPa (≈ 29000 ksi) with a coefficient of variation of 0.033.

8.4 CONCRETE STATISTICAL PROPERTIES

Although the statistical distribution of concrete compressive strength has been of interest for a long time [e.g. Julian, 1957; Freudenthal, 1956] it has a much smaller influence on structural strength and behaviour than do reinforcement properties. This is due entirely to the conventional design philosophy of attempting to achieve ductility in the structure.

A normal distribution is commonly adopted for the compressive strength of good quality concrete; a lognormal distribution appears more appropriate where control is poor. Based on test results for cast *in-situ* concrete [e.g. Entroy, 1960; Murdock, 1953; Rüsch *et al.*, 1969], it appears appropriate to suggest the values of the coefficient of variation or standard deviation given in Table 8.6. The values of the coefficient of variation are roughly halved for within-batch variation.

Table 8.6 — Variation for between-batch *in-situ* concrete compressive strength

Control	Coefficient of variation ($F'_c < 27$ MPa)	Standard deviation ($F'_c \geqslant 27$ MPa)
Excellent	0.10	2.7 MPa
Average	0.15	4.0 MPa
Poor	0.20	5.4 MPa

Note that 2.7 MPa $\approx$ ksi.

Of particular interest in reinforced concrete construction is dimensional variability [Mirza and MacGregor, 1979a]. In most cases it has been found that the actual thickness of slabs is greater than the nominal thickness by ratios varying up to about 1.06, with a coefficient of variation up to about

0.08, but with corresponding values 1.005 and 0.02 for high quality bridge decks. Similar values also apply to precast slabs.

In contrast, the effective depth to the reinforcement for *in-situ* slabs appears to be generally less than specified, in the range (actual/nominal) 0.93–0.99 with a coefficient of variation of around 0.08. There is some evidence that these values are considerably better in good-quality work and that in precast slabs the deviation and variability is almost negligible. Considerably fewer data are available for other concrete elements [Mirza and MacGregor, 1979a].

8.5 STATISTICAL PROPERTIES OF STRUCTURAL MEMBERS

8.5.1 Introduction

The statistical properties of structural members depend on the properties which describe the member, such as the cross-sectional dimensions and material strength properties. When member properties are derived from these using mathematical relationships, differences between the derived result(s) and field or experimental results would be expected. In part this is due to the inherent variability in experimental techniques and observations. The greater part of the difference, however, is the result of the simplification(s) introduced by the mathematical model which relates material and geometric parameters to structural element behaviour. For example, in deriving an expression for the ultimate moment capacity of a reinforced concrete beam section, it is well known that assumptions are made about the concrete compressive stress distribution and about the form of the stress–strain relationships for the reinforcement and that the concrete tensile strength is usually ignored, etc. These assumptions are usually conservative. However, they add a degree of uncertainty to the transition from individual parameters to member strength. This variability is known variously as the "modelling" error or the "professional factor". It does not arise, of course, if the statistical properties of a structural member are obtained directly from (extensive) experimental observations on the member itself. However, such tests are not always practical and recourse may have to be made to modelling the member behaviour mathematically and using data on material and geometric properties.

8.5.2 Methods of analysis

The random variable strength R of the member can be expressed as

$$R = R(\sigma_R, \mathbf{d}, m) \tag{8.1}$$

where σ_R is a vector of material strengths, $\mathbf{d}$ a vector of dimensions, cross-sectional areas, etc. (including those due to workmanship) and m is a model factor. If the relationship (8.1) is known explicitly and is of simple form, R can be evaluated relatively easily using second-moment techniques. Otherwise simulation may have to be used to obtain the probability distribution of R.

8.5.3 Second-moment analysis

In converting from parameter to member statistical properties, second-moment analysis can be used if the relationship between member strength and parameter properties is simple in form. This is the case for a number of important resistance properties of steel members. Typically, the relationship between test strength R and nominal strength R_n (as determined from a code rule, say) can be expressed as (Cornell, 1969)

$$R = P M F R_n \tag{8.2}$$

where P is the so-called "professional" or "modelling" factor, which accounts for the accuracy of the model (expression) used to predict the actual strength, M represents the material properties, such as yield strength, and F is the so-called "fabrication" factor, representing sectional properties, including the effect of fabrication variability. P, M and F are typically ratios of actual to nominal values and will have their own distributional properties. If it is assumed that each can be represented in second-moment format, then it follows that ((A.166) and (A.169))

$$\overline{R} \approx \overline{P}\,\overline{M}\,\overline{F}R_n \tag{8.3}$$

and

$$V_R^2 \approx V_P^2 + V_M^2 + V_F^2 \tag{8.4}$$

where the bar $(\overline{})$ denotes the sample mean of the quantity () and the V_i are coefficients of variation.

The nominal resistance R_n can be obtained directly from codes of practice, while the distributional properties of M and F have been discussed in the previous sections. It might be noted that the assumption of second-moment (i.e. normal distribution) applicability is usually not strictly valid for all the properties discussed.

To apply the simplified approach of expressions (8.2)–(8.4), information is required about the professional or modelling factor P. For the tensile strength of an element, no modelling error term is needed as this situation corresponds directly to the experimental observations used to derive the probability distribution for the material strength. For compact beam sections, with adequate lateral bracing, the resistance is given by the plastic moment based on the nominal plastic modulus, the nominal (specified) yield stress and the modelling factor. The latter can be obtained directly from tests on beams for which "simple plastic theory" was the basis for analysis [Yura *et al.*, 1978]; thus (in direct correspondence to (8.3))

$$R_{\text{test}} = \left(\frac{\text{test capacity}}{\text{nominal capacity}}\right)_{\text{mean}} \frac{\overline{f_y}}{f_{yn}} \frac{\overline{S}}{S_n} R_n \tag{8.5}$$

where $\bar{S}$ is the (sample) mean plastic section modulus, $\bar{f}_y$ is the (sample) mean yield stress, and S_n and f_{yn} are the corresponding nominal values. R_n is the nominal plastic moment. Typically, $\bar{F} = \bar{S}/S_n = 1.0$, $V_F = 0.05$, $\bar{M} = \bar{f}_y/f_{yn} = 1.05$ and $V_M = 0.10$, as can be seen from Tables 8.2–8.4.

The results for beam tests depend on the moment gradient; typical values of $\bar{P}$ equal to the mean of the test to normal values and V_p are given in Table 8.7. The value of V_p was obtained using (8.4) with V_R known from the scatter in the test results.

Table 8.7 — Typical ratios (of test to nominal) for beams in the plastic range
[Yura *et al.*, 1978]

Beam type and moment type	Mean of test/nominal	Coefficient of variation	Number of tests
Determinate; uniform	1.02	0.06	33
Determinate; gradient	1.24	0.10	43
Indeterminate (also frames)	1.06	0.07	41

Generally similar but rather more complex analyses can be performed for beams laterally unsupported, for which elastic or inelastic buckling load is critical, for beam-columns, for plate girders, etc. The models which might be used in conjunction with the present approach to predict actual strength and the relevant model errors have been described in the literature [e.g. Yura *et al.*, 1978; Bjorhovde *et al.*, 1978; Cooper *et al.*, 1978). Some typical values for $\bar{P}$ and V_p are shown in Table 8.8 [Ellingwood *et al.*, 1980]. The

Table 8.8 — Modelling statistics (professional factor P)

Element type	$\bar{P}$	V_p	Remark
Tension member	1.00	0 00	
Compact wide flange beams			
Uniform moment	1.02	0.06	M_p
Continuous beams	1.06	0.07	mechanism
Wide flange beams			
Elastic lateral torsional buckling	1.03	0.09	
Inelastic lateral torsional buckling	1.06	0.09	
Beam-columns	1.02	0.10	SSRC[a] column curves

[a]Structural Stability Research Council.

main limitation with the above approach is that a purely second-moment interpretation is taken of all the parameters. As shown earlier, this is not necessarily the case for steel yield strength, nor for other parameters. Nor is the probability distribution of the modelling error necessarily describable simply in second-moment terms, and separable as in equation (8.2).

8.5.4 Simulation

For the derivation of member properties from (non-normal) dimensional and material properties, with a complex relationship between them, simulation may be the only viable approach. Consider, for example, the strength of reinforced beam–columns. It is well known that according to ultimate strength theory the strength of a beam-column can be represented in terms of a locus of points in the P-M plane, where P represents axial thrust and M bending moment. Typically, in conventional reinforced concrete theory notation, the implicit relationship between P, M and various parameters is given by [Ellingwood, 1977]

$$P = \frac{M}{e} = 0.85 f'_c b(\beta_1 c) + A'_s(f'_s - 0.85 f'_c) - A_s f_s \tag{8.6a}$$

$$M = 0.85 f'_c b(\beta_1 c)\left(\frac{h}{2} - \tfrac{1}{2}\beta_1 c\right) + A'_s(f'_s - 0.85 f'_s)\left(\frac{h}{2} - d'\right) + A_s f_s\left(d - \frac{h}{2}\right) \tag{8.6b}$$

where e is the eccentricity measured from the centroid, c is the depth to neutral axis, $\beta_1 c$ is the depth to the resultant of compression stress "block", h is the depth of the section, b is the breadth of section, d and d' are the concrete covers for the tensile and compressive reinforcements respectively, A_s and A'_s are the areas of the tensile and compressive reinforcements respectively, and f_s and f'_s are the stresses in tensile and compressive reinforcement respectively.

Because there are two strength terms, P and M, the resistance of a beam-column may be expresssed in a variety of ways including:

(1) fixed P;
(2) fixed M;
(3) fixed eccentricity.

Approach (1) may be appropriate for earthquake-type situations, whereas approach (3) is probably appropriate for the usual design situation where axial thrust and moment increase roughly in proportion owing to the nature of the applied loading. The most appropriate choice must be made for any analysis; as will be evident, this choice may be difficult in code calibration work.

With known statistical properties for each of the variables in (8.6), conventional Monte Carlo simulation may be used to generate a sample distribution of resistance values. A typical result [Ellingwood, 1977] is shown in Fig. 8.4 for the case of fixed eccentricity. The simulation procedure

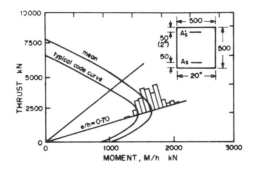

Fig. 8.4—Typical member strength simulation results for reinforced concrete beam-columns [Ellingwood, 1977].

is shown schematically in Fig. 8.5; it is evident that since the distribution of the resistance is desired, relatively few Monte Carlo simulations are required to obtained reasonable results, typically 100–500.

Unfortunately, equations (8.6) do not predict the actual (or experimental) strength of beam-columns accurately since they are nominal relationships used for code (or nominal) strength determination. If strengths predicted by (8.6) are compared with experimental data, it would be possible to obtain the ratio of test strength to calculated strength for each Monte Carlo sample and hence the mean and standard deviation for this ratio. Thus the "professional" factor of the previous section would be found. However, owing to the poor predictability of (8.6), there would be much scatter in the ratio, i.e. its coefficient of variation would be high.

A better relationship between P, M and e can be obtained by integrating the stress–strain curves for the steel and concrete to obtain moment–curvature relationships for given P and hence noting that failure occurs at the peak of the curve. This is a standard procedure for calculating beam-column strength [e.g. Tall, 1964] and has generally been found to agree reasonably well with experimental observations. From a series of simulations using such an accurate model to predict test observations, Grant et al. (1978) found that for reinforced beam-columns the theoretical approach slightly underestimated the available test data. The mean value of the ratio R_T/R_S (i.e. test strength to simulated strength) was found to be 1.007 with a coefficient of variation of 0.064. This factor can be further subdivided into:

(1) error due to the theory used to find the simulated strength, the so-called "modelling error";
(2) errors due to the test procedures used to determine the test strength;
(3) errors due to in-batch variabilities of concrete strength, reinforcement strength and dimensional variability.

For simplicity, the relationship might be expressed as

$$\frac{R_T}{R_S} = C_{T/S} = C_{model} C_{testing} C_{in\text{-}batch} \tag{8.7}$$

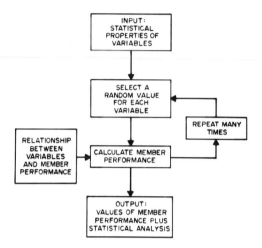

Fig. 8.5 — Simulation procedure for member strength statistical properties.

and, if a second-moment approach is used (see (A.169)),

$$V_{T/S}^2 \approx V_{model}^2 + V_{testing}^2 + V_{in\text{-}batch}^2 \tag{8.8}$$

From these expressions C_{model} and V_{model} can be determined if estimates for the other correction factors are available. Typically $V_{testing}$ is in the range 0.02–0.04 and $V_{in\text{-}batch}$ is about 0.04. (These could be determined experimentally or by Monte Carlo simulation). The respective means are approximately unity. It follows readily that $C_{T/S}$ has a mean of approximately 1.00 and a coefficient $V_{T/S}$ of variation of 0.03–0.046, the latter value corresponding to good-quality testing ($V_{testing} = 0.02$). A general similar correction factor has also been found to be appropriate for other reinforced (and prestressed) concrete elements, such as beams [Ellingwood *et al.*, 1980; Allen, 1970].

With an accurate means to predict actual (i.e. test) results, the ratio of actual to nominal strength, for an element, can now be determined. The nominal values of all the various parameters (stresses, dimensions, etc.) can be substituted into the code rules to obtain a nominal resistance. Then by using Monte Carlo simulation, and the probability distributions for each parameter, accurate theoretical predictions of strength can be obtained, each of which is then modified by the correction factor, $C_{T/S}$, to predict actual strengths. The ratio R_T/R_n (i.e. test strength to nominal strength) is found for each prediction, and, after a sufficient number of Monte Carlo trials, the mean and coefficient of variation can be determined. Hence $\bar{R}/R_n$ and V_R are available. Note that, by directly determining predicted test strength and nominal strength, no explicit mention need be made of the "professional factor" of section 8.5.3. Some typical values of $\bar{R}/R_n$ and V_R are given in Table 8.9 for US data and design codes [Ellingwood *et al.*, 1980].

Table 8.9 — Typical resistance statistics ($F_c' = 34$ MPa ≈ 4800 ksi)

Action	Description	$\overline{R}/R_n$	V_R
Bending	Continuous one-way slabs	1.22	0.15
Bending	Two-way slabs	1.12–1.16	0.14
Bending	Beams	1.05–1.16	0.08–0.14
Bending and axial	Short columns, compression failure	0.95	0.14
Bending and axial	Short columns, tension failure	1.05	0.12
Axial load	Slender columns, compression failure	1.10	0.17
Axial load	slender columns, tension failure	0.95	0.12
Shear	No stirrups	0.93	0.21
Shear	Minimum stirrups	1.00	0.19

8.6   CONNECTIONS

As might be expected, the data available for structural connections in both steel and reinforced concrete construction are rather limited. For steel connections it is usual to assume that the connection is sufficiently ductile to mobilize its full plastic strength. As a result, no "modelling" (or "professional") factor is needed.

Data quoted by Fisher et al. (1978) for electrodes matched to the parent steel indicate that for fillet welds in tension the actual strength is on average about 1.05 of that specified, with a coefficient of variation of 0.04.

For fillet welds in shear, the ratio of fillet weld shear strength to weld electrode tensile strength is typically about 0.84, with a standard deviation of 0.09 and coefficient of variation of 0.10 (whichever is maximum). The ratio of fillet weld shear strength to specified strength is then $0.84 \times 1.05 = 0.88$, with a coefficient of variation given by $V = (0.1^2 + 0.04^2)^{1/2} = 0.11$. As noted, there is no need to have a factor to account for modelling. However, fabrication of the weld will produce additional variability. $V_F = 0.15$ has been suggested [cf. Fisher et al., 1978].

Butt or "groove" welds may be considered to develop the strength of the parent metal provided that they are adequately fabricated and correctly specified [Fisher et al., 1978].

Data and probabilistic models for the strength of high-strength bolts in tension, in shear and in friction grip configuration have been described by Fisher and Struik (1974) and Fisher et al. (1978).

8.7   INCORPORATION OF MEMBER STRENGTH IN DESIGN

Because rolled steel sections and reinforcing bars are available only in discrete sizes, it will generally be the case that a greater section, or more bars, are provided at the end of the design process than strictly required according to the calculations. Therefore the actual resistance of elements will usually be greater than that determined thus far. Occasionally designers will select downwards, if the error is less than, say, about 5%, but generally an upward selection to the next largest section or next whole number of bars is made. Thus the ratio of strength provided to strength required, called the "discretization factor", would be expected to have a mean greater than 1.0 and the corresponding probability distribution function would be expected to be skewed. Figure 8.6 shows a typical relationship for reinforced concrete

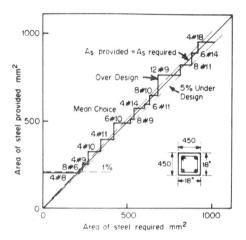

Fig. 8.6 — Effect of discrete sizes on actual strength provided for reinforced concrete columns [after Mirza and MacGregor, 1979a].

columns [Mirza and MacGregor, 1979a]; similar graphs can be drawn for structural steel sections. A typical probability density function obtained by simulation and not limiting over-design, is shown in Fig. 8.7. [Mirza and MacGregor, 1979a].

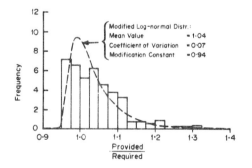

Fig. 8.7 — Typical probability density function for ratio of provided to required capacity for reinforced concrete beams [Mirza and MacGregor, 1979a].

Table 8.10 indicates a size effect for the discretization factor for rein-

Table 8.10 — Typical statistical properties for discretization

Element	Size (mm × mm) (in × in)	Mean of: provided/required	Coefficient of variation	Remarks	Reference
Flexural reinforcement in reinforced concrete beam	250 × 375 (10 × 30)	1.04	0.07	LN: c = 0.94	[Mirza and MacGregor, 1979a]
Flexural reinforcement in reinforced concrete beam	500 × 750 (20 × 30)	1.00	0.03	LN: c = 0.90	[Mirza and MacGregor, 1979a]
Stirrups (reinforced		1.03	0.06	LN: c = 0.93	[Mirza and MacGregor, 1979a]
Vertical steel	300 × 300 (12 × 12)	1.03	0.06	LN: c = 0.93	[Mirza and MacGregor, 1979a]
Reinforced concrete columns	900 × 900 (36 × 36)	1.01	0.04	LN: c = 0.91	[Mirza and MacGregor, 1979a]
Steel beam-columns		1.05	0.07		[Lind, 1976a]

LN = modified lognormal distribution in which [log(provided/required) − c] is normally distributed.

forced concrete elements; this is greatest for flexural reinforcement in beams for the parts of the probability distribution away from the lower tail. Since the lower tail is usually of interest in reliability calculations, size effects may often be ignored. For reinforced concrete design generally, however, a discretization factor with modified lognormal distribution, $c = 0.91$, a mean of 1.01 and coefficient of variation of 0.04 appears a reasonable choice.

It is important to note that the discretization factor does not take into account possible human error in the design process, nor does it consider self checking or other checking. It rests purely on an assumption of reasonable underdesign (and over) design.

8.8 CONCLUSION

Typical resistance properties for use in a structural reliability analysis were reviewed in this chapter, with particular reference to properties for hot-rolled steel members. Some concrete strength properties were also noted, together with dimensional information.

For structural members it may be necessary to integrate statistical properties of material strength, dimensions and design models to arrive at the member statistical properties. This may be done using a second-moment approach or a Monte Carlo analysis.

Finally some comments on connections and on systematic overdesign due to discrete sizing close this chapter.

9

Codes and structural reliability

9.1 INTRODUCTION

The design of structures using a structural design code, or an equivalent regulation, may be described as "structural proportioning by delegated professional authority" [Lind, 1969]. All designers required to use a particular design code are forced to follow one algorithm (usually) to achieve "their" design. Key steps in the structural proportioning process have been predetermined for the designer by code rules. Any structural design for which a particular code was used therefore reflects the "design" of the code itself. It follows that the formulation of a structural design code is the same as the bulk detail proportioning of all structures that will be designed using that code. The individual designer is left free to determine structural modelling, layout, main dimensions, connection types, support and loading conditions, etc., largely independent of the structural design code, but once such decisions have been made, the detail design is largely prescribed by code rules.

The objective of structural engineering design may be reasonably taken to be the maximization of the total expected utility of the structure, given a prescribed reference data set and availability of materials and labour. There are then two complementary aspects of the design task that must be considered:

(a) optimization of the total expected utility of the structure (by the designer);
(b) optimization of the structural design code (by code-writing committee).

The aim of this chapter will be to consider item (b), i.e. how a code should be formulated to be optimum for the range of structures for which it is likely to be used. In particular, attention will be confined to:

(1) safety-checking rules in code theory;
(2) the likely formats for so-called limit state design codes;
(3) the relationship of these codes to the theoretical reliability concepts
      introduced in the previous chapters.

Practical matters such as proper wording of requirements, interpretation, etc., will not be discussed. It will be assumed that physical models, such as those describing column strength, etc., are given and not subject to optimization, although that would clearly be a possibility. Fundamental aspects of code-writing philosophy have been discussed by Lind (1969, 1972), Legerer (1970), Veneziano (1976), Turkstra (1970) and others.

9.2 STRUCTURAL CODE OPTIMIZATION

A structural design code can be viewed as a predictive tool, in the sense that the designer, having followed the requirements of the code, would expect the resulting structure to be sufficiently safe and sufficiently serviceable during the period of its expected life. However, at the design stage, some matters are not known with any degree of certainty and can only be predicted (see Section 2.2). The way that a code is structured to allow for these uncertainties will affect the expected utility of any structure designed using it.

The degree of optimality of expected utility of a code will depend on the interests of those affected by it. The various requirements which a particular structure should satisfy will also depend on whose viewpoint is considered. Undoubtedly some of the requirements have the potential for conflict. Interested parties usually include the builder, the owner, the eventual user and various regulatory authorities.

Of these, regulatory authorities and structural engineers tend to rate structural safety as a very important requirement to be met. However, from the point of view of the other parties, safety might be seen as "necessary but not sufficient"; other matters such as serviceability and costs are also important. Mostly, in countries with well-established rules and traditions and a legacy of successful engineering, structural safety for other than major or unusual structures is largely a "non-issue" [Bosshard, 1979]. The functions of a structural design code tend, in these circumstances, to be seen more in economic and decision-making terms.

A further complication is that structural design codes must be applicable to whole families of structures each with details and with requirements which a code-writing committee can only guess at on the basis of previous experience. Again, prediction is involved. It is unlikely that in a single Code the differing requirements of the many structures and the many interested parties within its scope can be reconciled. As a result, it is very unlikely that optimization of a code *ab initio* will be successful.

An alternative approach is:

(1) to recognize that many structural design codes have been in existence for some considerable time;
(2) to assume that the codes represent the collective wisdom of the profession and reflect a degree of consensus arrived at over many years.

This process can be viewed as representing an alternative form of code optimization, now via refinement, or the gradual improvement of code provisions (if not necessarily their presentation and complexity) through accumulated experience and trade-offs.

Changes to a design code should be only gradually introduced since all the effects will not be known until the longer term. They should also be sufficiently small not to cause uncertainty and anxiety to code-writing committees, to code users and to other interested parties within its scope. Code revisions which result in safety level changes greater than about 10% are generally considered to alarm practitioners [Sexsmith and Lind, 1977].

Code changes which do not involve safety levels should also be treated carefully. Perceived code complexity increases dramatically with the introduction of changes of philosophy (e.g. from permissible stress to limit state design format), with the introduction of more comprehensive rules and with the change of measurement units (e.g. Imperial to SI units). All three changes were made at once in Britain's CP110 reinforced concrete code, with the result that "limit state design" was commonly held responsible for making design more complex.

The major inconsistency in traditional (non-limit state design) codes is in dealing with safety checking, and in particular the load combination rules for safety checking. It is typical for structural design codes each to deal with only one material or form of construction, such as steel, reinforced concrete, prestressed concrete and timber structures. By tradition, each of these codes has (had) a different set of load combination requirements. These are seldom consistent from one code to the next.

The main thrust of the introduction of limit state design code formats is to overcome the inconsistencies between various structural design codes and to employ reasonably rational rules for safety and serviceability checking. Such rules should require the designer to check all relevant limit states explicitly. A further, and important, aim is to make all code requirements "transparent" to the code user, so that it is clear which limit state is actually being checked.

9.3 IMPROVED SAFETY-CHECKING FORMATS

9.3.1 General

All the improved safety-checking formats are developments of existing deterministic formulations, rather than fully or partially probabilistic formulations. Hence they are all at Level 1 (see Table 2.10), involve no obvious probability calculations for the code user and will be largely similar in appearance to traditional safety-checking formats (even though there are substantial conceptual differences). Thus, the partial factors system already widely used in concrete codes is likely be retained as a basis for development.

In principle, codes could employ level 2 procedures for safety checking, provided that codes also prescribed the basic variables for each design situation, the statistical parameters and the appropriate distributions, as well as notional acceptable p_f values. Each limit state would also need to be prescribed in an appropriate form. In view of the difficulty likely to be faced on reaching decisions about these matters, the gaps in the data required, the general unpreparedness of practitioners and the need to use computers even for simpler design problems involving reliability calculations, it is at present unlikely that safety-checking methods other than at level 1 will be used for day-to-day design application in the foreseeable future.

9.3.2 Unified rules of the Comité Européen du Béton and others

Common design rules to form the basis of all codes have been proposed by the Joint Committee on Structural Safety (JCSS) acting for Comité Européen du Béton (CEB) and five other European structural engineering groups. The rules are modifications of earlier code formats. The safety-checking analysis is formulated in the space of the load effects (stress resultants) and has the following general form [CEB, 1976]:

$$g_R\left(\frac{f_k}{\gamma_{m1}\gamma_{m2}\gamma_{m3}}\right) \geq g_S(\gamma_{f1}\gamma_{f2}\gamma_{f3}Q_k) \qquad (9.1)$$

where g_R and g_S are resistance and load effect functions which convert the ( ) terms to resistance and load effects respectively, f_k and Q_k are characteristic material strengths and loads respectively (see section 1.4.4) and γ_{mi} and γ_{fi} are the respective partial factors.

The partial factors γ_{mi}, on material strength, take account of the following factors [CEB, 1976]:

(a) the possibility of unfavourable deviations of the strengths of materials or elements from the specified characteristic value;
(b) possible differences between the strength of the material or element in the structure from that derived from control test specimens;
(c) possible local weaknesses in the structural material or element arising principally from, or in, the construction process;
(d) the possible inaccurate assessment of the resistance of elements derived from the strength of the material including the variations of the dimensional accuracy achieved in construction as they affect the resistance.

The partial factors γ_{fi} on actions take account of the following factors:

γ_{f1} deviation of actions from their characteristic values;

γ_{f2} (load combination factor) allows for the reduced probability that all actions are at their characteristic value; it is given by three factors;

$$\psi_{pi} \leqslant 1 \qquad (i = 1, 2, 3) \ ;$$

γ_{f3} possible inaccurate assessment of the action effects, including dimensional inaccuracies.

In addition, either γ_m or γ_f may be modified to allow for low consequences of failure and/or the possibility of brittle failure.

A particular form of expression (9.1) adopted by CEB (1976)] is

$$g_R\left(\frac{f_{k1}}{\gamma_{m1}}, \frac{f_{k2}}{\gamma_{m2}}, \ldots\right) \geqslant$$

$$g_S\left[\gamma_{f1_{max}}\gamma_{f3}\sum Q_{max} + \gamma_{f1_{min}}\gamma_{f3}\sum Q_{min} + \gamma_{f1_{mean}}\gamma_{f3}\sum Q_{mean}\right.$$

$$\left. + \gamma_{f1}\gamma_{f3}\left(\psi_{p1}Q_{k1} + \sum_{1=2}^{n}\psi_{pi}Q_{ki}\right)\right] \qquad (9.2)$$

where $\psi_{pi} \leqslant 1$ are load combination factors with $\psi_{p1}Q_{k1}$ denoting the most unfavourable loading. The ψ_{pi} may be considered as the ratio of the arbitrary-point-in-time loading (see Chapter 7) to the characteristic value of that loading. Q_{max} and Q_{min} denote the dead load action (and other permanent actions) acting in the most unfavourable manner for the limit state being considered. Values for the parameters in expression (9.2) for various limit states are given in Tables 9.1 and 9.2. In this particular implementation, γ_{f3} has been incorporated into γ_{f1}.

Values for γ_{mi} would be fixed by code committees dealing with specific materials. For reinforced concrete, for example, typical values are $\gamma_m = 1.5$ for concrete and $\gamma_m = 1.15$ for steel reinforcement, given normal control, average inspection and "normal" consequences of failure.

The characteristic values of the loads and resistances are considered as their 95 or 5 percentile values as appropriate, or their currently accepted values in lieu.

9.3.3 National Building Code, Canada

A somewhat less complex safety-checking format [National Research Council of Canada (NRCC) (1977), Canadian Standards Association

Table 9.1 — Partial factors for CEB code format [CEB, 1976]

| Partial factor | Limit states for the following combinations | | | | |
| | ultimate | | | serviceability | |
	Funda-mental loads	Accidental loads	Infrequent loads	Quasi-permanent loads	Frequent loads
$\gamma_{f1\,max}$	1.2	1.0	0	0	0
$\gamma_{f1\,min}$	0.9	1.0	0	0	0
$\gamma_{f1\,mean}$	0	0	1.0	1.0	1.0
γ_{f1}	1.4	1.0	1.0	1.0	1.0
ψ_{p1}	1.0	B	1.0	C	B
ψ_{pi}, $i \geqslant 2$	A	C	B	C	C

See Table 9.2 for A, B and C.

Table 9.2 — Load combination factors [CEB, 1976]

| | Load combination factors ψ_p | | |
	A	B	C
Domestic buildings	0.5	0.7	0.4
Office buildings	0.5	0.8	0.4
Retail premises	0.5	0.9	0.4
Parking garages	0.6	0.7	0.6
Wind	0.55	0.2	0
Snow	0.55		
Wind and snow	0.55 and 0.4		

(CSA) (1974), [Allen (1975)] has been adopted for use in Canadian building codes. Typically it takes the form

$$\phi R \geqslant g_S[\gamma_D D_n + \psi(\gamma_L L_n + \gamma_W W_n + \gamma_T T_n + \ldots)] \qquad (9.3)$$

The left hand-side represents factored resistance, composed of the resistance R, based on characteristic strengths and material properties, dimensions, etc., and a partial factor ϕ. The right-hand side represents factored load effects. The function $g_S[\ ]$ converts the loads to load effects.

The partial factors γ_D, γ_L, γ_W and γ_T apply to the nominal dead load D_n, live load L_n, wind load W_n, etc. The partial factors are related to an "important factor" γ_I which is a measure of the consequences of failure. For most ordinary structures, $\gamma_I = 1.0$. For structures, such as hospitals, which must survive a disaster γ_I would be set at greater than unity. The partial factors are then related to γ_I, e.g. $\gamma_D = 1.25\gamma_I$ normally, and $0.85\gamma_I$ when D_n counteracts L_n, etc. For structures of lesser importance, or for structures subject mainly to deadload, $\gamma_I = 0.8$–1.0.

The load combination factor ψ accounts for the reduced probability that L_n, W_n and T_n reach their nominal values simultaneously. Typically it takes on values $1.0, 0.7$ and 0.6 respectively for one, two or three loadings to which it applies acting concurrently.

The NBC safety-checking format is derived partly from the earlier (non-probabilistic) safety-checking format of the American Concrete Institute (ACI) [e.g. MacGregor, 1976]. It is similar to the CEB format for load combinations, although the numerical values are somewhat different. However, the major difference lies in the treatment of resistance calculations. Whereas the CEB format uses factored material strengths, factored dimensions, etc., in design to allow for the possibility that the material, dimensions, etc., may be less than anticipated, the NBC format (and the load and resistance factor design format of section 9.3.4) combines all the member understrength and geometrical error terms into the factor ϕ. This factor is intended to reflect the probability that the member as a whole is understrength.

The main disadvantage of the NBC safety-checking format is that it does not always allow adequately for the strength variance of members composed of different materials. If the axial load and moment are calculated for a given eccentrically loaded column using a Monte Carlo approach, a spread of results will be obtained [Grant *et al.*, 1978]. As can be seen from Fig. 8.5, the spread of results is much smaller for columns failing in tension. This could be predicted using the CEB safety-checking format, but not the NBC (or load and resistance factor design) format. Other examples, indicating that partial factors on material strengths are preferable to those on member strength, have been given by Allen (1981b). However, the major advantage of the NBC safety-checking format is that of simplicity, i.e. ϕ needs to be considered only once in calculations.

9.3.4   Load and Resistance Factor Design [LRFD]

The load and resistance factor design safety-checking format was proposed by Ravindra and Galambos (1978) for use in US codes. It has the following form:

$$\phi R_n \geq \sum_{k=1}^{i} \gamma_k S_{km} \tag{9.4}$$

where ϕ and R are the "resistance factor" and "nominal resistance" as used already in US practice in the (non-probabilistic) ACI safety-checking format, the γ_k are the "load factors" or partial factors, and S_{km} the "mean load effects". Note that in this format the load effects rather than the loads themselves are combined. Where the relationship between loading and load effect is linear, such forms are, of course, equivalent.

Based on a series of calibration exercises involving only simple second-moment concepts (see also section 9.6) it was recommended that four specific versions of expression (9.4) would be sufficient for most design situations, i.e. only four load combinations need be examined [Ravindra and Galambos, 1978]. These are

$$\phi R_n \;\geqslant\; \gamma_D \overline{D} + \gamma_L \overline{L} \tag{9.5a}$$

$$\phi R_n \;\geqslant\; \gamma_D \overline{D} + \gamma_{apt} \overline{L}_{apt} + \gamma_W \overline{W} \tag{9.5b}$$

$$\phi R_n \;\geqslant\; \gamma_D \overline{D} + \gamma_{apt} \overline{L}_{apt} + \gamma_S \overline{S} \tag{9.5c}$$

$$\Phi R_n \;\geqslant\; \gamma_W \overline{W} - \gamma_D \overline{D} \tag{9.5d}$$

where $\overline{D}$ is the load effect due to the mean dead load, and $\overline{L}$, $\overline{W}$ and $\overline{S}$ are the means of the maximum lifetime live load, maximum lifetime wind load and maximum lifetime snow load respectively; γ_D, γ_L, γ_W and γ_S (each > 1.0) are the corresponding load factors and $(\ )_{apt}$ represents the arbitrary-point-in-time value (or "sustained" values; see Chapter 7). The load factor $\gamma_D < 1.0$ refers to the minimum dead load effect $\overline{D}$.

Expression (9.5b) is equivalent to the CEB format with $\gamma_{apt}\overline{L}_{apt}$ equivalent to $g_S(\gamma_{f1}\gamma_{f3}\psi_{pi}Q_{ki})$ in expression (9.2) for the case of wind and live and dead load, except that Q_{ki} refers to the characteristic load (a "maximum") whereas $\overline{L}_{apt}$ is in the nature of a mean load. Similar comparisons can also be made for the other formulations [Ellingwood *et al.*, 1980]. Compared with the CEB format, the LRFD format has the advantage of being very similar to safety-checking formats at present in use in many countries, as well as having fewer load combinations which need to be considered. Thus, for a situation with dead, live, wind and snow loading, the CEB safety-checking format requires checking of 32 load combinations, the NBC format requires 14 and the LRFD format requires four (including in each case load reversal due to wind uplift).

9.3.5 Some observations
The safety-checking formats sketched above show different complexities; this implies that code-writers have considerable freedom in choosing a format. If too many partial factors are chosen, it will not generally be possible to derive consistent values for them all; the ideal number of partial factors is dependent on the number of degrees of freedom in the problem. Conversely, if for simplicity and practicality only a few partial factors are used, it must be expected that they will not strictly be constant over all design

situations, since the safety-checking format then has to apply to too great a range of possibilities. In effect, each partial factor has to allow for a number of sources of uncertainty or variability. The net effect of simplicity is that conservative values for partial factors must be used in order to cover all likely design situations. This obviously has a cost penalty.

Finally, it is often considered desirable to have the same safety-checking format for all codes, in particular in terms of load combinations. In principle these should be independent of the material used for the structure. Variation in material strength and behaviour would then be included in the ϕ or γ_{mi} factors only.

9.4 RELATIONSHIP BETWEEN LEVEL 1 AND LEVEL 2 SAFETY MEASURES

Safety factors were traditionally selected largely on the basis of intuition and experience [e.g. Pugsley *et al.*, 1955]. However, the availability of level 2 probability methods has made it possible to relate probabilistic safety measures such as p_{fN} or β to the partial factors of the level 1 safety-checking formats, provided that some simplifications and approximations are accepted. As will become evident later, the relationships which will be developed below are not strictly necessary for practical code calibration, but they are useful in illustrating that there is a reasonable link between level 1 and level 2 safety measures.

The discussion to follow will be in terms of the time-integrated approach of section 6.2. This means that specific reference to time-dependent behaviour of loads and resistances, etc., may be neglected provided that the appropriate (extreme value) probability distributions are used for the variables, and an appropriate load combination rule is employed.

Also, as noted earlier, reliability assessments for code work are made without detailed knowledge of the specific loads, materials and workmanship which will be involved in the actual project. Thus all such properties are predictions of some possible future implementation, and conservative parameters and probability density functions must be used. This means that the nominal failure probability and the corresponding safety index so calculated do not necessarily bear a close relationship to the nominal failure probability and safety index which would be obtained on the basis of data for the completed structure.

Largely in order to keep in mind that it is a predicted value on the basis of quite uncertain information, the nominal failure probability and the corresponding safety index used in this chapter for code calibration will be denoted p_{fC} and β_C respectively. The distinction between p_{fN} and p_{fC} (and β and β_C) is not normally made in the structural reliability literature; however, failure to have done so might appear to be a reason for the common lack of understanding of the meaning of the nominal failure probability as used in code work.

9.4.1 Derivation from first-order second-moment theory

For level 2 reliability calculations using first-order second-moment (FOSM) theory, equations (4.3) and (4.56) give the coordinates in basic variable space X at the checking point P* (see Fig. 4.2):

$$x_i^* = F_{X_i}^{-1}[\Phi(y_i^*)] = \mu_{X_i}(1 - \alpha_i\beta_C V_{X_i}) \tag{9.6}$$

with α_i defined by (4.5) and with the sign convention as adopted in Chapter 4. The limit state function evaluated at the checking point x* is

$$G(x^*) = G\{F_X^{-1}[\Phi(y^*)]\} = G[\mu_{X_i}(1 - \alpha_i\beta_C V_{X_i})_{i=1,\dots,n}] = 0 \tag{9.7}$$

This limit state function in the basic variables (e.g. material strengths, dimensions, loads) must be compatible with the appropriate safety checking format chosen.

Let X_i, $i = 1, \dots, m$ be resistance basic variables. Converting from means to characteristic values by the use of equation (1.24), equation (9.6) becomes

$$x_i^* = \mu_{X_i}(1 - \alpha_i\beta_C V_{X_i}) = \frac{1 - \alpha_i\beta_C V_{X_i}}{1 - k_{X_i}V_{X_i}} x_{ki} \tag{9.8}$$

where k_{X_i} represents the appropriate coefficient corresponding to the characteristic fractile of the normal distribution (e.g. Table 1.3). Equation (9.8) may also be written as

$$x_i^* = \mu_{X_i}(1 - \alpha_i\beta_C V_{X_i}) = \frac{x_{ki}}{\gamma_{mi}} \tag{9.9}$$

where $1/\gamma_{mi} = (1 - \alpha_i\beta_C V_{X_i})/(1 - k_{X_i}V_{X_i}) = x_i^*/x_{ki}$ is defined as the partial factor on the basic variable X_i.

Similarly, if X_i, $i = m + 1, \dots, n$ represent loading basic variables,

$$x_i^* = \mu_{X_i}(1 - \alpha_i\beta_C V_{X_i}) = \frac{1 - \alpha_i\beta_C V_{X_i}}{1 + k_{X_i}V_{X_i}} x_{ki} \tag{9.10}$$

$$= \gamma_{fi} x_{ki} \tag{9.11}$$

where $\gamma_{fi} = x_i^*/x_{ki}^*$ represents the partial factor on the load X_i.

The limit state equation (9.7) now becomes

$$G\left(\frac{x_{ki}}{\gamma_{mi}}, \dots, \gamma_{fj}x_{kj}, \dots\right) = 0 , \qquad \begin{array}{l} i = 1, \dots, m \\ j = m+1, \dots, n \end{array} \tag{9.12}$$

which contains partial factors in a format similar to those of the various limit state design safety-checking formats described above.

Since expression (9.12) is also given by $G(\mathbf{x}^*) = 0$, it follows directly that general expressions for γ_i are

$$\gamma_{mi} = \frac{x_{ki}}{x_j^*} = \frac{x_{ki}}{F_{X_i}^{-1}[\Phi(y_i^*)]} \tag{9.13a}$$

$$\gamma_{fj} = \frac{x_j^*}{x_{kj}} = \frac{F_{X_i}^{-1}[\Phi(y_j^*)]}{x_{kj}} \tag{9.13b}$$

These expressions could be applied where X_i consists of a non-normal variable.

It is important to note that the γ_i values are not necessarily unique. In reduced variable space, selection of a point $\mathbf{y}^{(1)}$ different from the "checking point" $\mathbf{y}^*$ (see section 4.2) will lead to different α_i and hence to different γ_i values. Selecting different mean values for the basic variables (but leaving σ_{X_i} unchanged) will give, in general, a new set $\mathbf{y}^{(1)}$ for the checking point in the $\mathbf{y}$ space. As a result, the set γ will not necessarily be constant but may be a function of the mean or characteristic values selected for the basic variables. (Such a set of values will be called a "calibration point" below).

9.4.2 Special case: linear limit state function
If the limit state function is linear, and attention is confined to just one load case, the above results can be expressed in a particularly simple form. The limit state function is now

$$G(\mathbf{X}) = Z = R - S \tag{9.14}$$

so that

$$\mu_Z = \mu_R - \mu_S \tag{9.15}$$

and

$$\sigma_Z = (\sigma_R^2 + \sigma_S^2)^{1/2} \tag{9.16}$$

in the notation of section 1.4.3. Ravindra *et al.* (1969) noted that (9.16) could be approximately linearized to

$$\sigma_Z \approx \alpha(\sigma_R + \sigma_S) \tag{9.17}$$

where α was termed a "separation" function. For $\frac{1}{3} \leqslant \sigma_R/\sigma_S \leqslant 3$, $\alpha = 0.75 \pm 0.06$, with an error $< 10\%$ (Fig. 9.1). From (1.22) it follows that $\mu_R - \mu_S = \beta(\sigma_R^2 + \sigma_S^2)^{1/2}$ so that using (9.17)

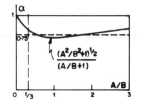

Fig. 9.1 — Variation in separation function α.

$$\mu_Z = \mu_R - \mu_S \geq \alpha\beta_C(\sigma_R + \sigma_S)$$
$$\geq \alpha\beta_C(\mu_R V_R + \mu_S V_S) \qquad (9.18)$$

The central safety factor is obtained from rearranging (9.18):

$$\lambda_0 = \frac{\mu_R}{\mu_S} \geq \frac{1 + \alpha\beta_C V_S}{1 - \alpha\beta_C V_R} \qquad (9.19)$$

Using equations (1.24), (1.25) and (1.26) the characteristic safety factor is then given by

$$\lambda_k = \frac{R_k}{S_k} = \frac{\mu_R(1 - k_R V_R)}{\mu_S(1 + k_S V_S)} \qquad (9.20)$$

so that

$$\frac{R_k}{S_k} \geq \frac{1 - k_R V_R}{1 - \alpha\beta_C V_R} \frac{1 + \alpha\beta_C V_S}{1 + k_S V_S} \qquad (9.21)$$

or

$$R_k \geq \gamma_R \gamma_S S_k \qquad (9.22)$$

Expression (9.22) is in a partial factor format similar to those described in section 9.3. It has partial factors $\gamma_R = (1 - k_R V_R)/(1 - \alpha\beta_C V_R)$ for resistance, and $\gamma_S = (1 + \alpha\beta_C V_S)/(1 + k_S V_S)$ for load effect. In particular for the load and resistance factor design format (9.4) it is immediately evident that $\phi = 1/\gamma_R$.

The partial factors γ may be evaluated by noting that, for agreed percentiles for resistance and loads, the values of k_R and k_S are known. It then remains to fix β_C to obtain γ_R and γ_S. The selection of an appropriate β_C value (or p_{fC}) is therefore central to the derivation of partial factors.

It will be seen that expressions (9.9) and (9.11) for the resistance and loading partial factors are identical with the forms given in expression (9.21) when it is recognized that the separation function α (see expression (9.17)) is a special case of the direction cosines (or sensitivity factors) α_i for the reliability problem with two basic variables. In fact, for two basic variables, $X_1 = R, X_2 = S$ with $G( ) = R - S = 0$, it is not difficult to show that $\alpha_R = \alpha$ and $-\alpha_S = \alpha$. (The sign change is necessary since $(\alpha_R, \alpha_S) = \mathbf{\alpha}$ is a vector, defined in Fig. 4.2, whereas α in (9.17) is a simple variable.)

9.5 SELECTION OF CODE SAFETY LEVELS

The selection of an appropriate value of β_C (or p_{fC}) in (9.21) is not an easy matter. As discussed in Chapter 2, appropriate *a priori* values for the structural probability p_F of failure either over the planned life span, or per annum, are not readily available. Moreover, for codified safety-checking rules using second-moment concepts, it is not p_f that is directly involved, but p_{fC}, the notional failure probability (and corresponding safety index B_C) and, as noted in section 9.2, this may bear very little relationship to any real or imputed p_f.

As also noted in section 9.2, it is generally agreed that codes should be "calibrated". That is, a new generation code is framed in such a way that, for particular design situations, it has average p_{fC} or β_C values not significantly different from those implicit in the previous generation code. Of course, the distribution of the variation in β_C across different design situations might well be different. The most common approach is to attempt to obtain a reasonable constant value of β_C in the new generation code, at least for given design situations most representative of practical design. As noted in section 2.5, such an approach has strict limitations in the way that it accounts for the probability of failure due to human variability and human error.

Event without the presence of human error and variability it would be expected that β_C need not be constant over all design, since, as also discussed in Chapter 2, a complete analysis should consider the risk and consequences of failure. Thus a cheap but critical component should have a higher β_C value, while a very low probability load combination could have a lower β_C value. Such allowance can seldom be rationally made owing to the impossibility of predicting the consequences of failure for every application of the structural design code. In principle, codes could be written to allow a trade-off between expected consequences of failure and partial factors, but to date this has been done only to a very limited extent (e.g. lower load factors for farm buildings).

9.6 CODE CALIBRATION PROCEDURE

A number of pioneering attempts to calibrate a new generation structural design code to an existing code have been reported, not all based, however, on precisely the second-moment reliability concepts described here [e.g.

Allen, 1975; Baker, 1976; CIRIA, 1977; Hawrenek and Rackwitz, 1976; Guiffre and Pinto, 1976; Skov, 1976; Ravindra and Galambos, 1978; Ellingwood *et al.* 1980]. Despite differences of detail, the general approach to code calibration is much the same irrespective of the refinement of the reliability theory used, and broad agreement exists on the essential steps in the process [Lind, 1976b; Baker, 1976].

Step 1: *Define scope*   Since it is still unrealistic to expect one structural design code format to represent all design situations, it is sometimes convenient to limit the scope of the code to be calibrated. Thus the material may be prescribed (e.g. steel structures), the structural type may be prescribed (e.g. building structures), etc.

Step 2: *Select calibration points*   A design space, consisting of all basic variables, such as beam lengths, cross-sectional areas and properties, nominal permitted yield stresses, range of applied loads and loading types, continuity conditions, is chosen. It is then divided into a set of approximately equal discrete zones (e.g. a simply supported 5 m mild steel beam, supporting 25 m^2 precast concrete flooring of given dead load and design for normal office loading). The resulting discrete valued points are used to calculate β_C values for the safety-checking format of the existing code (see below).

    It is extremely important that the effect of changes in the code format on all possible designs for which it will be used is considered. This means that realistic extensive calibration points must be selected.

Step 3: *Existing design code*   The existing structural design code(s) is now used to design the element (i.e. the 5 m beam). This is repeated for all appropriate combinations of calibration points within each discrete zone.

Step 4: *Define limit states*   Limit state functions for each limit state (usually corresponding with those already explicit (or implicit) in existing codes) are now defined. This might include, for example in the case of steel beams, limit states for bending strength, shear strength, local buckling, web buckling, flexural–torsional buckling, etc. Each of these limit states must be expressed in terms of the basic variables.

    The definition of limit states may involve the use of appropriate strength models of the type described in Chapter 8 to convert strength basic variables to member strength, etc. These models should be realistic rather than code-based conservative approximations.

    The definition of the limit states also requires a decision about the load combination model(s) to be employed (cf. Chapter 7). For practical code calibration, a simple load combination model such as Turkstra's rule (6.144) is usually preferred.

Step 5: *Determine statistical properties*   Appropriate statistical properties (distributions, means, variances, average-point-in-time values) for each of the basic variables are required for the determination of the β_C index. Data such as given in Chapters 7 and 8 for loads and resistances can be used.

Step 6: *Apply method of reliability analysis*   Using an appropriate method of reliability analysis, assumed here to be the FOSM method, together with the limit state functions (Step 4) and the statistical data (Step 5), each of the designs obtained in Step 3 is analysed to determine β_C for each calibration point within each zone. The results may be conveniently arranged such that the applied loading becomes the chief independent parameter. Typical results, obtained for the American National Standard A58 (Minimum Design Loads in Buildings Code) [Ellingwood *et al.*, 1980], using the FOSM algorithm of section 4.4.3.2 are given in Figs 9.2 and 9.3. Both different

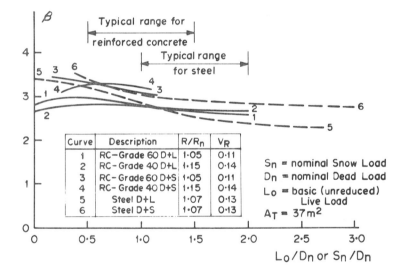

Fig. 9.2 — β index for steel and reinforced concrete beams in the existing code: gravity loads [Ellingwood *et al.*, 1980].

material combinations and loading situations for a given area supported (A_T) are illustrated.

It is evident from Fig. 9.3 that the existing code(s) safety checking rules do not produce uniform β_C values even in the one design situation (e.g. steel beam bending).

It should be noted that the β_C values obtained depend very much on the probability models and the parameters chosen for the basic variables. Other calibration exercises have found β_C values considerably different from those given in Figs. 9.2 and 9.3. This does not matter provided that the data used to obtain β_C are used consistently throughout the calibration process.

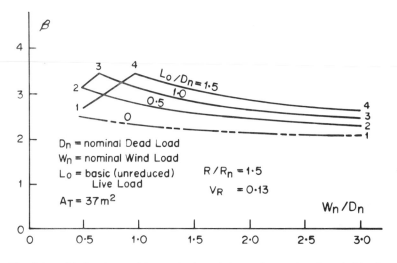

Fig. 9.3 — β index for steel beams in the existing code: gravity plus wind loads
[Ellingwood *et al.*, 1980].

Step 7: Select target β_C value   From repeated analyses such as Step 6 above, the variation of β_C in existing design practice becomes evident and from this information, a weighted-average β_C value could be determined and used as a target β_C ($= \beta_T$) [Baker, 1976]. Some allowance could also be made for consequences of failure, with higher β_T values for high consequence failures.

In practice, this is seldom feasible, owing to the lack of appropriate information. One approach is simply to note the complexity of the issue and to select β_T values on a semi-intuitive basis; for example $\beta_T = 3$ for dead and live load (snow load) combinations, $\beta_T = 2.5$ for dead, live and wind and $\beta_T = 1.75$ under earthquake loads [Ellingwood *et al.*, 1980]. These appear to correspond reasonably well with the range of values obtained from analysing existing (US) design codes, as illustrated in Figs 9.2 and 9.3. Exceptions are masonry structures, for which β_C ranges from 4 to 8, and for glue-laminated timber members, for which β_C ranges from 2.0 to 3.0 with a strong mean of 2.5. These divergences could (and should) be dealt with most appropriately by means of the partial factor(s) on material strength or resistance (e.g. through the ϕ factor).

For load combinations involving earthquakes β_T is significantly lower (i.e. p_f higher) than for more common load combinations. Although failure consequences for earthquakes are high, the forces involved in earthquake-resistant design are also high, so that the selected β_T value reflects a trade-off between cost of initial construction and cost of failure consequences.

Step 8: Observe partial factor format implicit in existing code   Although it is not essential, it is often useful to be able to see how the safety-checking format of the existing code converts to the partial factors in the safety-checking format of the new code, given that the target β_T value has been set. The process to obtain these partial factors is essentially the reverse of the process of Step 7. For a given calibration point, the β_C value is calculated for a given limit state, using the existing code safety-checking format to determine the resistance. If $\beta_C < \beta_T$ the required resistance is suitably increased until $\beta_C = \beta_T$ is obtained. The design, or checking point, as well as the direction cosines α_i are then calculated. From these, the partial factors γ_i can be computed using equations (9.8) and (9.10). This may then be repeated for different calibration points and the variation in partial factors noted. Obviously, the partial factors would not be expected to be constant over all calibration points since β_C is generally not constant. For the revision of the A58 code [Ellingwood *et al.*, 1980], typical results are given in Fig. 9.4.

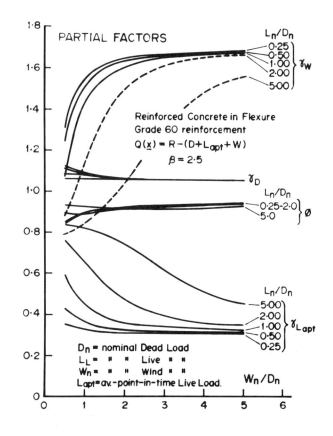

Fig. 9.4 — Variation of γ and ϕ factors for new safety-checking format for steel beam bending in the existing code rules [Ellingwood *et al.*, 1980].

Figure 9.4 shows that the ϕ factor (on resistances) is relatively constant over a wide range of L_n/D_n and W_n/D_n for a given material. (It is also rather insensitive to differences in material.) Although somewhat lower ϕ values occur for low W_n/D_n or L_n/D_n values, these are not in the commonly employed design range. Associated with these lower ϕ values are significant changes in γ_L and γ_W, although γ_D is much more uniform for wide ranges in applied loading.

It is therefore evident that uncoupling between resistance and loading partial factors, as used in section 9.4, is reasonable.

The values of γ_D obtained from analysing US standards in this way are rather lower than expected, being around 1.1. Evidently this is a result of the relatively lower variability in dead load. Further the value of γ_L in Fig. 9.4 is low since it applies to L_{apt} (cf. Chapter 7).

Step 9: Select partial factors As seen above, the partial factors are not constant for a given code safety-checking format and given target β_T value. If, for normal design convenience, the partial factors in the code safety-checking format are to be constant, at least over large groups of design situations, it is to be expected that this will involve some deviation from the target β_T value. Hence, the selection of appropriate partial factors involves a certain amount of subjective judgement.

For a given range $1, \ldots, m$ of calibration points, the partial factors which best approximate the uniform target reliability can, in principle, be obtained by minimizing the value of a measure of "closeness" to the target reliability. Appropriate measures are of the weighted least-squares fit type, such as:

$$S = \sum_{i=1}^{m} (\beta_T - \beta_{Ci})^2 w_i \tag{9.23}$$

where $p_{fC} = \Phi(-\beta_C)$ is the nominal probability of failure for a given calibration point i, $p_{fT} = \Phi(-\beta_T)$ is the target value and w_i is a weighting factor to account for the importance of the calibration point relative to design practice. Obviously $\sum_{i}^{m} w_i = 1$. Log p_{fC} rather than p_{fC} itself is sometimes used in expression (9.23) in an attempt to make S relatively more sensitive to very low values of the nominal failure probability. In addition, an upper limit such as $S \leqslant 0.25$ might be placed on S to limit the deviation of p_{fC} from p_{fT} [CEB, 1976]. Other forms of equation (9.23) have been suggested [Ellingwood *et al.*, 1980]. The weighting factors w_i must sometimes be selected subjectively.

To determine the partial factors for the new code checking format, trial values of the partial factors are used with the new generation code format to calculate β_C and hence p_{fC} for each calibration point (cf. Steps 3–6 above). This result is substituted into (9.23). By repeated trial and error, and perhaps by (arbitrarily) assigning values to one or more partial factors, the set of partial factors which minimizes (9.23) can be determined. These values will then be the partial factors in the new generation code safety-checking format. The complete process for calibration is summarized in Fig. 9.5.

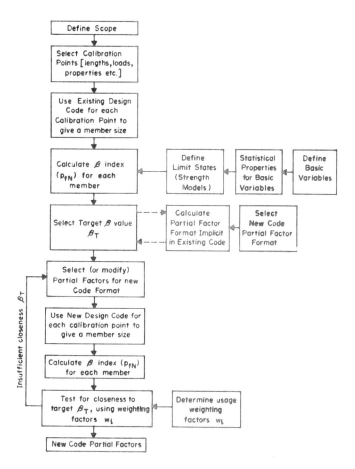

Fig. 9.5 — Flow chart for calibration of code safety-checking format.

Clearly, code calibration is by no means a wholly objective exercise. The choice of limit state format for which the calibration will be done, of minimal acceptable safety indices for various load combinations and materials, of some minimal acceptable partial factors, of the desirable degree of linearization of the safety margin over the design space, as well as of loading and resistance properties all require subjective assessment.

There is obviously some interplay between these and other factors. For example, if it is found that S in equation (9.23) cannot be reduced to a sufficiently low value, then there is probably an insufficient number of partial factors in the limit state design format. Alternatively, it may be necessary to reduce the range of the basic variables covered by the particular limit state under consideration.

9.7 EXAMPLE OF CODE CALIBRATION

The calibration process may be illustrated with the following example, based on steel beam bending. This defines the scope (Step 1) for calibration. For simplicity, it will be assumed that the calibration will be done in the load space and that no allowance for live load reduction will be made. Also only dead load plus live load will be considered and wind loading ignored. Let the existing code safety-checking format for this situation be

$$R_n = (LF)(D_n + L_n) \qquad (9.24)$$

where $(LF) = 1.7$ is the load factor, and all values are nominal values. By expressing L_n in terms of D_n, it will not be necessary to deal specifically with beam size, length, area supported, etc. Hence the calibration points have been selected. Equation (9.24) specifies the required resistance, given the nominal loads D_n and L_n which the resistance rules of the code must satisfy. In general, R_n will be a function of material and geometric properties as well as resistance modelling rules selected by code committees.

Let the new code format be of load and resistance factor design form $\phi R_n = \gamma_D D_n + \gamma_L L_n$ and let it be assumed, for simplicity, that the specification of D_n, L_n and R_n is not changed. This means that the code calibration exercise reduces essentially to one of finding a new load combination rule(s). It is now required to determine ϕ, γ_D and γ_L such that β_C is approximately constant and consistent with the existing code.

For beam bending, the limit state function formulated in the load space is (Step 4)

$$G(\mathbf{x}) = R - D - L = 0 \qquad (9.25)$$

where R is determined from codified beam-bending rules.

For Step 5, statistical properties for D, L and R are given in Chapters 7 and 8. For the present example let it be assumed that [cf. Ellingwood *et al.*, 1980]

$$\frac{\mu_R}{R_n} = 1.18 \text{ (beam bending)}, \qquad V_R = 0.13$$

$$\frac{\mu_D}{D_n} = 1.05, \qquad V_D = 0.10$$

$$\frac{\mu_L}{L_n} = 1.00, \qquad\qquad V_L = 0.25$$

where the subscript n refers to "nominal" values, i.e. values in the existing loading and resistance (material) codes. These values correspond roughly to "characteristic" values. For purposes of illustration in this example, each variable will be assumed normally distributed. Of course, more generally R is often lognormal, and L is extreme value type I; this can be easily treated using the first-order reliability method of section 4.4 but would destroy the simplicity of the example given here.

The reliability analysis, Step 6, may be performed using the FOSM method and working in the original basic variable space (since (9.25) is linear):

$$\beta_C = \frac{G(\mu_X)}{\sigma_G}$$

where

$$G(\mu_X) = 1.18R_n - 1.05D_n - 1.0L_n$$

and from equation (9.24)

$$R_n = (LF)\left(1 + \frac{L_n}{D_n}\right)D_n$$

so that

$$G(\mu_X) = 1.18(LF)\left(1 + \frac{L_n}{D_n}\right)D_n - 1.05D_n - \frac{L_n}{D_n}D_n$$

Further

$$\sigma_G^2 = (\mu_R V_R)^2 + (\mu_D V_D)^2 + (\mu_L V_L)^2$$

$$= \left[1.18(LF)\left(1 + \frac{L_n}{D_n}\right)D_n \times 0.13\right]^2 + (1.05D_n \times 0.1)^2$$

$$+ \left(0.25\frac{L_n}{D_n}D_n\right)^2$$

$$= D_n^2\left\{\left[0.153(LF)\left(1 + \frac{L_n}{D_n}\right)\right]^2 + (0.105)^2 + \left(\frac{0.25L_n}{D_n}\right)^2\right\}$$

Consider now the case $L_n/D_n = 1.0$; then

$$\beta_C = \frac{2.36(LF) - 2.05}{[(0.306(LF))^2 + 0.0735]^{1/2}}$$

and if $(LF) = 1.7$ (the existing code value), then $\beta = 3.34$. The above process can now be repeated for a number of other L_n/D_n values (i.e. calibration points). A plot similar to Fig. 9.2 would be expected. On this basis, a target safety index β_T might be selected (Step 7). Let this be $\beta_T = 3.0$ for purposes of illustration.

The partial factors implicit in the existing code formulation (9.24) may be determined as follows (Step 8), again using $L_n/D_n = 1.0$ and $(LF) = 1.7$ as example values.

Let the variables R, D and L be transformed to the "reduced" variable space defined by (cf. Chapter 4)

$$r = \frac{R - \mu_R}{\sigma_R} = \frac{R - 1.18(LF)(1 + L_n/D_n)D_n}{1.18(LF)(1 + L_n/D_n)D_nV_R}$$

$$or \quad \frac{R}{D_n} = 0.522r + 4.01$$

$$d = \frac{D - \mu_D}{\sigma_D} = \frac{D - 1.05D_n}{1.05D_nV_D}$$

$$or \quad \frac{D}{D_n} = 0.105d + 1.05$$

$$l = \frac{L - \mu_L}{\sigma_L} = \frac{L - L_n}{D_nV_L}$$

$$or \quad \frac{L}{D_n} = 0.25l + 1.0$$

The limit state equation $G(x) = R - D - L$ now becomes, in reduced space

$$g(y) = (0.52r - 0.105d - 0.25l + 4.0 - 1.05 - 1.0)D_n$$

The direction cosines α_i follow from (4.5):

$$c_r = \frac{\partial g}{\partial r} = 0.522D_n$$

$$c_d = -0.105D_n$$

$$c_l = -0.25D_n$$

and

$$l = \left[\sum_i \left(\frac{\partial g}{\partial y_i} \right)^2 \right]^{1/2} = 0.588 D_n$$

so that

$$\alpha_r = \frac{0.522}{0.588} = 0.89$$

$$\alpha_d = -\frac{0.105}{0.588} = -0.18$$

$$\alpha_l = -\frac{0.25}{0.588} = -0.43$$

The partial factors are given by equation (9.9) for resistance:

$$\phi = \frac{1}{\gamma_i} = \frac{1 - \alpha_i \beta_C V_{X_i}}{1 - k_{X_i} V_{X_i}} = \frac{x_i^*}{x_{ki}}$$

or

$$\phi = \frac{1}{\gamma_i} = (1 - \alpha_i \beta_C V_{X_i}) \frac{\mu_X}{x_{ki}}$$

where x_k is the "characteristic" value. This corresponds to the nominal value here. Substituting for R,

$$\phi = \frac{1}{\gamma_R} = [1 - 0.89(3)(0.13)] \, (1.18) = 0.77$$

Similarly for loads, the partial factors are given by equation (9.10) as

$$\gamma_i = \frac{x_i^*}{x_{ki}} = \frac{1 - \alpha_i \beta_C V_{X_i}}{1 + k_{X_i} V_{X_i}} = (1 - \alpha_i \beta_C V_{X_i}) \frac{\mu_X}{x_{ki}}$$

so that

$$\gamma_D = [1 + 0.18(3)(0.10)](1.05) = 1.11$$
$$\gamma_L = [1 + 0.43(3)(0.25)](1.0) = 1.32$$

Hence the partial factor format at $L_n/D_n = 1$ for $\beta_T = 3$, corresponding to the existing code format and the data given above, is

$$0.77R_n \geqslant 1.11D_n + 1.32L_n$$

Again, the above process may be repeated for a number of other L_n/D_n values (i.e. calibration points) and the results graphed (cf. Fig. 9.4).

For purposes of illustration, let it now be decided that, for the adopted code format $\phi R_n = \gamma_D D_n + \gamma_L L_n$, the values for ϕ and γ_D be (arbitrarily) selected at $\phi = 0.80$ and $\gamma_D = 1.2$. The remaining γ_L is sought. A trial-and-error procedure is required [Baker, 1976].

Let the first trial be $\gamma_L = 1.4$. Then

$$R_n = \frac{1}{0.8}(1.2D_n + 1.4L_n)$$

and

$$G(\mu_X) = \frac{1.18}{0.8}\left(1.2 + 1.4\frac{L_n}{D_n}\right)D_n - 1.05D_n - \frac{L_n}{D_n}D_n$$

and

$$\sigma_G^2 = \left[\frac{1.18}{0.8}\left(1.2 + 1.4\frac{L_n}{D_n}\right)D_n \times 0.13\right]^2 + (1.05D_n \times 0.1)^2 +$$
$$+ \left(0.25\frac{L_n}{D_n}\right)^2 D_n^2$$

so that

$$\beta_C = \frac{1.77(1 + 1.167L_n/D_n) - 1.05 - L_n/D_n}{[0.053(1 + 1.167L_n/D_n)^2 + 0.011 + 0.0625(L_n/D_n)^2]^{1/2}}$$

The β_C and p_{fC} values are given in Table 9.3 for various values of L_n/D_n

Table 9.3.

$\dfrac{L_n}{D_n}$	β_C	p_{fC}	$\log p_{fC}$	w_i	$S = (\log p_{fT} - \log p_{fC})^2 w_i$
1	3.14	0.84×10^{-3}	-3.07	0.2	0.0080
2	3.09	1.0×10^{-3}	-3.00	0.4	0.0068
3	3.05	1.14×10^{-3}	-2.94	0.3	0.0015
4	3.02	1.26×10^{-3}	-2.90	0.1	0.0001
	$\beta_C = 3.0$	1.35×10^{-3}	-2.87	$\Sigma = 1$	$S = 0.0164$

(calibration points), together with an illustrative set of weighting factors and the calculation of S (equation (9.23)) using the log p version.

These calculations can be repeated for other trial values of γ_L; say $\gamma_L =$ 1.3. In the expression for β_C the constant 1.167 is then replaced by 1.083 and a new value for S can be determined. Trial values for γ_L are chosen until the minimum value of S is found. Naturally, the procedure could also be repeated for the variables ϕ and γ_D if this was desired.

9.8 OBSERVATIONS

The principles outlined in the previous sections of this chapter have been successfully applied for determining partial factor safety-checking formats for codes dealing with the design of elements of building structures. For these, many of the required statistical data have been collected and the necessary theoretical models constructed. The application to structural systems as distinct from structural elements has not received much attention; yet it is well known that there is considerable reserve capacity in many framed structures (see also Chapter 5).

Calibration concepts have also been applied to bridge design codes [e.g. Nowak and Lind, 1979; Flint et al., 1981] although it appears to be a more difficult task owing to the nature of bridge loading, the influence of fatigue and the need to consider stability limit states, all of which are important for bridges. These phenomena have been more difficult to quantify in probabilistic terms and, in some cases, do not easily allow the resistance to be separated from the load effects.

There are also still some difficulties with extending the application of code calibration to some other materials, such as timber and brickwork. For these, calibration exercises have revealed quite high, but varied, values for the safety index implied by current design codes.

For timber design, it is generally the case that the ratio μ_R/R_n (mean to nominal resistance) is much higher (2–3) than for steel or reinforced concrete members, owing to the conservatism of strength rules to allow for the large variability in member properties. As a result, when reasonable target reliabilities are used with accepted loads for calibration purposes, the ϕ factor tends to be greater than unity (as R_n is so low). This is not a desirable result, and adjustment of the strength rules (i.e. R_n) to achieve $\phi \approx 0.8$ would be desirable. The ϕ factor can then be used as a parameter to express workmanship and quality of construction. An essentially similar situation exists for masonry structures [Ellingwood et al., 1980].

The calibration referred to in this chapter has been phrased in terms of "ultimate" limit states. In principle, the calibration concept is also applicable to "serviceability" limit states. For example, the limit state for deflection Δ might be written as

$$Z = \Delta_C - \Delta_A \leqslant 0$$

where Δ_C is a deflection criteria and Δ_A is the deflection due to applied loading. Δ_A can be obtained directly from structural analysis. It would be an uncertain quantity as the loading (and the structural behaviour) is uncertain. The deflection criterion Δ_C might be a constant or an uncertain quantity. The latter is possible since there are difficulties with specification of serviceability criteria because they are associated with subjective human reactions. Relatively few data are available in this area; in any case the variability is likely to be high, with coefficients of variation in the range 0.2–0.5.

In general, there are difficulties in defining appropriate limit states for serviceability requirements, as these are much more dependent on the consequences of exceeding the limit state [Leicester and Beresford, 1977].

9.9 CONCLUSION

This chapter has been concerned with applying the methods of reliability analysis outlined in the previous chapters to derive rational non-probabilistic safety-checking rules for use in so-called "limit state design" codes. The format of the safety-checking rules which has been adopted for such codes is generally of the partial factor format. This was shown in Chapter 1 to be not necessarily mechanically invariant. However, provided that the method of application of the safety-checking rules is clearly set out in the structural design code, the problem of lack of invariance of the safety measure can be largely ignored.

It was sufficient in the present chapter to confine attention to the nominal failure probability for purposes of sizing structural members. The justification for so doing was discussed by reference to Chapter 2. Further, a distinction was made between the nominal failure probability P_{fN} used elsewhere and p_{fC} used in the present chapter. This was considered to be appropriate because the latter probability is obtained to a large extent from predicted information. It therefore reflects a large amount of uncertainty. Such a nominal failure probability cannot rationally be compared with observed rates of failure of structures.

Appendix A: Summary of probability theory

A.1 PROBABILITY

Probability can be considered as a numerical measure of the likelihood that an event occurs relative to a set of alternative events that do not occur. The set of all possible events must be known.

The determination of the probability that an event occurs can be based on:

(1) *a priori* assumptions about the underlying mechanisms governing event(s);
(2) relative frequency of empirical observations in the past;
(3) intuitive or subjective assumptions.

The probability of an event E is denoted $P(E|X)$ where P is the probability operator, and $|X$ denotes the condition "subject to X", where X denotes whatever may be known or assumed in determining $P(E|X)$. Hence any probability depends on the state of knowledge (and ignorance) (or more generally "the state of nature") at the time that the probability is calculated. Seen in this way, all probabilities are conditional (as indicated by $|X$) [e.g. Tribus, 1969]. In many cases $P(E|X)$ will be denoted simply $P(E)$, the state of nature X being understood.

A.2 MATHEMATICS OF PROBABILITY

A.2.1 Axioms

(a) The probability $P(E)$ of an event E is a real non-negative number: $0 \leqslant P(E) \leqslant 1$.

(b) The probability of an inevitable event C is $P(C) = 1.0$. Hence the probability $P(0)$ of an impossible event equals 0.

(c) Addition rule. The probability that either or both of two events E_1 and E_2 occur is

$$P(E_1 \cup E_2) = P(E_1) + P(E_2) - P(E_1 \cap E_2) \tag{A.1}$$

Hence for two *mutually exclusive* events

$$P(E_1 \cup E_2) = P(E_1) + P(E_2) \tag{A.2}$$

A.2.2   Derived results

(d) Multiplication rule. The probability that both events E_1 and E_2 occur is

$$P(E_1 \cap E_2) = P(E_1|E_2)P(E_2) \tag{A.3}$$

If E_1 and E_2 are independent,

$$P(E_1|E_2) = P(E_1) \text{ and } P(E_1 \cap E_2) = P(E_1)\,P(E_2) \tag{A.4}$$

This provides the definition of *independent* events.

(e) If $\overline{E}$ is the event that E does not occur,

$$P(E \cup \overline{E}) = P(E) + P(\overline{E}) = P(C) = 1$$

Therefore

$$P(\overline{E}) = 1 - P(E) \tag{A.5}$$

(f) Conditional probability

$$P(E_1|E_2) = \frac{P(E_1 \cap E_2)}{P(E_2)} \tag{A.3a}$$

follows direction from (A.3).

(g) Total probability theorem. For events E_i, $i = 1, \ldots, n$, mutually exclusive and collectively exhaustive (i.e. covering all possiblities without overlap),

$$P(A) = P(A|E_1)P(E_1) + P(A|E_2)P(E_2) + \ldots + P(A|E_n)P(E_n) \tag{A.6}$$

by virtue of the multiplication rule.

(h) Bayes' theorem. From the multiplication theorem, for a particular event E_i and the event A,

$$P(A \cap E_i) = P(A|E_i)P(E_i) = P(E_i|A)P(A)$$

but $P(A)$ is given by the total probability theorem, so that

$$P(E_i|A) = \frac{P(A|E_i)P(E_i)}{\displaystyle\sum_{i=1}^{n} P(A|E_i)P(E_i)} = \frac{P(A|E_i)P(E_i)}{P(A)} \qquad (A.7)$$

A.3 DESCRIPTION OF RANDOM VARIABLES

In what follows, only the case of continuous random variables will be noted; for discrete random variables the results are exactly analogous with integration being replaced by summation. The probability that the random variable X takes on a value less than or equal to x (a specific value) is given by:

$$P(X \leq x) \equiv F_X(x) = \int_{-\infty}^{x} f_X(\varepsilon)\,d\varepsilon \qquad (A.8)$$

where $F_X(x)$ is the cumulative distribution function of X, and $f_X(x)$ is the probability density function. Obviously $f_X(x) = dF_X(x)/dx$; thus $f_X(x)$ is not a probability, but its local derivative. Specific cases of $f_X(x)$ and $F_X(x)$ are considered in section A.5. Any function satisfying $F_X(-\infty) = 0$, $F_X(+\infty) = 1.0$, $F_X(x) \geq 0$, $f_X(x) \geq 0$, and for which the derivative $dF_X/dx = f_X$ exists is a possible cumulative distribution function. However, in practice, attention is restricted to a restricted set

$$P(a < X \leq b) = \int_{-\infty}^{b} f_X(x)\,dx - \int_{-\infty}^{a} f_X(x)\,dx$$
$$= F_X(b) - F_X(a) \qquad (A.9)$$

A distribution may be described by a number of derived properties, commonly called "moments", without specific reference to either f_X or F_X. Also, for discrete functions, the probability density function is replaced by the probability mass function p_X.

A.4 MOMENTS OF RANDOM VARIABLES

A.4.1 Mean or expected value (first moment).
This is a "weighted average" of all the values that a random variable may take:

$$E(X) \equiv \mu_X = \sum_i x_i p_X(x_i) \; \overset{\text{or}}{=} \; \int_{-\infty}^{\infty} x f_X(x) \, dx \qquad (A.10)$$

It is called the "first moment" since it is the first moment of area of the probability density function. (The mean μ_X is analogous to the centroidal distance of a (beam) cross-section.)

Other central tendency measures are the *mode* which is most probable value, i.e. the value of x for which p_X or f_X is greatest and the *median* which is the value of x for which $F_X(x) = 0.5$, i.e. values above and below the median are equally likely.

A.4.2 Variance and standard deviation (second moment)

The variance of a random variable is a measure of the degree of randomness about the mean

$$E(X - \mu_X)^2 = \text{var}(X) = \sum_i (x_i - \mu_X)^2 p_X(x_i) =$$

$$\overset{\text{or}}{=} \int_{-\infty}^{\infty} (x - \mu_X)^2 f_X(x) \, dx$$

$$= E(X^2) - (\mu_X)^2 \qquad (A.11)$$

which is a useful result. The standard deviation is

$$\sigma_X = [\text{var}(X)]^{1/2} \qquad (A.12)$$

The coefficient of variation is

$$V_X = \frac{\sigma_x}{\mu_X} \qquad (A.13)$$

A.4.3 Bounds on deviations from the mean

For all discrete and continuous random variables there are bounds on the relationship between the standard deviation and the mean. The most well-known is due to Bienayme and Chebychev. For a random variable X, with mean μ_X and standard deviation σ_X,

$$P(|X - \mu_X| \geq k\sigma_X) \leq \frac{1}{k^2} \qquad (A.14)$$

where $k > 0$ is a real number. This is a (weak) general bound on the amount of deviation $|X - \mu_X|$ from the mean relative to the standard deviation. If it is known that $f_X(x)$ has a single peak and has so-called "high order" contact with the x axis at $x = \pm \infty$, the right-hand side of expression (A.14) may be replaced by $1/2.25k^2$. This is the "Camp–Meidall" inequality [Freeman, 1963, p. 30]

A.4.4   Skewness γ_1 (third moment)
A measure of skewness or lack of symmetry of a distribution is given by the third central moment about the mean:

$$E(X-\mu_X)^3 \;=\; \sum_i (x_i-\mu_X)^3 p_X(x_i) \;\overset{\text{or}}{=}\; \int_{-\infty}^{\infty} (x-\mu_X)^3 f_X(x)\,\mathrm{d}x$$

$$\text{(A.15)}$$

$E(X-\mu_X)^3$ will be positive if there is greater dispersion of values of $X \geqslant \mu_X$ than for values of $X < \mu_X$; the sign and magnitude of $E(X-\mu_X)^3$ governs the sign and degree of skewness

$$\gamma_1 \;=\; \frac{E(X-\mu_X)^3}{\sigma_X^3}$$

Positive skewness is indicated by the larger "tail" of the distribution in the positive direction.

A.4.5   Coefficient γ_2 of kurtosis (fourth moment)
The fourth central moment gives a measure of the "flatness" of the distribution. The greater the moment, the "flatter" (less peaked) is the distribution.

$$E(X-\mu_X)^4 \;=\; \sum_i (x_i-\mu_X)^4 p_X(x_i) \;\overset{\text{or}}{=}\; \int_{-\infty}^{\infty} (x-\mu_X)^4 f_X(x)\,\mathrm{d}x$$

$$\text{(A.16)}$$

For a standard normal distribution,

$$\gamma_2 \;=\; \frac{E(X-\mu_X)^4}{\sigma_X^4} \;=\; 3.0$$

The measure is used in statistics only for large samples.

A.4.6   Higher moments
Higher-order moments can be developed. A systematic way of developing moments indirectly employs the "moment-generating function" [e.g. Ang and Tang, 1975]. In general, the set of all moments of a probability density function describes the function exactly; any subset of moments represents an approximation to the probability density function. For some probability density functions, a limited set of moments is sufficient to describe the

function completely. Thus, the normal, or Gaussian distribution, is completely described by its first two moments.

A.5 COMMON UNIVARIATE PROBABILITY DISTRIBUTIONS

A.5.1 Binomial B(*n*, *p*)

It gives the probability of exactly x "successes" in n trials.

The probability mass function is

$$P(X=x) = p_X(x) = \binom{n}{x}p^x(1-p)^{n-x}, x = 1, 2, \ldots, n \quad (A.17)$$

The cumulative distribution function is

$$P(X \leqslant x) = F_X(x) = \sum_{\mu=0}^{x} \binom{n}{\mu}p^\mu(1-p)^{n-\mu}, x = 1, 2, \ldots, n$$

$$(A.18)$$

where

$$\binom{n}{x} = \frac{n!}{x!(n-x)!} \quad (A.19)$$

is the binomial coefficient.

The parameters are the number n of independent trials and the probability p of success per trial ($=$ constant).

The moments are

$$E(X) = \mu_X = np \quad (A.20)$$

$$\text{var}(X) = \sigma_x^2 = np(1-p) \quad (A.21)$$

It is applicable where there are only two discrete alternatives per
independent trial, with probability p and $1-p$ respectively, with $p =$
constant. The "multinomial" distribution applies where there are more than
two discrete alternatives possible.

It has the useful property that

$$B(n_1,p) + B(n_2,p) = B(n_1+n_2,p) \quad (A.22)$$

A.5.2 Geometric G(*p*)

It gives the probability that the nth trial is a success given that the first $n-1$
trials are failures.

The probability mass function is

$$P(N=n) = p_N(n) = (1-p)^{n-1}p \quad n = 1, 2, 3, .. \quad (A.23)$$

The cumulative distribution function is

$$P(N \leqslant n) = F_N(n) = \sum_{i=1}^{n}(1-p)^{i-1}p = 1-(1-p)^n \qquad \text{(A.24)}$$

The parameters are the number n of independent trials and the probability p of success per trial ($=$ constant).
The moments are

$$E(N) = \mu_N = \frac{1}{p} \qquad \text{(A.25)}$$

$$\text{var}(N) = \sigma_N^2 = \frac{1-p}{p^2} \qquad \text{(A.26)}$$

For applications, see those for the binomial distribution. It is assumed that the trials are independent. A case in which this is not so occurs in sampling without replacement. The corresponding distribution is the hypergeometric $HG(p)$ [Ang and Tang, 1975, p. 127; Freeman, 1963, p. 113].

A.5.3 Negative binomial $NB(k, p)$
It gives the probability that the kth occurrence of a success occurs at the tth trial.
The probability mass function is

$$P(T=t) = p_T(t) = \binom{t-1}{k-1}(1-p)^{t-k}p^k, \, t = k, k+1, \ldots \text{ (A.27)}$$

The parameters are the number k of successes, the probability p of success per trial ($=$ constant) and the number t of trials before k successes ($t \geqslant k$).
The moments are

$$E(T) = \mu_T = \frac{k}{p} \qquad \text{(A.28)}$$

$$\text{var}(T) = \sigma_T^2 = \frac{k(1-p)}{p^2} \qquad \text{(A.29)}$$

The number of trials can also be interpreted as the number of time units. It is also known as the Pascal distribution. Other forms also exist.

A.5.4 Poisson $PN(\upsilon t)$
It gives the probability of a number of occurrences of a random event in a given (time) interval given that the mean rate of occurrences is known.

The probability mass function is

$$P(X_t = x) = p_X(x) = \frac{(vt)^x}{x!} s^{-vt} \qquad\qquad (A.30)$$

The cumulative distribution function is

$$P(X_t \leq x) = F_X(x) = \sum_{r=0}^{x} \frac{(vt)^r}{r!} e^{-vt} \qquad\qquad (A.31)$$

The parameters are the mean occurrence rate (per unit time or space units) v, the time or space interval t, the average number λ ($\equiv vt$) of events in t and the number X_t of occurrences in t.

The moments are

$$E(X) = \mu_x = vt \qquad\qquad (A.32)$$

$$\text{var}(X) = \sigma_X^2 = vt \qquad\qquad (A.33)$$

The Poisson distribution results from (approximates) the Binomial distribution $B(n, P)$ as $n \to \infty$ and p is small. The Poisson distribution replaces the time interval for the number of trials. The relationship is given by $np = vt$ in the limit as $n \to \infty$, $p \to 0$. In practice, this is satisfied if, say, $n = 50$ for $p < 0.10$, or $n = 100$ for $p < 0.05$. It has wide application in its own right and not merely as an approximation to the binomial distribution. When the Poisson distribution is used in terms of time or space units, with independence of events, a Poisson process results (see Chapter 6). Note that $(vt)^0/0! = 1$.

It has the useful property that

$$PN(\lambda_1) + PN(\lambda_2) = PN(\lambda_1 + \lambda_2) \qquad\qquad (A.34)$$

A.5.5 Exponential $EX(v)$

It gives, for events occurring according to a Poisson process, the probability of the time to the first occurrence of an event.

The probability density function is

$$P(T = t) = f_T(t) = v e^{-vt}, \qquad\qquad t \geq 0 \quad (A.35)$$

The cumulative distribution function is

$$P(T \leq t) = F_T(t) = 1 - e^{-vt}, \qquad\qquad t \geq 0 \quad (A.36)$$

or

$$P(T \geqslant t) = e^{-\upsilon t}$$

The parameters are the mean occurrence rate υ and the time or space interval t.

The moments are

$$E(T) = \mu_T = \frac{1}{\upsilon} = \overline{\Delta t} \qquad (A.37)$$

where $\overline{\Delta t}$ is the average time between arrivals or mean life

$$\text{var}(T) = \sigma_T^2 = \left(\frac{1}{\upsilon}\right)^2 = (\overline{\Delta t})^2 \qquad (A.38)$$

The exponential distribution is the continuous analogue of the geometric distribution. Since a Poisson process is stationary by definition, any starting time can be used, and hence T can refer to "inter-arrival time", which is therefore exponentially distributed.

A.5.6   Gamma $GM(k, \upsilon)$ [Fig. A.1]

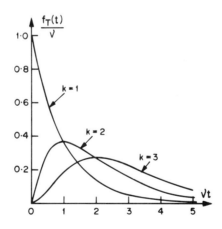

Fig. A.1 — Gamma probability density function.

It gives, for events occurring according to a Poisson process, the probability of the time T to the kth occurrence of an event. More generally, k need not be an integer.

The probability density function is

$$P(T=t) = f_T(t) = \frac{\upsilon(\upsilon t)^{k-1}}{\Gamma(k)} e^{-\upsilon t}, \qquad t \geqslant 0 \tag{A.39}$$

The cumulative distribution function is

$$P(T<t) = F_T(t) = 1 - \sum_{x=0}^{k-1} \frac{(\upsilon t)^x}{x!} e^{-\upsilon t}, \ t \geqslant 0, \ k \text{ integer} \tag{A.40}$$

$$= \frac{\Gamma(k, \upsilon t)}{\Gamma(k)} \qquad\qquad t \geqslant 0, \text{ all } k$$

where

$$\Gamma(k) = \int_0^\infty e^{-u} u^{k-1} du = (k-1)! \qquad \text{for } k \text{ integer} \tag{A.41}$$

and

$$\Gamma(k,x) = \int_0^x e^{-u} u^{k-1} du \tag{A.42}$$

The parameters are the mean occurrence rate υ and the time or space interval t.

The moments are

$$E(T) = \mu_T = \frac{k}{\upsilon} \tag{A.43}$$

$$\text{var}(T) = \sigma_T^2 = \frac{k}{\upsilon^2} \tag{A.44}$$

$$E(T-\mu_T)^3 = 2k^{-1/2}\sigma_T^{-3} \text{ (skewness)} \tag{A.45}$$

The gamma function $\Gamma(k)$ is the generalization of the factorial for non-integer k. It is tabulated as the "incomplete gamma function" $\Gamma(k, \upsilon t)$ [Ang and Tang, 1975, p. 127; Benjamin and Cornell, 1970, p. 247].

For k integer, the distribution is also known as the "Erlang" distribution, and is the continuous analogue of the negative binomial distribution.

It has the useful property that, if X_i is $GM(k_i, \upsilon)$, then

$$\sum_i^m X_i = GM\left(\sum_i^m k_i, \upsilon\right) \tag{A.46}$$

A.5.7 Normal (Gaussian) $N(\mu, \sigma)$ (Fig. A.2)

It gives a distribution which arises frequently in practice as a limiting case of other probability distributions. It is also a reasonable description for many physical observations.

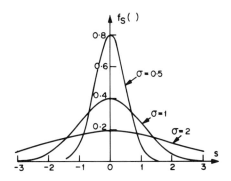

Fig. A.2 — Normal probability density function.

The probability density function is

$$f_X(x) = \frac{1}{(2\pi)^{1/2}\sigma_X} \exp\left[-\frac{1}{2}\left(\frac{x-\mu_X}{\sigma_X}\right)^2\right], \quad -\infty \leqslant x \leqslant \infty \quad (A.47)$$

The cumulative distribution function is

$$P(X \leqslant x) = F_X(x) =$$

$$= \frac{1}{(2\pi)^{1/2}} \int_{-\infty}^{s} e^{-\frac{1}{2}v^2} dv, \quad -\infty \leqslant x \leqslant \infty \quad (A.48)$$

where $s = (x-\mu_X)/\sigma_X$.

There is no simple expression for $F_X(x)$. However, there are many approximations (see below).

The parameters are the mean value μ_x and the standard deviation σ_x. The moments are

$$E(X) = \mu_X \qquad (A.49)$$

$$\text{var}(X) = \sigma_X^2 \qquad (A.50)$$

$$E(X-\mu_X)^3 = 0 \text{ (no skewness)} \qquad (A.51)$$

$$E(X-\mu_X)^4 = \frac{3}{\sigma_X^4} \qquad (A.52)$$

The standard normal distribution $N(0,1)$ is usually tabulated for both $f_X(x)$ and $F_X(x)$ (see Appendix C). The variate is then $s = (x-\mu_X)/\sigma_X$ and $F_S(s)$ is commonly denoted $\Phi(s)$, while $f_S(s)$ is sometimes denoted $\phi(s)$. Note that, if X is $N(\mu_X, \sigma_X)$, then

$$f_X(x) = \frac{1}{\sigma_X} \phi\left(\frac{x - \mu_X}{\sigma_X}\right), \quad F_X(x) = \Phi\left(\frac{x - \mu_X}{\sigma_X}\right)$$

It has the useful properties that

$$\Phi(-s) = 1 - \Phi(s) \qquad\qquad\qquad\qquad\qquad (A.53)$$

$$s = \Phi^{-1}(p) = -\Phi^{-1}(1 - p) \qquad\qquad\qquad (A.54)$$

$$P(a < x \leq b) = \Phi\left(\frac{b - \mu_X}{\sigma_X}\right) - \Phi\left(\frac{a - \mu_X}{\sigma_X}\right) \qquad (A.55)$$

If $Y = \sum_i X_i$ where X_i are independent $N(\mu_{Xi}, \sigma_{Xi})$,

$$\mu_Y = \sum_i \mu_{Xi} \qquad\qquad\qquad\qquad\qquad\qquad (A.56)$$

$$\sigma_Y^2 = \sum_i \sigma_{Xi}^2 \quad (\text{cf. section A.11.1(a)}) \qquad\qquad (A.57)$$

Approximate expressions [Abramowitz and Stegun, 1966; Hastings, 1955] are as follows:

(i) $\Phi(-\beta) \approx \dfrac{1}{\beta(2\pi)^{1/2}} e^{-\frac{1}{2}\beta^2}$ (A.58)

(ii) $\Phi(s) = P(s \leq s) = 1 - \dfrac{1}{(2\pi)^{1/2}} e^{-\frac{1}{2}s^2}\left[\displaystyle\sum_{i=1}^{5} b_i t^i + \varepsilon(s)\right]$ (A.59a)

where
$$t = (1 + 0.231\,641\,9s)^{-1}$$
$$|\varepsilon(s)| < 7.5 \times 10^{-8}$$
$$b_1 = 0.319\,381\,530$$
$$b_2 = 0.356\,563\,782$$
$$b_3 = 1.781\,477\,937$$
$$b_4 = 1.821\,255\,978$$
$$b_5 = 1.330\,274\,429$$

(iii) $\Phi(s) = P(S \leq s) = 1 - 0.5\left(1 + \displaystyle\sum_{i=1}^{6} d_i x^i\right)^{-16} + \varepsilon(s)$ (A.59b)

where

$$|\varepsilon(s)| < 1.5 \times 10^{-7}$$
$$d_1 = 0.049\,867\,3470$$
$$d_2 = 0.021\,141\,0061$$
$$d_3 = 0.003\,277\,6263$$
$$d_4 = 3.800\,36 \times 10^{-5}$$
$$d_5 = 4.889\,06 \times 10^{-5}$$
$$d_6 = 0.538\,30 \times 10^{-5}$$

(iv) for $\beta \geqslant 1$ [Rosenblueth, 1985b],

$$\Phi(-\beta) \approx \left[\frac{\beta}{1+\beta^2} + \left(\sum_{i=0}^{5} a_i \beta^i\right)^{-1}\right]\phi(\beta) + \varepsilon(\beta) \qquad (A.59c)$$

where

$$|\varepsilon(\beta)| < 5 \times 10^{-5}$$

$$a_0 = \left(\frac{2}{\pi}\right)^{0.5}$$

$$a_1 = 1.280$$
$$a_2 = 1.560$$
$$a_3 = 1.775$$
$$a_4 = 0.584$$
$$a_5 = 0.427$$

A.5.8 Central limit theorem

This famous theorem states that the sum of a large number of random variables approaches the normal distribution, irrespective of the individual distributions of the random variables [Freeman, 1963, p. 181; Benjamin and Cornell, 1970, p. 251].

A.5.9 Lognormal $LN(\lambda, \xi)$ (Fig. A.3)

In this distribution, the natural logarithm of the random variable X, rather than X itself, has a normal distribution.

The probability density function is

$$f_X(x) = \frac{1}{(2\pi)^{1/2}x\varepsilon}\exp\left[-\frac{1}{2}\left(\frac{\ln x - \lambda}{\varepsilon}\right)^2\right], \qquad 0 \leqslant x < \infty \qquad (A.60)$$

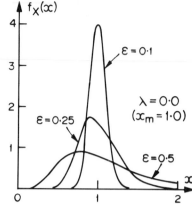

Fig. A.3 — Lognormal probability density functions.

The cumulative distribution function is

$$F_X(x) = \int_{-\infty}^{u} f_X(u)\,du = \Phi\left(\frac{\ln x - \lambda}{\varepsilon}\right)$$
$$\text{(no explicit form)} \qquad \text{(A.61)}$$

The parameters are

$$\lambda = E(\ln X) = \text{mean of } \ln(X) = \mu_{\ln X} \qquad \text{(A.61a)}$$
$$\xi^2 = \text{var}(\ln X) = \sigma_{\ln x}^2 \qquad \text{(A.61b)}$$

The moments [e.g. Ang and Tang, 1975, p. 105] are

$$E(X) = \mu_X = \exp(\lambda + \tfrac{1}{2}\xi^2) \qquad \text{(A.62)}$$
$$\text{var}(X) = \sigma_X^2 = \mu_X^2(e^{\xi^2} - 1) \qquad \text{(A.63)}$$

Because

$$P(a < x \leq b) = \Phi\left(\frac{\ln b - \lambda}{\xi}\right) - \Phi\left(\frac{\ln a - \lambda}{\xi}\right) \qquad \text{(A.64)}$$

this allows use of standard normal tables for evaluation of probabilities with X lognormal.

The lognormal distribution has the following useful properties.

(1) From equation (A.62),

$$\lambda = \mu_{\ln X} = \ln \mu_X - \tfrac{1}{2}\xi^2$$

or

$$\lambda = \ln x_{\mathrm{m}} = \ln\left[\frac{\mu_X}{(1 + V_X^2)^{1/2}}\right]$$

or

$$x_{\mathrm{m}} = \mu_X e^{\frac{1}{2}\varepsilon^2}$$

θηερε x_{m} is the median of x $(P(X \leqslant x_{\mathrm{m}}) = 0.5)$.

(2) From equation (A.63),

$$\xi^2 = \sigma_{\ln X}^2 = \ln\left(1 + \frac{\sigma_X^2}{\mu_X^2}\right) = \ln(1 + V_X^2) \approx V_X^2 \qquad \text{for } V \leqslant 0.30$$

where V_X is the coefficient of variation.

(3) Also

$$x_{\mathrm{m}} = \frac{\mu_X}{(1 + V_X^2)^{1/2}}$$

so that $x_{\mathrm{m}} \leqslant \mu_X$.

(4) *If* $Y = \prod_{i=1}^{n} X_i$ *where* X_i *are lognormal,* $LN(\lambda_i, \xi_i)$, *then*

$$\mu_{\ln Y} = \sum_{i=1}^{n} \mu_{\ln X_i} \quad \text{or} \quad y_{\mathrm{m}} = \prod_{i=1}^{n} x_{\mathrm{m}_i} \quad \text{and} \quad \sigma_{\ln Y}^2 = \sum_{i=1}^{n} \sigma_{\ln X_i}^2 \qquad \text{(A.65)}$$

A.5.9.1 *Central limit theorem for lognormal*

Because of the relationship between the normal and lognormal distributions, the central limit theorem extended to the lognormal distribution states that the *product* of a large number of random variables approaches the lognormal distribution, irrespective of the individual distributions of the random variables.

A.5.10 Beta *BT(a, b, q, r)* (Fig. A.4)

Although the beta distribution may arise from physical considerations, its chief advantage is its great flexibility in being able to be fitted to observed data. It is given in various, essentially equivalent forms.

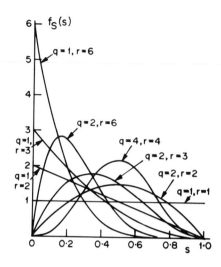

Fig. A.4 — Beta distribution probability density functions for different parameters.

The probability density function is

$$f_X(x) = \frac{1}{\beta(q,r)} \frac{(x-a)^{q-1}(b-x)^{r-1}}{(b-a)^{q+r-1}}, \qquad a \leqslant x \leqslant b \quad \text{(A.66)}$$

$$= 0, \qquad\qquad\qquad\qquad \text{elsewhere}$$

where the Beta function $\beta(q,r)$ is given by

$$\beta(q,r) = \int_0^1 x^{q-1}(1-x)^{r-1}dx$$

$$= \frac{\Gamma(q)\Gamma(r)}{\Gamma(q+r)}$$

$$= \frac{(q-1)!(r-1)!}{(q+r-1)!} \qquad \text{for } q,r \text{ integer}$$

If $a = 0$, $b = 1$ (standard beta distribution),

$$f_S(s) = \frac{1}{\beta(q,r)}s^{q-1}(1-s)^{r-1}, \qquad 0 \leqslant s \leqslant 1 \quad \text{(A.67)}$$

$$= 0, \qquad\qquad\qquad\qquad \text{elsewhere}$$

The cumulative distribution function

$$F_S(s) = \frac{\beta_s(q,r)}{\beta(q,r)}, \qquad\qquad 0 \leqslant s \leqslant 1 \quad (A.68)$$

where the incomplete beta function $\beta_s(q,r)$ is given by

$$\beta_s(q,r) = \int_0^s y^{q-1}(1-y)^{r-1}\,dy$$

The parameters a and b describe the intervals for the general beta distribution.

The moments are

$$E(x) = \mu_x = a + \frac{q(b-a)}{q+r} \qquad\qquad (A.69)$$

$$\mathrm{var}(X) = \sigma_x^2 = \frac{qr(b-a)^2}{(q+r)^2(q+r+1)} \qquad\qquad (A.70)$$

The coefficient of skewness is

$$\gamma_1 = \frac{2(r-q)}{(q+r)(q+r+2)\sigma_x} \qquad\qquad (A.71)$$

The *incomplete Beta function ratio* $\beta_s(q,r)/\beta(q,r)$ has been tabulated [Pearson and Johnson, 1968]. If q,r are both integral, $BT(0,1,q,r)$ is binomially distributed such that

$$f_S(s) = (q+r-1)p_X(x) \qquad\qquad (A.72)$$

where $p_X(x)$ is binomially distributed as $B(q+r-2,s)$ with $x = q-1$.

A special case of the general beta distribution is the *rectangular* or *uniform* distribution $BT(a,b,1,1) = R(a,b)$:

$$F_X(x) = \frac{1}{b-a} \qquad\qquad a < x < b \quad (A.73)$$

$$= 0, \qquad\qquad\qquad \text{elsewhere}$$

$$F_X(x) = \frac{x-a}{b-a}, \qquad\qquad a < x < b \quad (A.74)$$

$$= 0, \qquad\qquad\qquad x \leqslant a$$

$$= 1, \qquad\qquad\qquad x \geqslant b$$

with moments

$$\mu_X = \frac{(a+b)}{2}, \qquad \sigma_X^2 = \frac{(b-a)}{12} \qquad\qquad (A.75)$$

A.5.11   Extreme value distribution type I $EV\text{-}I(\mu, \alpha)$ (Fig. A.5)

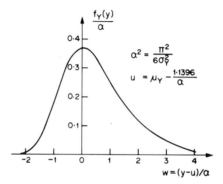

Fig. A.5 — Extreme value distribution type I (Gumbel).

The limiting (asymptotic) distribution of the largest (smallest) of n random variables X_i as $n \to \infty$. The distribution of the X_i must be of the form $F_X(x) = 1 - e^{-g(x)}$, $f_X(x) = e^{-g(x)}$, with $dg/dx > 0$. The normal, gamma and exponential distributions are of this type. The distribution Y, the *largest* of many independent X_i, is given asymptotically [Gumbel, 1958, pp. 159, 169], for the probability density function, by

$$f_Y(y) = \alpha \exp[-\alpha(y - u) - e^{-\alpha(y - u)}], \quad -\infty < y < \infty \qquad (A.76)$$

and, for the cumulative distribution function, by

$$F_Y(y) = \exp[-e^{-\alpha(y - u)}], \qquad\qquad -\infty < y < \infty \qquad (A.77)$$

The parameters are the mode u of the distribution and α which is a measure of the dispersion of the distribution. α^{-1} is sometimes known as the "slope" of the distribution (obtained when plotting the distribution on so-called "Gumbel" paper).

Both u and α may be obtained, via the moments, from curve fitting to observed data.

The moments are

$$E(Y) = \mu_Y = u + \frac{\gamma}{\alpha} \tag{A.78}$$

where $\gamma = 0.577\,215\quad 664\quad 9\ldots$ is Euler's constant.

$$\mathrm{var}(Y) = \sigma_Y^2 = \frac{\pi^2}{6\alpha^2} \tag{A.79}$$

The skewness is

$$\gamma_1 = 1.1396 \tag{A.80}$$

(i.e. independent of u and α).

The following points should be noted in applications using this distribution.

(1) In practice, the X_i of the underlying population need not be completely independent nor completely identical [Gumbel, 1958, p. 166]. Also it may be difficult to determine the appropriate underlying distribution of the X_i, and convergence to the asymptotic distribution may be slow. Nevertheless extreme value distributions are useful for fitting to experimental data even where the underlying mechanisms are not fully understood.

(2) The distribution is tabulated by the National Bureau of Standards (NBS) (1953) in terms of a reduced variate $W = (Y - u)\alpha$ for which $u = 0$, $\alpha = 1$ and $F_W(w) = \mathrm{e}^{-\mathrm{e}^{-w}}$. The probability density function and cumulative distribution function in terms of Y are recovered from

$$f_Y(y) = \alpha f_W[(y - u)\alpha] \tag{A.81}$$

$$F_Y(y) = F_W[(y - u)\alpha] \tag{A.82}$$

(3) This distribution is also termed the "double exponential", "Gumbel" or "Fisher–Tippett Type I" distribution.

The complementary result is as follows. The distribution Y^S of the *smallest* of many independent X_i is given asymptotically by, for the probability density function,

$$f_{Y^S}(y^S) = \alpha \exp[\alpha(y^S - u) - \mathrm{e}^{\alpha(y^S - u)}], \quad -\infty < y^S < \infty \tag{A.83}$$

and, for the cumulative distribution function,

$$F_{Y^S}(y^S) = 1 - \exp[-\mathrm{e}^{\alpha(y^S - u)}], \quad -\infty < y^S < \infty \tag{A.84}$$

with

$$\mu_{Y^S} = u - \frac{\gamma}{\alpha} \tag{A.85}$$

$$\sigma^2_{Y^S} = \frac{\pi^2}{6\alpha^2} \tag{A.86}$$

$$\gamma_1 = -1.1396 \tag{A.87}$$

The distribution for Y^S is related to that for the reduced variable W above, by:

$$f_{Y^S}(y^S) = \alpha f_W[-(y^S - u)\alpha] \tag{A.88}$$

$$F_{Y^S}(y^S) = 1 - F_W[-(y^S - u)\alpha] \tag{A.89}$$

Hence, the tabulated results for the reduced variable W can be applied.

The extreme value distribution Y^S for the minimum has less practical application than that for Y; the Weibull distribution (type III) is more commonly used for smallest values.

A.5.12 Extreme value distribution type II $EV\text{-}II(u, k)$ (Fig. A.6)

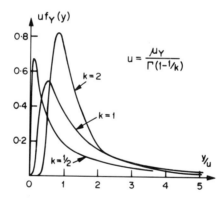

Fig. A.6 — Extreme value distribution type II (Frechet).

The limiting (asymptotic) distribution of the largest of n random variables X_i as $n \to \infty$. The distribution of the X_i must be of the form $F_X(x) = 1 - Ax^{-k}$, $x \geqslant 0$, $A = $ constant [Gumbel, 1958, pp. 160, 255–272]. Typically of this form is the Pareto distribution, and the Cauchy distribution for $x \geqslant 0$.

The probability density function is

$$f_Y(y) = \frac{k}{y}\left(\frac{u}{y}\right)^k e^{-(u/y)^k}, \qquad\qquad y \geqslant 0 \quad \text{(A.90)}$$

The cumulative distribution function is

$$F_Y(y) = e^{-(u/y)^k}, \qquad\qquad y \geqslant 0 \quad \text{(A.91)}$$

The parameters are the characteristic value u of the distribution (median $> u >$ mode; median $\approx u \approx$ mode for $k > 4$) and k which is a dimensionless inverse measure of the dispersion of the distribution.
The moments are

$$E(Y) = \mu_Y = u\Gamma\left(1 - \frac{1}{k}\right), \qquad\qquad k > 1 \quad \text{(A.92)}$$

$$\text{var}(Y) = \sigma_Y^2 = u^2\left[\Gamma\left(1 - \frac{2}{k}\right) - \Gamma^2\left(1 - \frac{1}{k}\right)\right], \quad k > 2 \quad \text{(A.93)}$$

so that

$$V_Y^2 = \frac{\mu_Y^2}{\sigma_y^2} = \frac{\Gamma(1 - 2/k)}{\Gamma^2(1 - 1/k)} - 1 \qquad\qquad \text{(A.94)}$$

Moments of order $l \geqslant k$ do not exist; this complicates the estimation of parameters u and k.
The following points should be noted in applications using this distribution.

(1) If it is known that $k > 2$, equation (A.94) for V_Y^2 may be used to evaluate k, and u may be evaluated from equation (A.92) for μ_Y.
(2) The type II distribution for Y ($EV\text{-}II(u, k)$) may be transformed to the type I for Z ($EV\text{-}I(u, \alpha)$) by letting $Z = \ln Y$. Then

$$f_Y(y) = \frac{1}{y} f_Z(\ln y) \qquad\qquad \text{(A.95)}$$

$$F_Y(y) = F_Z(\ln y) \qquad\qquad \text{(A.96)}$$

$$\alpha = k \qquad\qquad \text{(A.97)}$$

Hence, in terms of the reduced variable W, which is tabulated (see section A.5.11),

$$f_F(y) = \frac{k}{y} f_W[(\ln y - \ln u)k] \qquad\qquad \text{(A.98)}$$

$$F_Y(y) = F_W[\ln y - \ln u)k] \tag{A.99}$$

(3) The above properties hold for $y \geq 0$. A more general result, for $y \geq \varepsilon$, $\varepsilon \geq 0$, can be obtained by linear transformation by writing $u - \varepsilon$ for u and $y - \varepsilon$ for y.
(4) This distribution is also termed the "Frechet" distribution.
(5) The distribution for the smallest extreme values is of no practical interest.

The distribution has a useful property in that the underlying distributions X_i for the type II distributions typically have longer tails ($x \geq 0$) than those for the type I distribution. This is also evident for the corresponding asymptotic distributions.

A.5.13   Extreme value distribution type III $EV\text{-}III(\varepsilon,u,k)$ (Fig. A.7)

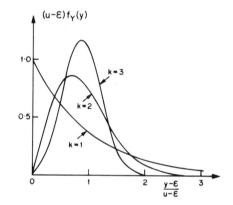

Fig. A.7 — Extreme value distribution type III (Weibull).

The limiting (asymptotic) distribution of the largest (smallest) of n random variables X_i as $n \rightarrow \infty$, with X_i limited *in the tail of interest* to some maximum (minimum) value w (or ε), and X_i having a distribution of general form

$$F_X(x) = 1 - A(w - x)^k, \qquad x \leq w, \ k > 0, \ A = \text{constant}$$

The rectangular ($k = 1$), triangular ($k = 2$) and the Gamma distribution ($\varepsilon = 0$) are of this form. The distribution Y^L *of the largest* of many independent X_i is given by [Gumbel, 1958, p. 275], for the probability density function,

$$f_{Y^L}(y^L) = \frac{k}{w-u}\left(\frac{w-y^L}{w-u}\right)^{k-1} F_{Y^L}(y^L), \qquad y \leqslant w \quad \text{(A.100)}$$

and, for the cumulative distribution function,

$$F_{Y^L}(y^L) = \exp\left[-\left(\frac{w-y^L}{w-u}\right)^k\right], \qquad y \leqslant w \quad \text{(A.101)}$$

More useful is the distribution Y of the *smallest* of many (independent) X_i [Gumbel, 1958, p. 277]; the probability density function is

$$F_Y(y) = P(Y \leqslant y) = 1 - P_Y(y). \qquad y \geqslant \varepsilon \quad \text{(A.102)}$$

where

$$P_Y(y) = \exp\left[-\left(\frac{y-\varepsilon}{u-\varepsilon}\right)^k\right], \qquad y \geqslant \varepsilon \quad \text{(A.103)}$$

which equals the probability of a value Y larger than y, i.e. $P(Y > y)$. Also,

$$f_Y(y) = \frac{dF_Y(y)}{dy} = \frac{k}{u-\varepsilon}\left(\frac{y-\varepsilon}{u-\varepsilon}\right)^{k-1} P_Y(y), \qquad y \geqslant \varepsilon \quad \text{(A.104)}$$

The parameters are the minimum value ε of X_i (and hence Y), the characteristic value u of the distribution (which converges to μ_Y as $k \to \infty$) and the "scale parameter" $1/k$ (usually $k > 1$).

The moments are

$$E(y) = \mu_Y = \varepsilon + (u-\varepsilon)\Gamma\left(1 + \frac{1}{k}\right) \qquad \text{(A.105)}$$

$$\text{var}(y) = \sigma_Y^2 = (u-\varepsilon)^2\left[\Gamma\left(1 + \frac{2}{k}\right) - \Gamma^2\left(1 + \frac{1}{k}\right)\right] \qquad \text{(A.106)}$$

The following point should be noted in applications using this distribution.

(1) Estimation of the parameters ε, u and k is not generally straightforward. If the underlying distribution in X_i is known, k is known, and ε and u can be estimated from the estimates for μ_Y and σ_Y^2. Otherwise, k may be estimated from sample skewness [Gumbel 1958, p. 289] or u may be estimated from order statistics [Gumbel 1958, p. 291]. If the lower limit ε

is known, or is zero, then u and k can be evaluated from equations (A.105) and (A.106) for μ_Y and σ_Y^2 by writing y for $y - \varepsilon$ and hence

$$\mu_Y = u\Gamma\left(1 + \frac{1}{k}\right)$$

$$\sigma_Y^2 = u^2\left[\Gamma\left(1 + \frac{2}{k}\right) - \Gamma^2\left(1 + \frac{1}{k}\right)\right]$$

and

$$1 + V_Y^2 = \frac{\Gamma(1 + 2/k)}{\Gamma^2(1 + 1/k)} \quad \text{or} \quad k \approx V_Y^{-1.09}$$

all of which can be estimated from sample data [Gumbel, 1958, p. 293]. However, the procedure is cumbersome [see also Mann et al., 1974, p. 189].

(2) The distribution $F_Y(y)$ is pseudo-symmetric for $3.2 < k < 3.7$.

(3) If Y is $EV\text{-}III(\varepsilon, u, k)$ for smallest values, then $Z = \ln(Y - \varepsilon)$ is $EV\text{-}I[\ln(u - \varepsilon), k]$ for smallest values. This enables the third extreme value distribution to be evaluated using the tables for $EV\text{-}I$ (largest) in terms of the reduced variable W:

$$F_Y(y) = 1 - F_W\{-k[\ln(y - \varepsilon) - \ln(u - \varepsilon)]\}, \quad y \geqslant \varepsilon \quad \text{(A.107)}$$

$$f_Y(y) = \frac{k}{y - \varepsilon} f_W\left[-k\ln\left(\frac{y - \varepsilon}{u - \varepsilon}\right)\right], \quad y \geqslant \varepsilon \quad \text{(A.108)}$$

(4) The distribution $P_Y(y)$ is also known as the Weibull distribution.

(5) If $\varepsilon = 0$, $k = 2$, the distribution is also known as the Rayleigh distribution:

$$f_Y(y) = \frac{y}{\sigma_Y^2} \exp\left(-\frac{y^2}{2\sigma_Y^2}\right) \quad \text{(A.102a)}$$

$$F_Y(y) = 1 - \exp\left(-\frac{y^2}{2\sigma_Y^2}\right) \quad \text{(A.103a)}$$

A.6   JOINTLY DISTRIBUTED RANDOM VARIABLES

A6.1   Joint probability distribution.

If an event is the result of two (or more) continuous random variables, X_1 and X_2 say, the probabilities that the event occurs for given values of x_1 and x_2 are described by the joint cumulative distribution function

$$F_{X_1 X_2}(x_1, x_2) = P[(X_1 \leqslant x_1) \cap (X_2 \leqslant x_2)] \geqslant 0$$

$$= \int_{-\infty}^{x_1} \int_{-\infty}^{x_2} f_{X_1 X_2}(u, v) \, du \, dv \qquad (A.109)$$

where $f_{X_1 X_2}(x_1, x_2) \geqslant 0$ is the joint probability density function.
Evidently, if the partial derivatives exist,

$$f_{X_1 X_2}(x_1, x_2)$$

$$\equiv \lim_{\delta x_1, \delta x_2 \to 0} \{(P[x_1 < X_1 \leqslant x_1 + \delta x_1) \cap (x_2 < X_2 \leqslant x_2 + \delta x_2)]\}$$

$$= \frac{\partial^2 F_{X_1 X_2}(x_1, x_2)}{\partial x_1 \partial x_2} \qquad (A.110)$$

Also,

$$F_{X_1 X_2}(-\infty, -\infty) = 0 \qquad (A.111)$$
$$F_{X_1 X_2}(-\infty, y) \quad = 0 \text{ and vice versa in } (x_1, x_2) \qquad (A.112)$$
$$F_{X_1 X_2}(\infty, y) \quad = F_{X_2}(y) \text{ and vice versa in } (x_1, x_2) \qquad (A.113)$$
$$F_{X_1 X_2}(\infty, \infty) \quad = 1.0 \qquad (A.114)$$

The last expression is equivalent to stating that the volume under $f_{X_1 X_2}$ is unity.

For discrete random variables, analogous expressions apply.

A.6.2 Conditional probability distributions

If the probability that $(x_1 < X_1 \leqslant x_1 + \delta x_1)$ is a function of X_2, we can write

$$\lim_{\delta x_1 \delta x_2 \to 0} \{P[(x_1 < X_1 \leqslant x_1 + \delta x)|(x_2 < X_2 \leqslant x_2 + \delta x_2)]\}$$

$$\equiv f_{X_1 | X_2}(x_1 | x_2) \qquad (A.115)$$

and according to equation (A.3), reading f as a probability over the infinitesimal region $(\delta x_1, \delta x_2)$,

$$f_{X_1 | X_2}(x_1 | x_2) = \frac{f_{X_1 X_2}(x_1 | x_2)}{f_{X_2}(x_2)} \qquad (A.116)$$

By analogy to equation (A.4), if X_1 and X_2 are independent,

$$f_{X_1 X_2}(x_1, x_2) = f_{X_1}(x_1) f_{X_1}(x_2) \qquad (A.117)$$

A.6.3 Marginal probability distributions

A marginal probability density function may be obtained from the joint density function by integrating over the other variables, i.e. by invoking the total probability theorem (A.6):

$$f_{X_1}(x_1) = \int_{-\infty}^{\infty} f_{X_1|X_2}(x_1|x_2) f_{X_2}(x_2)\,dx_2$$

$$= \int_{-\infty}^{\infty} f_{X_1X_2}(x_1, x_2)\,dx_2 \qquad (A.118)$$

by virtue of (A.116).

If X_1 and X_2 are *independent*, the conditional and marginal distributions are identical, so that $f_{X_1|X_2} = f_{X_1}, f_{X_1X_2} = f_{X_1}f_{X_2}$, etc. In general, the relationship between $f_{X_1X_2}, f_{X_1|X_2}$ and f_{X_1}, etc., takes the form shown in Fig. A.8.

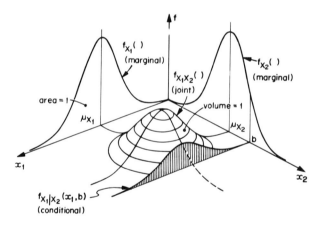

Fig. A.8 — Joint, marginal and conditional probability density functions.

Multivariate distributions and concepts follow directly by extending the above relationships.

A.7   MOMENTS OF JOINTLY DISTRIBUTED RANDOM VARIABLES

The concept of moments may be extended to an event which depends on two (or more) random variables. Let these variables be the jointly distributed random variables X_1 and X_2.

A.7.1   Mean

$$\mu_{X_1} \equiv E(X_1) = \int_{-\infty}^{\infty} \int_{-\infty}^{\infty} x_1 f_{X_1 X_2}(x_1, x_2) \, dx_1 \, dx_2$$

$$= \int_{-\infty}^{\infty} \mu_{X_1|X_2} f_{X_2}(x_2) \, dx_2 \qquad (A.119)$$

This is the first (marginal) moment of X_1, or, equivalently, the mean value of X_1 over all X_2. The fact that this is so can also be seen from the first double-integral term, which can be rewritten as

$$\int_{-\infty}^{\infty} x_1 \left[\int_{-\infty}^{\infty} f_{X_1 X_2}(x_1, x_2) \, dx_2 \right] dx_1 = \int_{-\infty}^{\infty} x_1 f_{X_1}(x_1) \, dx_1 = \mu_{X_1}$$

since the term [ ] is the marginal distribution f_{X_1} according to equation (A.118).

The term

$$\mu_{X_1|X_2} = E(X_1|X_2 = x_2) = \int_{-\infty}^{\infty} x_1 f_{X_1|X_2}(x_1|x_2) \, dx_1$$

is the conditional mean of X_1, given that $X_2 = x_2$.

A.7.2   Variance
Extending as before

$$\text{var}(X_1) \equiv E[(X_1 - \mu_{X_1})^2] = \int_{-\infty}^{\infty} \int_{-\infty}^{\infty} (x_1 - \mu_{X_1})^2 f_{X_1 X_2}(x_1, x_2) \, dx_1 \, dx_2$$

$$\qquad (A.120)$$

$$= \int_{-\infty}^{\infty} \text{var}(X_1|x_2) f_{X_2}(x_2) \, dx_2 = \text{var}(X_1)$$

$$\qquad (A.121)$$

Here the marginal variance of X_1

$$\text{var}(X_1|x_2) \equiv \text{var}(X_1|X_2 = x_2) = E[(X_1 - \mu_{X_1|X_2})^2|X_2 = x_2]$$

$$= \int_{-\infty}^{\infty} (x_1 - \mu_{X_2|X_2})^2 f_{X_1|X_2}(x_1|x_2) \, dx_1 \qquad (A.122)$$

A.7.3   Covariance and correlation

The above expressions for the mean and the variance are symmetrical in (X_1, X_2). A further elementary moment exists involving both X_1 and X_2; this is the covariance. It has the same dimension as variance.

$$\text{cov}(X_1, X_2) \equiv E[(X_1 - \mu_{X_1})(X_2 - \mu_{X2})]$$

$$= \int_{-\infty}^{\infty} \int_{-\infty}^{\infty} (x_1 - \mu_{X1})(x_2 - \mu_{X2}) f_{X_1 X_2}(x_1, x_2) \, dx_1 \, dx_2$$

(A.123)

The correlation coefficient

$$\rho_{X_1 X_2} \equiv \frac{\text{cov}(X_1, X_2)}{(+)[\text{var}(X_1) \, \text{var}(X_2)]^{1/2}} = \frac{\text{cov}(X_1, X_2)}{\sigma_{X_1} \sigma_{X_2}},$$
$$1 \leqslant \rho \leqslant +1$$

(A.124)

is a measure of *linear* dependence between two random variables; if $\rho_{X_1 X_2} = 0$, it follows only that X_1, X_2 are not *linearly* related, but they may be related in some other (non-linear) way. Correlation makes no statement about cause or effect [Benjamin and Cornell, 1970, p. 165; Freeman, 1963, p. 51]. A summary of the significance of ρ is given in Fig. A.9. Higher-order moments may also be developed; these have little practical interest.

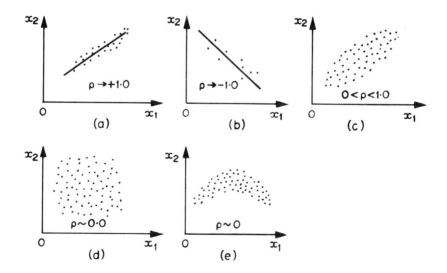

Fig. A.9 — Linear dependence between two variables as a function of correlation coefficient ρ.

A.8   BIVARIATE NORMAL DISTRIBUTION $BiN(\mu_{X_1}, \mu_{X_2}, \sigma_{X_1}^2, \sigma_{X_2}^2, \rho)$

The bivariate normal distribution describes the joint behaviour of two random variables for which the marginal distributions are normally distributed.

The probability density function is

$$f_{X_1X_2}(x_1, x_2, \rho) = \frac{1}{2\pi\sigma_{X_1}\sigma_{X_2}(1-\rho^2)^{1/2}} \exp\left[\frac{-\frac{1}{2}(h^2 + k^2 - 2\rho h\, k)}{1-\rho^2}\right]$$

$$(A.125)$$

for $-\infty < x_i < \infty$, $i = 1, 2$ and where $h = (x_1 - \mu_{X_1})/\sigma_{X_1}$ and $k = (x_2 - \mu_{X_2})/\sigma_{X_2}$. The "standard" form, with zero means and unit standard deviations, is

$$\phi_2(h, k, \rho) = \frac{1}{2\pi(1-\rho^2)^{1/2}} \exp\left[\frac{-\frac{1}{2}(h^2 + k^2 - 2\rho hk)}{(1-\rho^2)}\right]$$

$$-\infty < h < \infty, \quad -\infty < k < \infty \qquad (A.126)$$

with h, k as before and

$$f_{X_1X_2}(\ ) = \frac{1}{\sigma_{X_1}\sigma_{X_2}} \phi_2(h, k, \rho)$$

The cumulative distribution function is

$$F_{X_1X_2}(x_1, x_2, \rho) = \int_{-\infty}^{x_1} \int_{-\infty}^{x_2} f_{X_1X_2}(u, v, \rho)\, du\, dv$$

There is no simple expression for $\Phi_2(\ )$; however, values here have been tabulated (see below) for the standard form

$$F_{X_1X_2}(x_1, x_2, \rho) = \Phi_2(h, k, \rho) = \int_{-\infty}^{h} \int_{-\infty}^{k} \phi_2(u, v, \rho)\, du\, dv$$

$$(A.127)$$

The parameters are the mean value μ_{X_i} of X_i, $i = 1, 2$, the variance $\sigma_{x_i}^2$ of X_i, $i = 1, 2$, and the correlation coefficient ρ between X_1 and X_2.

The conditional moments (see section A.7) are

$$E(X_2|X_1 = x_1) = \mu_{X_2|X_1} = \mu_{X_2} + \rho\frac{\sigma_{X_2}}{\sigma_{X_1}}(x_1 - \mu_{X_1}) \qquad (A.128)$$

Equation (A.128) is also termed the "regression function" of X_2 on x_1.

$$\text{var}(X_2|X_1 = x_1) = \sigma^2_{X_2|X_1} = \sigma^2_{X_2}(1 - \rho^2) \tag{A.129}$$

$$\text{cov}(X_1, X_2) = \rho\sigma_{X_1}\sigma_{X_2} \tag{A.130}$$

where μ_{X_2}, the marginal mean, and $\sigma^2_{X_2}$, the marginal variance, are defined in the marginal density function

$$f_{X_i}(x_i) = \int_{-\infty}^{\infty} f_{X_1X_2}(x_1, x_2)\,dx_i, \qquad i = 1.2$$

$$= \frac{1}{(2\pi)^{1/2}\sigma_{x_i}}\exp\left\{-\frac{1}{2}\left[\frac{x_i - \mu_{X_i}}{\sigma_{X_i}}\right]^2\right\} \tag{A.131}$$

This represents the normal distribution $N(\mu_{X_i}, \sigma^2_{X_i})$.

The properties of the bivariate normal distribution are as follows [Owen, 1956; Johnson and Kotz, 1972].

(1) The marginal distributions are normal (see above). However, the converse may *not* be true; if $f_{X_1}(x_1)$ and $f_{X_2}(x_2)$ are normal, the joint density function $f_{X_1X_2}(x_1, x_2)$ is not necessarily bivariate normal.

(2) If $\rho = 0$, i.e. if X_1 and X_2 are uncorrelated normal random variables, they are also independent. Then $f_{X_1X_2}(x_1, x_2) = f_{X_1}(x_1)f_{X_2}(x_2)$ with f_{X_i} normally distributed.

(3) A commonly used expression is

$$L(h, k, \rho) \equiv$$

$$\equiv \frac{1}{2\pi(1-\rho^2)^{1/2}}\int_h^{\infty}\int_k^{\infty}\exp\left[\frac{-\frac{1}{2}(u^2 + v^2 - 2\rho uv)}{(1-\rho^2)}\right]du\,dv \tag{A.132}$$

for which the following properties hold.

(a) $L(h, k, \rho) = L(k, h, \rho)$ (A.133)

(b) $L(h, k, 0) = \frac{1}{4}[1 - \alpha(h)][1 - \alpha(k)] = [1 - \Phi(h)][1 - \Phi(k)]$ (A.134)

(c) $L(h, k, -1) = 0$,   if $h + k \geqslant 0$

                    $= 1 - \Phi(h) - \Phi(k)$,   if $h + k \leqslant 0$ (A.135)

(d) $L(h, k, 1) = \frac{1}{2}[1 - \alpha(t)]$ where $t = \max[h, k] \geqslant 0$ (A.136)

(e) $L(-h, k, \rho) = -L(h, k, -\rho) + \frac{1}{2}[1 - \alpha(h)]$ (A.137)

(f) $L(-h, -k, \rho) = L(h, k, \rho) + \frac{1}{2}[\alpha(k) + \alpha(h)] = \Phi_2(h, k, \rho)$ (A.138)

(g) $L(0, 0, \rho) = \frac{1}{4} + \frac{\arcsin\rho}{2\pi}$ (A.139)

(h) $L(h,k,\rho) = \dfrac{1}{2\pi} \displaystyle\int_{\text{arcos}\,\rho}^{\pi} \exp[-\tfrac{1}{2}(h^2+k^2-2hk\cos\theta)\csc^2\theta]\,d\theta,$

$$h,k \geqslant 0 \quad (A.140a)$$

(i) $\Phi_2(h,k,\rho) = \dfrac{1}{2\pi} \displaystyle\int_0^\rho \dfrac{1}{(1-z^2)^{1/2}} \exp\left[-\dfrac{\tfrac{1}{2}(h^2+9k^2-2hkz)}{1-z^2}\right]dz + \Phi(h)\Phi(k)$

$$h,k \geqslant 0 \quad (A.140b)$$

where

$$\Phi_2(h,k,\rho) = 1 - L(h,-\infty,\rho) - L(-\infty,k,\rho) + L(h,k,\rho) \quad (A.141a)$$

$$\alpha(v) = \dfrac{1}{(2\pi)^{1/2}} \int_{-v}^{v} e^{-\frac{1}{2}t^2}\,dt \quad (A.141b)$$

$$\Phi(v) = \dfrac{1}{(2\pi)^{1/2}} \int_{-\infty}^{v} e^{-\frac{1}{2}t^2}\,dt \quad (A.141c)$$

and

$$\tfrac{1}{2}[1-\alpha(v)] = 1 - \Phi(v) \quad (A.141d)$$

Charts for $L(h,0,\rho)$, $0 \leqslant h \leqslant 1$, $-1 \leqslant \rho \leqslant 1$, and $L(h,0,\rho)$, $h \geqslant 1$, $-1 \leqslant \rho \leqslant 1$, are given in Abramowitz and Stegun, [1966]. Using the relations (A.132)–(A.141) above, the probabilities over rectangular regions can be determined.

Regions of polygonal shapes can be broken down into triangular regions. For these, tabulations exist [NBS, 1959] in terms of the circular normal distribution, obtained by transforming the variables into an orthogonal set. Since triangles transform to triangles under rotation and scale change, the transformed variables may be used to obtain the required probability. Such a transformation is

$$\begin{pmatrix} u \\ v \end{pmatrix} = \dfrac{\pm 1}{(2\pm 2\rho)^{1/2}} \left(\dfrac{x_1 - \mu_{X_1}}{\sigma_{x_1}} \pm \dfrac{x_2 - \mu_{X_2}}{\sigma_{X_2}} \right), \quad \rho \neq \pm 1 \quad (A.142)$$

or alternatively

$$u = \dfrac{1}{(1-\rho^2)^{1/2}}\left[\dfrac{x_1-\mu_{X_1}}{\sigma_{x_1}} - \dfrac{\rho(x_2-\mu_{X_2})}{\sigma_{X_2}}\right], \quad v = \dfrac{x_2-\mu_{X_2}}{\sigma_{X_2}}, \neq \pm 1$$

$$(A.143)$$

The simpler function $V(h,k)$, defined by

$$V(h,k) = \dfrac{1}{2\pi} \int_0^h \int_0^{kx/h} e^{-\frac{1}{2}(x^2+y^2)}\,dy\,dx \quad (A.144)$$

may then be employed. It is extensively tabulated [NBS, 1959]. $V(h,k)$ is the probability content (in the transformed circular normal distribution) of the triangular region contained within the points $(0,0)$, $(h,0)$ and (h,k).

The relationship between L and V is

$$L(h,k,\rho) = V\left[h, \frac{k-\rho h}{(-\rho^2)^{1/2}}\right] + V\left[k, \frac{h-\rho k}{(1-\rho^2)^{1/2}}\right] + \frac{\arcsin \rho}{2\pi} +$$

$$\tfrac{1}{4}[1 - \alpha(h) - \alpha(k)] \qquad (A.145)$$

where $\alpha(\ )$ is defined above.

A.9 TRANSFORMATION OF RANDOM VARIABLES

A.9.1 Transformation of a single random variable.

If $Y = g(X)$ and $X = g^{-1}(Y)$, where g is a monotonic function, and X and Y are continuous random variables, it can be shown [Ang and Tang, 1975, p. 170; Freeman, 1963, pp. 70–77] that the probability density function for the dependent random variable Y in terms of that for X is

$$f_Y(y) = f_X(x)\left|\frac{\mathrm{d}x}{\mathrm{d}y}\right| \qquad (A.146)$$

where $x = g^{-1}(y)$. A physical meaning is evident. Equation (A.146) is equivalent, in the case of x and y increasing monotonically, to

$$f_Y(y)\,\mathrm{d}y = f_X(x)\,\mathrm{d}x \qquad (A.147)$$

which indicates that the infinitesimal area $f_X(x)\,\mathrm{d}x$ under the f_X curve at x is equal to the corresponding infinitesimal area $f_Y(y)\,\mathrm{d}y$ at $y(=g(x))$. Hence respective probabilities over $\mathrm{d}x$ and $\mathrm{d}y$ are maintained when f_Y is transformed from f_X by equation (A.146). See also Benjamin and Cornell (1970) (p. 107) for a good description of this situation.

If g is not a monotonic function, equation (A.146) does not apply, since y (or x) may take on more than one value. A general procedure, requiring adaptation to each individual problem, is to derive $F_Y(y)$ directly [Benjamin and Cornell, 1970, p. 110]:

$$F_Y(y) \equiv P(Y \leqslant y) = P[X \text{ has a value } x \text{ for which } g(x) \leqslant y]$$

$$= \int_{R_y} f_X(x)\,\mathrm{d}x \qquad (A.148)$$

where R_y is the region in which $g(x) \leqslant y$.

A particular transformation of use in Monte Carlo work is that given by

$y = F_X(x)$, $0 \leqslant y \leqslant 1$. If F_X is differentiable, $dy/dx = dF_X/dx = f_X(x)$. Hence, substituting into equation (A.146),

$$f_Y(y) = f_X(x)\left|\frac{1}{f_X(x)}\right| = 1, \qquad \text{for } 0 \leqslant y \leqslant 1 \qquad \text{(A.149)}$$

This is a rectangular distribution, irrespective of the form of $F_X(x)$.

A.9.2 Transformation of two or more random variables

Consider two random variables X_1 and X_2, with known joint density function $f_{X_1 X_2}(x_1, x_2)$ which are related to two random variables Y_1 and Y_2 by known functions [Freeman, 1963, p. 79]

$$y_1 = y_1(x_1, x_2), \qquad y_2 = y_2(x_1, x_2)$$

having unique (i.e. one-to-one) inverses

$$x_1 = x_1(y_1, y_2), \qquad x_2 = x_2(y_1, y_2)$$

Then

$$f_{Y_1 Y_2}(y_1, y_2) = f_{X_1 X_2}(x_1, x_2)|\mathbf{J}| \qquad \text{(A.150)}$$

where the Jacobian $\mathbf{J}$ defined by

$$\mathbf{J} \equiv \begin{bmatrix} \dfrac{\partial x_1}{\partial y_1} & \dfrac{\partial x_2}{\partial y_1} \\ \dfrac{\partial x_1}{\partial y_2} & \dfrac{\partial x_2}{\partial y_2} \end{bmatrix} \equiv \frac{\partial(x_1, x_2)}{\partial(y_1, y_2)} \qquad \text{(A.151)}$$

As in the case of the single random variable, an elemental volume defined by $f_{X_1 X_2}(x_1, x_2)\, dA(x_1, x_2)$ in the (x_1, x_2) variables remains invariant under the transformation (A.150), after which it is defined by $f_{Y_1 Y_2}(y_1, y_2)\, dA(y_1, y_2)$.

Transformation (A.150) can be directly extended to general random variables, provided the uniqueness of transformations remains valid, i.e. provided that there is a one-to-one relationship between X_i and Y_j. This is guaranteed "locally" if J does not change sign with "small" changes of X_i. The guarantee "in the large" is less straightforward [Freeman, 1963, p. 83].

A.9.3   Linear and orthogonal transformations
For the linear transformation

$$y_i = \sum_{j=1,n} a_{ij}x_j \qquad j = 1, 2, \ldots, n \tag{A.152}$$

the Jacobian is given by

$$\mathbf{J} = \begin{bmatrix} a_{11} & a_{1n} \\ \vdots & \vdots \\ a_{n1} & a_{nn} \end{bmatrix} = \mathbf{A} \tag{A.153}$$

In matrix theory, equation (A.152) expresses a linear transformation which may be expressed as $\mathbf{y} = \mathbf{A}\mathbf{x}$ where $\mathbf{A}$ is the transformation matrix. It is well known in matrix theory that the transformation (A.152) is orthogonal (i.e. the y_i are independent) if $\mathbf{A}\mathbf{A}^T = \mathbf{I}$, the identity matrix. It follows readily that in this case $|\mathbf{J}| = |\mathbf{A}| = (\mathbf{A}\mathbf{A}^T)^{1/2} = \pm 1$ (see also Appendix B).

A.10   FUNCTIONS OF RANDOM VARIABLES

A.10.1   Function of a single random variable
If $Y = g(X)$, then Y is simply X transformed through $g(\ )$, and the transformation results of section A.9 apply directly, equations (A.146) and (A.147) for monotonic functions, or else (A.148)

A.10.2   Function of two or more random variables
If a function $Y = Y(X_1, X_2)$ is sought when the function Y and its inverse are unique in the sense of section A.9, then the results of that section can be applied directly. Let $Y = Y_1$ and let, say, $Y_2 = X_2$ (or X_1), an entirely dummy relationship. From the joint density function $f_{Y_1Y_2}$, the density function f_Y can be obtained by integrating over Y_2, using equation (A.118) but with the limits of integration changed to $a \leqslant Y_2 \leqslant b$; thus

$$f_Y(y) = F_{Y_1}(y_1) = \int_a^b f_{Y_1Y_2}(y_1, y_2)\, dy_2$$

$$= \int_a^b f_{X_1X_2}(x_1, x_2) \left| \frac{\partial(x_1, x_2)}{\partial(y_1, y_2)} \right| dx_2 \tag{A.154}$$

[see Wadsworth and Bryan, 1974, p. 222, for an example]. This approach to finding what is essentially the marginal density function $f_Y = f_{Y_1}$ is an application of convolution. The resulting integrations are seldom straightforward.

If the function $Y = Y(X_1, X_2)$ and its inverse are not unique in the sense of section A.9, the above procedure is not valid. In parallel to the method of section A.9, a direct procedure is to establish $F_Y(y)$ directly. In particular, we may write

$$F_Y(y) \equiv P(Y \leqslant y) = P[X_1, X_2 \text{ have values } x_1, x_2 \text{ for which } Y(x_1,x_2) \leqslant y]$$

$$\int_{R_y} \int f_{X_1 X_2}(x_1, x_2) \, dx_1 \, dx_2 \tag{A.155}$$

where R_y is the region over which $Y(x_1,x_2) \leqslant y$ [see Benjamin and Cornell, 1970, pp. 114–117, for examples]. The integration in equation (A.155) is again seldom straightforward, particularly for multiple integration with multiple random variables.

A.10.3   Some special results
A.10.3.1 $Y = X_1 + X_2$
According to the convolution integral (A.154) and applying (A.117) for X_1 and X_2 independent

$$f_Y(y) = \int_{-\infty}^{\infty} f_{X_1}(x_1) f_{X_2}(y - x_1) \, dx_1 \tag{A.156}$$

If X_1 and X_2 are statistically independent, and if f_{X_1} and f_{X_2} are Poisson distributions, then f_Y will also be Poisson (see (A.34)). Similarly the sum of gamma distributions is given by (A.46). If X_1 and X_2 are normally distributed, then f_Y is normally distributed, with the mean and variance given by

$$\mu_Y = \mu_{X_1} + \mu_{X_2}, \qquad \text{var}(Y) = \text{var}(X_1) + \text{var}(X_2) \tag{A.157}$$

A.10.3.2 $Y = X_1 X_2$
By convolution,

$$f_Y(y) = \int_{-\infty}^{\infty} \left| \frac{1}{x_2} \right| f_{X_1 X_2}\left(\frac{y}{x_2}, x_2 \right) dx_2 \tag{A.158}$$

If $Y = X_1/X_2$, the same form holds with $|1/x_2|$ and y/x_2 replaced by x_2 and yx_2 [Ang and Tang, 1975, p. 183].

A.11   MOMENTS OF FUNCTIONS OF RANDOM VARIABLES

Because the joint density function for a function of several random variables is generally not easily obtainable, there is an interest in studying the moments of functions of several random variables. For many applications a knowledge of the first and second moments may be sufficient. This is particularly so where the distribution of variables X_i is normal, since the latter distribution is uniquely described by its first two moments.

The basic expressions for moments, equations (A.10) and (A.11), can be extended to a general function $Y = Y(X_1, X_2, \ldots, X_n)$ as follows

$$E(Y) = \int_{-\infty}^{\infty} \cdots \int_{-\infty}^{\infty} Y(x_1, x_2, \ldots, x_n) f_{X_1 X_2 \ldots X_n}(x_1, x_2, \ldots, x_n) \, dx_1 dx_2 \ldots dx_n$$

$$(A.159)$$

where Y is the function for which the moment is sought.

A.11.1   Some special results
A.11.1.1   *Linear functions*

If $Y = \displaystyle\sum_{i=1}^{n} a_i X_i$, then

$$E(Y) = \mu_Y = \sum_{i=1}^{n} a_i E(X_i) = \sum_{i=1}^{n} a_i \mu_{X_i} \qquad (A.160)$$

$$E[(Y - \mu_Y)^2] = \operatorname{var}(Y) = \sum_{i=1}^{n} a_i^2 \operatorname{var}(X_i) + \sum_{i \neq j}^{n} \sum^{n} a_i a_j \operatorname{cov}(X_i, X_j)$$

$$(A.161)$$

or, more compactly,

$$\operatorname{var}(Y) = \sum_{j}^{n} \sum_{j}^{n} a_i a_j \rho_{ij} \sigma_{X_i} \sigma_{X_j} \qquad (A.162)$$

where σ_{X_i} is the standard deviation of X_i, ρ_{ij} the correlation coefficient between X_i and X_j and $\rho_{ii} = 1$.

Further, if in addition $z = \displaystyle\sum_{i=1}^{n} b_i X_i$, then [Ang and Tang, 1975, p. 193]

$$\operatorname{cov}(Y, Z) = \sum_{i}^{n} \sum_{i}^{n} a_i b_j \rho_{ij} \sigma_{X_i} \sigma_{X_j} \qquad (A.163)$$

In both expressions, $\rho_{ii} = 1$ and, if the X_i are *independent*, $\rho_{ij} = 0$ if $i \neq j$.

A.11.1.2   Product of Variates

If $Y = \prod_{i=1}^{n} X_i$, then we have the following.

(a) If $n = 2$,

$$E(Y) = \mu_Y = E(X_1 X_2) = E(X_1)E(X_2) + \mathrm{cov}(X_1, X_2)$$
$$= \mu_{X_1}\mu_{X_2} + \rho\sigma_{X_1}\sigma_{X_2} \qquad (A.164)$$

and

$$\mathrm{var}(Y) = \sigma_Y^2 = [(\mu_{X_1}\sigma_{X_2})^2 + (\mu_{X_2}\sigma_{X_1})^2 + (\sigma_{X_1}\sigma_{X_2})^2](1 + \rho^2) \qquad (A.165)$$

where ρ is the correlation coefficient between X_1 and X_2.

(b) If $n = 2$ and X_1, X_2 are *independent* (i.e. $\rho = 0$), then equation (A.165) reduces to

$$V_Y^2 = V_{X_1}^2 + V_{X_2}^2 + V_{X_1}^2 V_{X_2}^2 \qquad (A.166)$$

where $V = \sigma_K/\mu_K$ is the coefficient of variation.

Further, $V_Y^2 \approx V_{X_1}^2 + V_{X_2}^2$ for V_{X_1} and V_{X_2} small, say <0.30. This result is of considerable practical importance.

(c) If $n \geqslant 2$ and the X_i are *independent*,

$$E(Y) = E\left(\prod_{i=1}^{n} X_i\right) = \prod_{i=1}^{n} E(X_i) = \prod_{i=1}^{n} \mu_{X_i} \qquad (A.167)$$

and from equation (A.11)

$$\mathrm{var}(Y) = E(Y^2) - [E(Y)]^2$$

$$= \prod_{i=1}^{n} \mu_{X_i^2} - \left(\prod_{i=1}^{n} \mu_{X_i}\right)^2 \qquad (A.168)$$

An *approximate* result, which ignores second-order terms is [Benjamin and Cornell, 1970, p. 185]

$$\mathrm{var}(Y) \approx \sum_{i=1}^{n} \left(\prod_{\substack{j=1 \\ j \neq i}}^{n} \mu_{X_j}^2\right) \sigma_{X_i}^2 \qquad (A.169)$$

This may be obtained using the procedure outlined in section A.12.

A.11.1.3 Division of variates

Excluding division by zero, some approximate results valid for random variables with reasonably small variance, are [Haugen, 1968, p. 121] as follows.

If $Y = X_1/X_2$,

$$E(Y) = \mu_Y = \frac{\mu_{X_1}}{\mu_{X_2}}\left[1 + \frac{\sigma_{X_1}}{\mu_{X_1}}\left(\frac{\sigma_{X_1}}{\mu_{X_1}} - \rho\frac{\sigma_{X_2}}{\mu_{X_2}}\right)\left(1 + \frac{\sigma_{X_1}^2}{\mu_{X_1}^2} + \ldots\right)\right] \tag{A.170}$$

$$= \frac{\mu_{X_1}}{\mu_{X_2}} \tag{A.171}$$

as a first order approximation.

$$\mathrm{var}(Y) = \sigma_Y^2 \approx \left(\frac{\mu_{X_1}}{\mu_{X_2}}\right)^2\left(\frac{\sigma_{X_2}^2}{\mu_{X_1}^2} - 2\rho\frac{\sigma_{X_1}\sigma_{X_2}}{\mu_{X_1}\mu_{X_2}} + \frac{\sigma_{X_1}^2}{\mu_{X_2}^2} + \text{ higher order terms}\right) \tag{A.172}$$

$$\approx \left(\frac{\mu_{X_1}}{\mu_{X_2}}\right)^2 (V_{X_1}^2 - 2\rho V_{X_1}V_{X_2} + V_{X_2}^2) \tag{A.173}$$

A.11.1.4 Moments of a Root [Haugen, 1968, p. 128]

If $Y = X^{\frac{1}{2}}$,

$$\mu_Y = (\mu_X^2 - \tfrac{1}{2}\sigma_X^2)^{\frac{1}{4}} \tag{A.174}$$

$$\sigma_Y^2 = \mu_X - (\mu_X^2 - \tfrac{1}{2}\sigma_X^2)^{\frac{1}{2}} \tag{A.175}$$

A.11.1.5 Moments of a quadratic form [Haugen, 1968, p. 128]

If $Y = aX^2 + bX + c$, then using results already established

$$\mu_Y = a(\mu_X^2 + \sigma_X^2) + b\mu_X + c \tag{A.176}$$

$$\sigma_Y^2 = \sigma_X^2(2a\mu_X + b)^2 + 2a^2\sigma_X^4 \tag{A.177}$$

A.12 APPROXIMATE MOMENTS FOR GENERAL FUNCTIONS

The mean and variance of general functions are not usually easily obtained owing to the integrations required in equation (A.159). Nor may more information that the first two moments of each X_i be available. A useful approach is therefore to calculate approximate moments by expanding $Y = Y(X_1, X_2, \ldots, X_n)$ in a Taylor series about the means $\mu_{X_1}\mu_{X_2}, \ldots, \mu_{X_n}$ [Ang and Tang, 1975, p. 199]. By truncating the series at linear terms, the first-order mean and variance are

$$E(Y) \approx Y(\mu_{X_1}, \mu_{X_2}, \ldots, \mu_{X_n}) \tag{A.178}$$

$$\text{var}(Y) \approx \sum_i^n \sum_j^n c_i c_j \text{cov}(X_i, X_j) \tag{A.179}$$

where

$$c_i \equiv \frac{\partial Y}{\partial X_i}\bigg|_{\mu_{X_1}, \mu_{X_2}, \ldots, \mu_{X_n}}$$

If the X_i are independent, $\text{cov}(X_i, X_j) = 0$ if $i \neq j$, $\text{cov}(X_i, X_j) = \text{var}(X_i)$ if $i = j$.

Similarly, the second order approximation is given by

$$E(Y) = Y(\mu_{X_1}, \mu_{X_2}, \ldots, \mu_{X_n}) + \frac{1}{2}\sum_{i=1}^n \sum_{j=1}^n \frac{\partial^2 Y}{\partial X_i \partial X_j} \text{cov}(X_i, X_j) \tag{A.180}$$

where the $\partial^2 Y / \partial X_i \partial X_j$ are evaluated at $\mu_{X_1}, \mu_{X_2}, \ldots$. The second term is negligibly small if V_{X_i} (the coefficient of variation) is small and the function Y does not depart greatly from non-linearity.

Appendix B: Rosenblatt and other transformations

B.1 ROSENBLATT TRANSFORMATION

A dependent random vector $\mathbf{X} = \{X_1, X_2, \ldots, X_n\}$ may be transformed to the independent uniformly distributed random vector $\mathbf{R} = \{R_1, \ldots, R_n\}$ through the Rosenblatt (1952) transformation $\mathbf{R} = T\mathbf{X}$ given by

$$
\begin{aligned}
r_1 &= P(X_1 \leqslant x_1) = F_1(x_1) \\
r_2 &= P(X_2 \leqslant x_2 | X_1 = x_1) = F_2(x_2 | x_1)
\end{aligned}
\tag{B.1}
$$

$$
\vdots
$$

$$
r_n = P(X_n \leqslant x_n | X_1 = x_1, \ldots, X_{n-1} = x_{n-1}) = F_n(x_n | x_1, \ldots, x_{n-1})
$$

where $F_i(\ )$ is shorthand for the conditional cumulative distribution function $F_{X_i | X_{i-1}, \ldots, x_1}(\ )$.

If the joint probability density function $f_{\mathbf{X}}(\ )$ is known then $F_i(\ )$ can be determined as follows. From section A.6.2, the conditional probability density function $f_i(\ )$ is given by

$$
f_i(x_i | x_1, \ldots, x_{n-1}) = \frac{f_{\mathbf{X}_i}(x_1, \ldots, x_i)}{f_{\mathbf{X}_{i-1}}(x_1, \ldots, x_{i-1})}
\tag{B.2}
$$

where $f_{\mathbf{X}_j}(x_1, \ldots, x_j)$ is a marginal probability density function obtained from

$$
f_{\mathbf{X}_j}(x_1, \ldots, x_j) = \int_{-\infty}^{\infty} \cdots \int_{-\infty}^{\infty} f_{\mathbf{X}}(x_1, \ldots, x_n) \, dx_{j+1}, \ldots, dx_n
\tag{B.3}
$$

$F_i(\ )$ is then obtained by integrating $f_i(\ )$ given by (B2) over x_i;

$$F_i(x_i|x_1, \ldots ,x_{n-1}) = \frac{\int_{-\infty}^{\infty} f_{\mathbf{X}_i}(x_1, \ldots ,x_{i-1}, t)\, dt}{f_{\mathbf{X}_{i-1}}(x_1, \ldots ,x_{i-1})} \qquad (B.4)$$

With all the conditional cumulative distribution functions $F_i(\ )$ determined in this way, (B.1) may be inverted successively to obtain

$$\begin{aligned}
x_1 &= F_1^{-1}(r_1) \\
x_2 &= F_2^{-1}(r_2|x_1)
\end{aligned} \qquad (B.5)$$

.

.

.

$$x_n = F_n^{-1}(r_n|x_1, \ldots ,x_{n-1})$$

It follows immediately that (B.5) can be used to generate the random vector **X** with probability density function $f_{\mathbf{X}}(\ )$ from **R**. A practical difficulty is that, unless $F_i(\ )$ is simple in form, the inversion will need to be done numerically.

As noted by Rosenblatt (1952), there are $n!$ possible ways in which expressions (B.1) can be written, depending on the numbering adopted for the variables in **X**. Equally, of course, there are $n!$ possible ways of conditioning the X_i in expression (B.1), as seen for the simple case $n = 2$ [Rubinstein, 1981]:

$$F_{X_1X_2}(x_1, x_2) = f_{X_1}(x_1)\, f_{X_2|X_1}(x_2|x_1) = f_{X_2}(x_2)\, f_{X_1|X_2}(x_1|x_2)$$

It is probably obvious that this freedom can lead to considerable differences in the difficulty of solving for **X**, i.e. in solving (B.5).

Unfortunately, the above method is not always useful for practical problems since $F_{\mathbf{X}}$, or the conditional probability density functions $f_i(\ )$, which characterize the dependence structure of the problem, are not always known. More commonly only some estimate of correlation may be available from the data. A special case arises when the X_i are independent. All the condition requirements in (B.5) then disappear and each transformation takes the form $x_i = F_i^{-1}(r_i)$ independent of all other x_j ($j \neq i$).

The Rosenblatt transformation may be used to transform from one distribution to another by applying (B.1) twice, using R as a transmitter, e.g.

$$F_1(u_1) = r_1 = F_1(x_1)$$

$$F_2(u_2|u_1) = r_2 = F_2(x_2|x_1) \qquad (B.6)$$

.

.

.

A particular case of interest is where **U** in (B.6) is standard normal distributed, with **X**, say, a vector of correlated random variables and **U** uncorrelated (independent). Then (B.6) may be written as

$$x_1 = F_1^{-1}[\Phi(u_1)]$$

$$x_2 = F_2^{-1}[\Phi(u_2)|x_1]$$

.

.

.

etc.

(B.7)

In practice solution of (B.7) requires multiple integration (cf. (B.4)). This technique is used in section 4.4.3.1 to convert non-normal distributed random variables to equivalent normal random variables. Where both **U** and **X** are normal vectors, an easier approach to finding the transformation implied by (B.7) is to make direct use of the special properties of the normal distribution. This is considered in the next section.

B.2   ORTHOGONAL TRANSFORMATION OF NORMAL RANDOM VARIABLES

Let **X** be a correlated vector of basic variables, with mean

$$E(\mathbf{X}) = [E(X_1), E(X_2), \ldots , E(X_n)]$$ (B.8)

and covariance matrix (cf. (A.123))

$$\mathbf{C}_X = \text{cov}(X_i, X_j)_{n \times m} = (\sigma_{ij})_{n \times m}$$ (B.9)

with $\text{cov}(X_i, X_i) = \text{var}(X_i)$. The covariance matrix will be strictly diagonal if the **X** are uncorrelated. Recall that "correlation" is a measure of linear dependence (see section A.7.3). This is a sufficient measure of dependence for normal distributions.

An uncorrelated vector **U**, and a linear transformation matrix **A**, is now sought, such that

$$\mathbf{U} = \mathbf{AX} \quad \text{or} \quad \mathbf{A}^{-1}\mathbf{U} = \mathbf{X}$$ (B.10)

It is clearly desirable that the transformation (B.10) is also orthogonal, i.e. the vector represented by **X** is unchanged in length under the transformation **A**. From well-known matrix theory this implies that $\mathbf{A}^T = \mathbf{A}^{-1}$. Further, the vector **U** will be uncorrelated if there exists a covariance matrix $\mathbf{C}_U$ which is strictly diagonal.

Under the linear transformation (B.10) the covariance matrix (B.9) is also transformed. It becomes the covariance matrix $\mathbf{C}_U$ for $\mathbf{U}$:

$$\mathbf{C}_U = \mathrm{cov}(U_i, U_j) = \mathrm{cov}(\mathbf{U}, \mathbf{U}^T) \tag{B.11}$$

$$= \mathrm{cov}(\mathbf{AX}, \mathbf{AX}) = \mathrm{cov}(\mathbf{AX}, \mathbf{X}^T\mathbf{A}^T)$$

$$= \mathbf{A}\,\mathrm{cov}(\mathbf{X}, \mathbf{X}^T)\,\mathbf{A}^T$$

or

$$\mathbf{C}_U = \mathbf{A}\mathbf{C}_X\mathbf{A}^T \tag{B.12}$$

To obtain $\mathbf{U}$ as an uncorrelated vector, the matrix $\mathbf{A}$ which makes (B.12) diagonal is sought. Thus, it is required that the off-diagonal terms (i.e. $i \neq j$) in $\mathbf{C}_U$ are zero. This can be done by finding the characteristic values (eigen-values) of $\mathbf{C}_X$ using well known methods. In particular, consider a matrix $\mathbf{D}$, having only diagonal coefficients λ_{ii}, defined by

$$\mathbf{C}_X\mathbf{A} = \mathbf{AD} \quad \text{or} \quad \mathbf{C}_X = \mathbf{ADA}^T \tag{B.13}$$

(provided that $\mathbf{A}$ is non-singular). Equation (B.13) may be written as a system of linear equations

$$\sum_i c_{ij}\, a_{jk} = a_{ik}\lambda_{ii}, \qquad i, k = 1, 2, \ldots, n \tag{B.14}$$

Equation (B.14) represents a system of n linear equations with $j = 1, \ldots, n$ terms of unknown coefficients a_{ij} on the left-hand side; typically

$$c_{11}\,a_{11} + c_{12}\,a_{21} + c_{13}\,a_{31} + \ldots = a_{11}\lambda_{11} \tag{B.15}$$

$$c_{11}\,a_{12} + c_{12}\,a_{22} + c_{13}\,a_{32} + \ldots = a_{12}\lambda_{22}$$

.

.

.

This system of equations is homogeneous and may be written using the Kronecker delta δ_{ij} ($= 1$ if $i = j$, otherwise $= 0$):

$$\sum_i (c_{ij} - \lambda_{ii}\delta_{ij})a_{jk} = 0, \qquad i, k = 1, \ldots, n \tag{B.16}$$

A system such as equation (B.16) has a non-trivial solution only if, for any value of k,

$$|c_{ij} - \lambda_{ii}\delta_{ij}| = 0 \qquad (B.17)$$

or, more generally,

$$|\mathbf{C}_X - \lambda\mathbf{I}| = 0$$

where $\mathbf{I}$ is the identity matrix. Equation (B.17) is the well-known "characteristic" equation. Its solution, the "characteristic" values $\lambda = \lambda_{ii} i = 1, \ldots, n$, of matrix $\mathbf{D}$ are obtained by expanding (B.17) and solving the determinant

$$\begin{vmatrix} c_{11} - \lambda_{11} & c_{12} & c_{13} \\ c_{21} & c_2 - \lambda_{22} & c_{23} \\ c_{31} & c_{32} & c_{33} - \lambda_{33} \\ \vdots & & \end{vmatrix}_{n \times n} = 0 \qquad (5.57e)$$

There are standard procedures available to obtain λ.

For each λ_{ii}, equations (B.15) have a solution $(a_{i1}, a_{i2}, \ldots, a_{ik}, \ldots, a_{in})$. This is termed the "characteristic vector" and yields the ith row of matrix $\mathbf{A}$. It represents the components of one of the uncorrelated vectors U_i in terms of the correlated vectors X_i, (and vice versa since $\mathbf{A}$ is symmetric). Thus solution for all characteristic values λ_{ii} and hence the characteristic vectors produces the complete matrix $\mathbf{A}$.

With $\mathbf{A}$ known, the uncorrelated vector $\mathbf{U}$ is then defined by

$$\mathbf{U} = \mathbf{AX} \qquad (B.10)$$

with

$$E(\mathbf{U}) = \mathbf{A}E(\mathbf{X}) \qquad (B.18)$$

$$\mathbf{C}_U = \mathbf{A}\mathbf{C}_X\mathbf{A}^T \qquad (B.12)$$

and hence

$$\mathbf{C}_U^{\frac{1}{2}} = (\mathbf{A}\mathbf{C}_X\mathbf{A}^T)^{\frac{1}{2}} \qquad (B.19)$$

where $\mathbf{C}_U^{\frac{1}{2}}$ is the matrix of standard deviations of $\mathbf{U}$. This comes about as follows. Since $\mathbf{U}$ is uncorrelated, $\mathbf{C}_U$ is strictly diagonal. For a diagonal matrix, such as $\mathbf{D}$, say, it is well known that $\{D_{ij}\}^2 = \{D_{ij}^2\}$. Hence $\mathbf{C}_U^{\frac{1}{2}}$ consists only of terms $C_{ii}^{\frac{1}{2}} = \sigma_{x_i} = \lambda_{ii}^{\frac{1}{2}}$ long the leading diagonal, where the λ_{ii} are the characteristic values.

A special case arises if vector $\mathbf{U}$ consists of only one term, $\mathbf{Z}$, say. Then (B.11) is given by

$$\mathbf{Z} = \mathbf{AX} \tag{B.20}$$

where the matrix $\mathbf{A}$ has become just one row of coefficients (see sections 4.2 and A.11.1). The covariance matrix for $\mathbf{Z}$ is then one term (the variance) given by

$$\mathbf{C_Z} = \text{var}(\mathbf{Z}) = \mathbf{AC}_X\mathbf{A}^{\mathsf{T}} \tag{B.21}$$

Mathematical aside. The standard deviations $\lambda_{ii}^{\frac{1}{2}}$ of $\mathbf{C}_U$ must all be positive (or zero), since they have no physical meaning otherwise. This means that the transformation (B.12) must have special properties; in particular it is said that $\mathbf{C}_X$ must be a positive (semi)definite matrix. In matrix terminology, this means that the determinant of any minor,

$$\det M_i = \begin{vmatrix} C_{11} & \cdots & c_{i1} \\ \cdot & & \\ \cdot & & \\ \cdot & & \\ c_{1i} & \cdots & c_{ii} \end{vmatrix} \tag{B.22}$$

is defined as positive definite if $\det M_i > 0$ or positive semidefinite if $\det M_i \geqslant 0$. A matrix is non-negative definite if it is either of these.

B.3 GENERATION OF DEPENDENT RANDOM VECTORS

In some situations it may be necessary to generate dependent variables (but see Section 3.4.6.2 for importance sampling). As will be seen, with the exception of normally distributed variables, this may not be easy in practice.

If the random variables in vector $\mathbf{X}$ are independent of each other, the joint probability density function can be decomposed as (cf. (A.117))

$$f_{\mathbf{X}}(x) = \sum_{i=1}^{n} f_{X_i}(x_i) \tag{B.23}$$

where $f_{X_i}(x_i)$ is the marginal probability density function of the random variable X_i. The generation of the vector $\mathbf{X}$ follows directly from the inverse transform method (see section 3.4.3) for each random variable X_i separately.

If there is dependence between the random variables constituting $\mathbf{X}$, (B.23) is replaced by (cf. (A.116) and A.3)

$$f_{\mathbf{X}}(\mathbf{x}) = f_{X_1}(x_1)f_{X_2|X_1}(x_2|x_1) \cdots f_{X_n|X_1, \ldots ,X_{n-1}}(x_n|x_1, \ldots ,x_{n-1}) \tag{B.24}$$

where $f_{X_k|X_i, \ldots, x_n}$ is the conditional probability density function of X_k given that $X_1 = x_1, \ldots$, and where $f_{X_1}(x_1)$ is the marginal probability density function of X_1.

Provided that the joint probability density function $f_{\mathbf{X}}(\mathbf{x})$ is known in terms of the conditional probability density functions as in (B.24), the inverse transform method of section 3.4.3 to generate sample values $\hat{x}_i$, can be extended to generate a sample vector $\hat{\mathbf{x}}$, such that $\hat{\mathbf{x}}$ is drawn from a distribution with mean vector $\boldsymbol{\mu}_X$ and covariance matrix $\mathbf{C}_X$ as defined by the joint probability density function $f_{\mathbf{X}}$ for the multivariate normal distribution (3.10):

$$f_{\mathbf{X}}(\mathbf{x}, \mathbf{C}_X) = \frac{1}{(2\pi)^{n/2}|\mathbf{C}_X|^{\frac{1}{2}}} \exp[-\tfrac{1}{2}(\mathbf{x} - \boldsymbol{\mu}_X)^T \mathbf{C}_X^{-1}(\mathbf{x} - \boldsymbol{\mu}_\mathbf{x}) \qquad \text{(B.25)}$$

The vector $\mathbf{X}$ of correlated normal (Gaussian) distributed random variables with given covariance matrix $\mathbf{C}_X$ can be generated from a vector of independent standardized normal random variables such as $\mathbf{Y} = \{Y_i\}$ using either the orthogonal transformation of section B.2 or the Rosenblatt transformation of section B.1.

To be specific, let the n correlated random variables be $\mathbf{X}$, with known mean vector $\boldsymbol{\mu}_x$, and known covariance matrix $\mathbf{C}_X$. Further, let U be the vector of independent random variables with strictly diagonal covariance matrix $\mathbf{C}_U$. Then the sample (i.e. generated) dependent values of $\mathbf{X}$ can be obtained from the sampled independent U by the orthogonal transformation (cf. B.11)):

$$\mathbf{X} = \mathbf{A}^T \mathbf{U} \qquad \text{(B.26)}$$

where, since $\mathbf{C}_X$ is positive definite and symmetric, the orthogonal transformation matrix $\mathbf{A}^T = \mathbf{A}^{-1}$ can be defined through the transformation of the covariance matrix (cf. B.12)):

$$\mathbf{C}_U = \mathbf{A}\mathbf{C}_X\mathbf{A}^T \text{ or } \mathbf{C}_X = \mathbf{A}^T \mathbf{C}_U \mathbf{A} \qquad \text{(B.27)}$$

If $\mathbf{B}$ is substituted for $\mathbf{A}^T = \mathbf{A}^{-1}$, then

$$\mathbf{C}_X = \mathbf{B} \, \mathbf{C}_U \cdot \mathbf{B}^T \qquad \text{(B.28)}$$

The elements b_{ij} of the square matrix $\mathbf{B}$ are given by

$$c_{ij} = \sum_{k=1}^{n} b_{ik} \, u_{kk} \, b_{jk} \qquad \text{(B.29)}$$

where c_{ij} is known from $\mathbf{C}_X$ and, since $\mathbf{U} = \mathbf{Y}$, the vector of dependent standardized normal variables, $u_{kk} = 1$ for all k, $u_{jk} = 0, j \neq k$.

The transformation (B.26) then becomes, after allowing for the normalization of the mean of $\mathbf{X}$,

$$\mathbf{X} - \boldsymbol{\mu}_x = \mathbf{BY} \quad \text{or} \quad \mathbf{X} = \mathbf{BY} + \boldsymbol{\mu}_x \qquad (B.30)$$

from which the correlated variables $\mathbf{X}$ can be generated for samples of $\mathbf{Y}$. Further (B.27) reduces to

$$\mathbf{C}_X = \mathbf{BB}^T = \mathbf{A}^T\mathbf{A} \qquad (B.31)$$

where the matrix $\mathbf{B}$ is square lower triangular and obtained directly from $\mathbf{A}$ as indicated for (B.28).

Alternatively, a recursive formula may be derived to obtain the elements b_{ij} of $\mathbf{B}$ (cf. Rosenblatt, 1952). Noting that $\mathbf{B}$ is lower triangular, consider for instance the following example of (B.31):

$$\begin{bmatrix} \sigma_{11} & \sigma_{12} & \sigma_{13} \\ \sigma_{21} & \sigma_{22} & \sigma_{23} \\ \sigma_{31} & \sigma_{32} & \sigma_{33} \end{bmatrix} = \begin{bmatrix} b_{11} & 0 & 0 \\ b_{21} & b_{22} & 0 \\ b_{31} & b_{32} & b_{33} \end{bmatrix} \begin{bmatrix} b_{11} & b_{21} & b_{31} \\ 0 & b_{22} & b_{32} \\ 0 & 0 & b_{33} \end{bmatrix} \qquad (B.32)$$

from which

$$\sigma_{11} = b_{11}^2 \qquad (B.32a)$$

$$\sigma_{22} = b_{21}^2 + b_{22}^2 \qquad (B.32b)$$

$$\sigma_{33} = b_{31}^2 + b_{32}^2 + b_{33}^2 \qquad (B.32c)$$

or generalizing

$$\sigma_{ii} = \sum_{k=1}^{n} b_{ik}^2 \qquad k \leq i$$

and

$$\sigma_{21} = \sigma_{12} = b_{21}b_{11} \qquad (B.32d)$$

$$\sigma_{31} = \sigma_{13} = b_{31}b_{11} \qquad (B.32e)$$

$$\sigma_{32} = \sigma_{23} = b_{31}b_{21} + b_{32}b_{22} \qquad (B.32f)$$

or generalizing

$$\sigma_{ij} = \sum_{k=1}^{n} b_{ik}b_{jk}, \; k \leq j < i$$

The recursive result now follows directly:
from (B.32a)

$$b_{11} = \sigma_{11}^{\frac{1}{2}}$$

from (B.32d)

$$b_{21} = \frac{\sigma_{21}}{b_{11}} = \frac{\sigma_{21}}{\sigma_{11}^{\frac{1}{2}}}$$

from (B.32b)

$$b_{22}^2 = \sigma_{22} - b_{21}^2$$

i.e. $$b_{22} = \left(\sigma_{22} - \frac{\sigma_{21}^2}{\sigma_{11}} \right)^{\frac{1}{2}}$$

from (B.32e)

$$b_{31} = \frac{\sigma_{31}}{b_{11}} = \frac{\sigma_{31}}{\sigma_{11}^{\frac{1}{2}}}$$

from (B.32f)

$$b_{32} = \frac{\sigma_{32} - b_{31}b_{21}}{b_{22}}$$

etc. From thus it can be readily verified that the recursive formula

$$b_{ij} = \frac{\sigma_{ij} - \sum_{k=1}^{j-1} b_{ik}b_{jk}}{\left(\sigma_{jj} - \sum_{k=1}^{j-1} b_{jk}^2 \right)^{\frac{1}{2}}} \qquad (B.33)$$

is valid provided that $\sum_{k=1}^{0}$ is interpreted as zero [Rubenstein, 1981].

Example B.1

It is desired to generate correlated normal variates X_1 and X_2 having means 10 and 12, respectively, and a covariance matrix C_X (which must be non-negative definite):

$$C_X = \begin{bmatrix} 2 & 2\sqrt{2} \\ 2\sqrt{2} & 7 \end{bmatrix}$$

From equations (B.33)

$$b_{11} = \sigma_{11}^{\frac{1}{2}} = \sqrt{2}$$

and

$$b_{21} = \frac{\sigma_{21}}{b_{11}} = \frac{2\sqrt{2}}{\sqrt{2}} = 2$$

$$b_{22} = \left(\sigma_{22} - \frac{\sigma_{21}^2}{\sigma_{11}} \right)^{\frac{1}{2}} = \left(7 - \frac{8}{2} \right)^{\frac{1}{2}} = \sqrt{3}$$

Hence

$$B = \begin{bmatrix} \sqrt{2} & 0 \\ 2 & \sqrt{3} \end{bmatrix}$$

and, from (3.61),

$$X_1 = \sqrt{2}Y_1 + 10$$
$$X_2 = 2Y_1 + \sqrt{3}Y + 12$$

where Y_1 and Y_2 are independent standardized normal variables generated through expressions (3.19).

Appendix C:
Complementary standard normal table

Table C.1 gives the $N(0,1)$ distribution; $\Phi(-\beta) = 1 - \Phi(\beta)$ (see section A.5.7).

Table C.1

β	$\Phi(-\beta)$	β	$\Phi(-\beta)$	β	$\Phi(-\beta)$
0.00	0.5000	0.40	0.3446	0.80	0.2119
0.01	0.4960	0.41	0.3409	0.81	0.2090
0.02	0.4920	0.42	0.3372	0.82	0.2061
0.03	0.4880	0.43	0.3336	0.83	0.2033
0.04	0.4841	0.44	0.3300	0.84	0.2005
0.05	0.4801	0.45	0.3264	0.85	0.1977
0.06	0.4761	0.46	0.3228	0.86	0.1949
0.07	0.4721	0.47	0.3192	0.87	0.1922
0.08	0.4681	0.48	0.3156	0.88	0.1894
0.09	0.4642	0.49	0.3121	0.89	0.1867
0.10	0.4602	0.50	0.3085	0.90	0.1841
0.11	0.4562	0.51	0.3050	0.91	0.1814
0.12	0.4522	0.52	0.3015	0.92	0.1788
0.13	0.4483	0.53	0.2981	0.93	0.1762
0.14	0.4443	0.54	0.2946	0.94	0.1736
0.15	0.4404	0.55	0.2912	0.95	0.1711
0.16	0.4364	0.56	0.2877	0.96	0.1685
0.17	0.4325	0.57	0.2843	0.97	0.1660
0.18	0.4286	0.58	0.2810	0.98	0.1635
0.19	0.4247	0.59	0.2776	0.99	0.1611
0.20	0.4207	0.60	0.2743	1.00	0.1587
0.21	0.4168	0.61	0.2709	1.01	0.1563
0.22	0.4129	0.62	0.2676	1.02	0.1539
0.23	0.4091	0.63	0.2644	1.03	0.1515
0.24	0.4052	0.64	0.2611	1.04	0.1492
0.25	0.4013	0.65	0.2579	1.05	0.1469
0.26	0.3974	0.66	0.2546	1.06	0.1446
0.27	0.3936	0.67	0.2514	1.07	0.1423
0.28	0.3897	0.68	0.2483	1.08	0.1401
0.29	0.3859	0.69	0.2451	1.09	0.1379
0.30	0.3821	0.70	0.2420	1.10	0.1357
0.31	0.3783	0.71	0.2389	1.11	0.1335
0.32	0.3745	0.72	0.2358	1.12	0.1314
0.33	0.3707	0.73	0.2327	1.13	0.1292
0.34	0.3669	0.74	0.2297	1.14	0.1271
0.35	0.3632	0.75	0.2266	1.15	0.1251
0.36	0.3594	0.76	0.2236	1.16	0.1230
0.37	0.3557	0.77	0.2207	1.17	0.1210
0.38	0.3520	0.78	0.2177	1.18	0.1190
0.39	0.3483	0.79	0.2148	1.19	0.1170

β	$\Phi(-\beta)$	β	$\Phi(-\beta)$	β	$\Phi(-\beta)$
1.20	0.1151	1.80	0.3593E-01	2.40	0.8198E-02
1.21	0.1131	1.81	0.3515E-01	2.41	0.7976E-02
1.22	0.1112	1.82	0.3438E-01	2.42	0.7760E-02
1.23	0.1094	1.83	0.3363E-01	2.43	0.7550E-02
1.24	0.1075	1.84	0.3289E-01	2.44	0.7344E-02
1.25	0.1057	1.85	0.3216E-01	2.45	0.7143E-02
1.26	0.1038	1.86	0.3144E-01	2.46	0.6947E-02
1.27	0.1020	1.87	0.3074E-01	2.47	0.6756E-02
1.28	0.1003	1.88	0.3005E-01	2.48	0.6569E-02
1.29	0.9853E-01	1.89	0.2938E-01	2.49	0.6387E-02
1.30	0.9680E-01	1.90	0.2872E-01	2.50	0.6210E-02
1.31	0.9510E-01	1.91	0.2807E-01	2.51	0.6037E-02
1.32	0.9342E-01	1.92	0.2743E-01	2.52	0.5868E-02
1.33	0.9176E-01	1.93	0.2680E-01	2.53	0.5703E-02
1.34	0.9013E-01	1.94	0.2619E-01	2.54	0.5543E-02
1.35	0.8851E-01	1.95	0.2559E-01	2.55	0.5386E-02
1.36	0.8692E-01	1.96	0.2500E-01	2.56	0.5234E-02
1.37	0.8535E-01	1.97	0.2442E-01	2.57	0.5085E-02
1.38	0.8380E-01	1.98	0.2385E-01	2.58	0.4940E-02
1.39	0.8227E-01	1.99	0.2330E-01	2.59	0.4799E-02
1.40	0.8076E-01	2.00	0.2275E-01	2.60	0.4661E-02
1.41	0.7927E-01	2.01	0.2222E-01	2.61	0.4527E-02
1.42	0.7781E-01	2.02	0.2169E-01	2.62	0.4397E-02
1.43	0.7636E-01	2.03	0.2118E-01	2.63	0.4269E-02
1.44	0.7494E-01	2.04	0.2068E-01	2.64	0.4145E-02
1.45	0.7353E-01	2.05	0.2018E-01	2.65	0.4025E-02
1.46	0.7215E-01	2.06	0.1970E-01	2.66	0.3907E-02
1.47	0.7078E-01	2.07	0.1923E-01	2.67	0.3793E-02
1.48	0.6944E-01	2.08	0.1876E-01	2.68	0.3681E-02
1.49	0.6811E-01	2.09	0.1831E-01	2.69	0.3573E-02
1.50	0.6681E-01	2.10	0.1786E-01	2.70	0.3467E-02
1.51	0.6552E-01	2.11	0.1743E-01	2.71	0.3364E-02
1.52	0.6426E-01	2.12	0.1700E-01	2.72	0.3264E-02
1.53	0.6301E-01	2.13	0.1659E-01	2.73	0.3167E-02
1.54	0.6178E-01	2.14	0.1618E-01	2.74	0.3072E-02
1.55	0.6057E-01	2.15	0.1578E-01	2.75	0.2980E-02
1.56	0.5938E-01	2.16	0.1539E-01	2.76	0.2890E-02
1.57	0.5821E-01	2.17	0.1500E-01	2.77	0.2803E-02
1.58	0.5706E-01	2.18	0.1463E-01	2.78	0.2718E-02
1.59	0.5592E-01	2.19	0.1426E-01	2.79	0.2635E-02
1.60	0.5480E-01	2.20	0.1390E-01	2.80	0.2555E-02
1.61	0.5370E-01	2.21	0.1355E-01	2.81	0.2477E-02
1.62	0.5262E-01	2.22	0.1321E-01	2.82	0.2401E-02
1.63	0.5155E-01	2.23	0.1287E-01	2.83	0.2327E-02
1.64	0.5050E-01	2.24	0.1255E-01	2.84	0.2256E-02
1.65	0.4947E-01	2.25	0.1222E-01	2.85	0.2186E-02
1.66	0.4846E-01	2.26	0.1191E-01	2.86	0.2118E-02
1.67	0.4746E-01	2.27	0.1160E-01	2.87	0.2052E-02
1.68	0.4648E-01	2.28	0.1130E-01	2.88	0.1988E-02
1.69	0.4552E-01	2.29	0.1101E-01	2.89	0.1926E-02
1.70	0.4457E-01	2.30	0.1072E-01	2.90	0.1866E-02
1.71	0.4363E-01	2.31	0.1044E-01	2.91	0.1807E-02
1.72	0.4272E-01	2.32	0.1017E-01	2.92	0.1750E-02
1.73	0.4182E-01	2.33	0.9903E-02	2.93	0.1695E-02
1.74	0.4093E-01	2.34	0.9642E-02	2.94	0.1641E-02
1.75	0.4006E-01	2.35	0.9387E-02	2.95	0.1589E-02
1.76	0.3921E-01	2.36	0.9138E-02	2.96	0.1538E-02
1.77	0.3836E-01	2.37	0.8894E-02	2.97	0.1489E-02
1.78	0.3754E-01	2.38	0.8657E-02	2.98	0.1441E-02
1.79	0.3673E-01	2.39	0.8424E-02	2.99	0.1395E-02

β	Φ(− β)	β	Φ(− β)	β	Φ(− β)
3.00	0.1350E-02	3.60	0.1591E-03	5.00	0.2859E-06
3.01	0.1306E-02	3.61	0.1531E-03	5.05	0.2203E-06
3.02	0.1264E-02	3.62	0.1473E-03	5.10	0.1694E-06
3.03	0.1223E-02	3.63	0.1417E-03	5.15	0.1299E-06
3.04	0.1183E-02	3.64	0.1363E-03	5.20	0.9935E-07
3.05	0.1144E-02	3.65	0.1311E-03	5.25	0.7582E-07
3.06	0.1107E-02	3.66	0.1261E-03	5.30	0.5772E-07
3.07	0.1070E-02	3.67	0.1212E-03	5.35	0.4384E-07
3.08	0.1035E-02	3.68	0.1166E-03	5.40	0.3321E-07
3.09	0.1001E-02	3.69	0.1121E-03	5.45	0.2510E-07
3.10	0.9676E-03	3.70	0.1077E-03	5.50	0.1892E-07
3.11	0.9354E-03	3.71	0.1036E-03	5.55	0.1423E-07
3.12	0.9042E-03	3.72	0.9956E-04	5.60	0.1067E-07
3.13	0.8740E-03	3.73	0.9569E-04	5.65	0.7985E-08
3.14	0.8447E-03	3.74	0.9196E-04	5.70	0.5959E-08
3.15	0.8163E-03	3.75	0.8837E-04	5.75	0.4436E-08
3.16	0.7888E-03	3.76	0.8491E-04	5.80	0.3293E-08
3.17	0.7622E-03	3.77	0.8157E-04	5.85	0.2438E-08
3.18	0.7363E-03	3.78	0.7836E-04	5.90	0.1800E-08
3.19	0.7113E-03	3.79	0.7527E-04	5.95	0.1325E-08
3.20	0.6871E-03	3.80	0.7230E-04	6.00	0.9716E-09
3.21	0.6636E-03	3.81	0.6943E-04	6.10	0.5220E-09
3.22	0.6409E-03	3.82	0.6667E-04	6.20	0.2778E-09
3.23	0.6189E-03	3.83	0.6402E-04	6.30	0.1463E-09
3.24	0.5976E-03	3.84	0.6147E-04	6.40	0.7636E-10
3.25	0.5770E-03	3.85	0.5901E-04	6.50	0.3945E-10
3.26	0.5570E-03	3.86	0.5664E-04	6.60	0.2018E-10
3.27	0.5377E-03	3.87	0.5437E-04	6.70	0.1023E-10
3.28	0.5190E-03	3.88	0.5218E-04	6.80	0.5130E-11
3.29	0.5009E-03	3.89	0.5007E-04	6.90	0.2549E-11
3.30	0.4834E-03	3.90	0.4804E-04	7.00	0.1254E-11
3.31	0.4664E-03	3.91	0.4610E-04	7.10	0.6107E-12
3.32	0.4500E-03	3.92	0.4422E-04	7.20	0.2946E-12
3.33	0.4342E-03	3.93	0.4242E-04	7.30	0.1407E-12
3.34	0.4189E-03	3.94	0.4069E-04	7.40	0.6654E-13
3.35	0.4040E-03	3.95	0.3902E-04	7.50	0.3116E-13
3.36	0.3897E-03	3.96	0.3742E-04	7.60	0.1445E-13
3.37	0.3758E-03	3.97	0.3588E-04	7.70	0.6636E-14
3.38	0.3624E-03	3.98	0.3441E-04	7.80	0.3017E-14
3.39	0.3494E-03	3.99	0.3298E-04	7.90	0.1359E-14
3.40	0.3369E-03	4.00	0.3162E-04	8.00	0.6056E-15
3.41	0.3248E-03	4.05	0.2557E-04	8.10	0.2673E-15
3.42	0.3131E-03	4.10	0.2062E-04	8.20	0.1169E-15
3.43	0.3017E-03	4.15	0.1659E-04	8.30	0.5058E-16
3.44	0.2908E-03	4.20	0.1332E-04	8.40	0.2167E-16
3.45	0.2802E-03	4.25	0.1067E-04	8.50	0.9197E-17
3.46	0.2700E-03	4.30	0.8524E-05	8.60	0.3864E-17
3.47	0.2602E-03	4.35	0.6794E-05	8.70	0.1608E-17
3.48	0.2507E-03	4.40	0.5402E-05	8.80	0.6623E-18
3.49	0.2415E-03	4.45	0.4285E-05	8.90	0.2701E-18
3.50	0.2326E-03	4.50	0.3391E-05	9.00	0.1091E-18
3.51	0.2240E-03	4.55	0.2677E-05	9.10	0.4363E-19
3.52	0.2157E-03	4.60	0.2108E-05	9.20	0.1728E-19
3.53	0.2077E-03	4.65	0.1656E-05	9.30	0.6773E-20
3.54	0.2000E-03	4.70	0.1298E-05	9.40	0.2629E-20
3.55	0.1926E-03	4.75	0.1015E-05	9.50	0.1011E-20
3.56	0.1854E-03	4.80	0.7914E-06	9.60	0.3847E-21
3.57	0.1784E-03	4.85	0.6158E-06	9.70	0.1450E-21
3.58	0.1717E-03	4.90	0.4780E-06	9.80	0.5408E-22
3.59	0.1653E-03	4.95	0.3701E-06	9.90	0.1998E-22

Appendix D:
Random numbers

Table D.1 gives a short list of random numbers generated for use with examples in the text only. Random numbers for use in realistic applications can be generated on a computer, or published tables may be used [Rand Corporation, 1955].

Table D.1.

0.9311	0.4537
0.7163	0.1827
0.4626	0.2765
0.7895	0.6939
0.8184	0.8189
0.3008	0.9415
0.3989	0.4967
0.0563	0.2097
0.1470	0.4575
0.2036	0.4950
0.6624	0.8463
0.2825	0.2812
0.9819	0.6504
0.1527	0.8517
0.0373	0.0716
0.2131	0.8970
0.4812	0.1217
0.7389	0.2333
0.7582	0.6336
0.8675	0.5620

References

Abramowitz, M., and Stegun, I. A. (eds) (1966). *Handbook of Mathematical Functions*, Applied Mathematics Series No. 55. National Bureau of Standards, Washington, DC.

Allen, D. E. (1968)., "Discussion of Turkstra, C. J., 'Choice of Failure Probabilities'", *J. Struct. Div.*, *ASCE*, **94**, pp. 2169–2173.

Allen, D. E. (1970). "Probabilistic Study of Reinforced Concrete in Bending", *J. Amer. Conc. Inst.*, **67**, No. 12, pp. 989–993.

Allen, D. E. (1975). "Limit States Design — A Probabilistic Study", *Can. J. Civ. Eng.*, **2**, No. 1, pp. 36–49.

Allen, D. E. (1981a). "Criteria for Design Safety Factors and Quality Assurance Expenditure", (in) Moan, T., and Shinozuka, M. (eds), *Structural Safety and Reliability*. Elsevier, Amsterdam, pp. 667–678.

Allen, D. E. (1981b). "Limit States Design: What do we really want?", *Can. J. Civ. Eng.*, **8**, pp. 40–50.

Alpsten, G. A. (1972). "Variations in Mechanical and Cross-Sectional Properties of Steel", (in) *Proc. Int. Conf. on Planning and Design of Tall Buildings*, Vol. Ib. Lehigh University, Bethlehem, PA, pp. 775–805.

Ang, A. H.-S., and Tang, W. H. (1975). *Probability Concepts in Engineering Planning and Design*, Vol. I, *Basic Principles*. John Wiley.

ASCE (Committee on Fatigue and Fracture Reliability) (1982). "Fatigue Reliability" (a series of papers), *J. Struct. Div.*, *ASCE*, **108**, No. ST1, pp. 3–88.

Augusti, G., and Baratta, A. (1973). "Theory of Probability and Limit Analysis of Structures under Multiparameter Loading", (in) Sawczuk, A. (ed.), *Foundations of Plasticity*. Noordhoff, Leyden, pp. 347–364.

Augusti, G., Baratta, A., and Casciati, F. (1984). *Probabilistic Methods in Structural Engineering*. Chapman and Hall, London.

Ayyab, B. M., and Haldar, A. (1984). "Practical Structural Reliability Techniques", *J. Struct. Engg.*, *ASCE*, **110**, No. 8, pp. 1707–1724.

Baker, M. J. (1969). "Variations in the Mechanical Properties of Structural

Steels", (in) *Final Report, Symposium on Concepts of Safety of Structures and Methods of Design*. IABSE, London, pp. 165–174.

Baker, M. J. (1976). "Evaluation of Partial Safety Factors for Level I Codes — Example of Application of Methods to Reinforced Concrete Beams", Bulletin d'Information No. 112. Comite Européen due Béton, Paris, pp. 190–211.

Baker, M. J. (1985). "The Reliability Concept as an Aid to Decision Making in Offshore Engineering", (in) *Behaviour of Offshore Structures*. Elsevier, Amsterdam, pp. 75–94.

Baker, M. J., and Wyatt, T. (1979). "Methods of Reliability Analysis for Jacket Platforms", (in) *Proc. 2nd Int. Conf. on Behaviour of Offshore Structures*, London. British Hydromechanics Research Association, Cranfield, Berks., pp. 499–520.

Basler, E. (1961), "Untersuchungen über den Sicherheitsbegriff von Bauwerken", *Schweiz. Arch.*, **27**, No. 4, pp. 133–160.

Batts, M. E., Russell, L. R., and Simiu, E. (1980). "Hurricane Wind Speed in the United States", *J. Struct. Div.*, ASCE, **106**, No. ST10, pp. 2001–2016.

Belyaev, Y. K. (1968). "On the Number of Exists Across the Boundary of a Region by a Vector Stochastic Process", *Theory Prob. Appl.*, **13**, No. 2, pp. 320–324.

Belyaev, Y. K., and Nosko, V. P. (1969). "Characteristics of Excursions Above a High Level for a Gaussian Process and its Envelope", *Theory Prob. Appl.*, **14**, pp. 296–309.

Benjamin, J. R. (1970). "Reliability Studies in Reinforced Concrete Design", (in) Lind, N. C. (ed.) *Structure Reliability and Codified Design*, SM Study No. 3, University of Waterloo, Waterloo, Ontario.

Benjamin, J. R., and Cornell, C. A. (1970). *Probability, Statistics and Decisions for Civil Engineers*. McGraw-Hill, New York.

Bennett, R. M., and Ang. A. H.-S. (1983). *Investigation of Methods for Structural Systems Reliability*, Structural Research Series No. 510. University of Ilinois, Urbana, IL.

Berthellamy, J., and Rackwitz, R. (1979). "Multiple Point Checking in System Reliability", SFB 96. Technical University, Munich.

Beveridge, G. S. G., and Schechter, R. S. (1970). *Optimization; Theory and Practice*. McGraw-Hill, New York.

Bjorhovde, R., Galambos, T. V., and Ravindra, M. K. (1978). "LRFD Criteria for Steel Beam-columns", *J. Struct. Div.*, ASCE, **104**, No. ST9, pp. 1371–1387.

Blockley, D. I. (1980). *The Nature of Structural Design and Safety*. Ellis Horwood, Chichester.

Blockley, D. I. (1985). "Discussion of Ditlevsen, O., 'Fundamental Postulate in Structural Safety'", *J. Engg. Mech.*, ASCE, **111**, No. 1, p. 108.

Bonferroni, C. E. (1936). "Teoria statistica classi e calcolo della proababilita", *Pubbl. R. Ist. Super Sci. Econ. Comm.*, Firenze, **8**, pp. 1–63.

Borgman, L. E. (1963), "Risk Criteria", *J. Waterways Harbours Div.*, ASCE, **89**, No. WW3, 1–35.

Borgman, L. E. (1967). "Spectral Analysis of Ocean Wave Forces on Piling", *J. Waterways Harbors Div.*, *ASCE*, **93**, No. WW2, pp. 129–156.

Bosshard, W. (1975). "On Stochastic Load Combinations", Technical Report No. 20. Department of Civil Engineering, Stanford University, Stanford, CA.

Bosshard, W. (1979). "Structural Safety — A matter of Decision and Control", IABSE Surveys No. S-9/1979, May. IABSE, London, pp. 1–27.

Box, G. E. P., and Muller, M. E. (1958). "A Note on the Generation of Normal Deviates", *Ann. Math. Stat.*, **29**, p. 610–611.

Breitung, K., and Rackwitz, R. (1982). "Non-linear Combination Load Processes", *J. Struct. Mech.*, **10**, No. 2, pp. 145–166.

Broding, W. C., Diederich, F. W., and Parker, P. S. (1964). "Structural Optimization and Design based on a Reliability Design Criterion", *J. Spacecraft*, **1**, No. 1, pp. 56–61.

CEB (1976). "Common Unified Rules for Different Types of Construction and Material" (3rd draft), Bulletin d'Information No. 116-E. Comité Européen du Béton, Paris.

Chalk, P. L., and Corotis, R. B. (1980). "Probability Model for Design Live Loads", *J. Struct. Div.*, *ASCE*, **106**, No. ST10, pp. 2107–2033.

Chen, X., and Lind, N. C. (1983). "Fast Probability Integration by Three-parameter Normal Tail Approximation", *Struct. Safety*, **1**, No. 4, pp. 269–276.

Cibula, E. (1971). "The Structure of Building Control — An International Comparison", Current Paper No. CP 28/71. Building Research Station, Garston, Herts.

CIRIA (1977). "Rationalization of Safety and Serviceability Factors in Structural Codes", Report No. 63. Construction Industry Research and Information Association, London.

Clough, R. W., and Penzien, J. (1975). *Dynamics of Structures*. McGraw-Hill, New York.

Cooper, P. B., Galambos, T. V., and Ravindra, M. K. (1978). "LRFD Criteria for Plate Girders", *J. Struct. Div.*, *ASCE*, **104**, No. ST9, pp. 1389–1407.

Cornell, C. A. (1967), "Bounds on the Reliability of Structural Systems", *J. Struct. Div.*, *ASCE*, **93**, No. ST1, 171–200.

Cornell, C. A. (1969). "A Probability Based Structural Code", *J. Amer. Conc. Inst.*, **66**, No. 12, pp. 974–985.

Corotis, R. B., and Doshi, V. A. (1977). "Probability Models for Live Load Survey Results", *J. Struct. Div.*, *ASCE*, **103**, No. ST6, pp. 1257–1274.

Cramer, H., and Leadbetter, M. R. (1967). *Stationary and Related Stochastic Processes*. John Wiley, New York.

Crandall, S. H., and Mark, W. D. (1963). *Random Vibration in Mechanical Systems*. Academic Press, New York.

CSA (1974). "Steel Structures for Buildings — Limit States Design", CSA Standard No. S16.1 — 1974. Canadian Standards Association.

Culver, C. G. (1976). "Survey Results for Fire Loads and Live Loads in Office Buildings", NBS Building Science Series Report No. 85. Center for Building Technology, National Bureau of Standards, Washington, DC.

Curnow, R. N., and Dunnett, C. W. (1962). "The Numerical Evaluation of Certain Multivariate Normal Integrals", *Ann. Math. Stat.*, **33**, No. 2, pp. 571–579.

Dahlquist, G., and Björck, A. (1974). *Numerical Methods*. Prentice-Hall, Englewood Cliffs, NJ.

Daley, D. J. (1974). "Computation of Bi- and Tri-variate Normal Integrals", *Appl. Stat.*, **23**, No. 3, pp. 435–438.

Daniels, H. E. (1945). "The Statistical Theory of the Strength of Bundles of Threads", *Proc. Roy. Soc.*, *Ser. A*, **183**, pp. 405–435.

Davenport, A. G. (1961). "The Application of Statistical Concepts to the Wind Loading of Structures", *Proc. Inst. Civ. Engrs.*, **19**, pp. 449–472.

Davenport, A. G. (1967), "Gust Loading Factors", *J. Struct. Div.*, *ASCE*, **93**, No. ST3, pp. 11–34.

Davenport, A. G. (1983). "The Reliability and Synthesis of Aerodynamic and Meterological Data for Wind Loading", (in) Thoft-Christensen, P. (ed.), *Reliability Theory and Its Application in Structural and Soil Mechanics*, NATO Advanced Study Institute Series E, No. 70. Martinus Nijhoff, The Hague, pp. 314–335.

Davis, P. J., and Rabinowitz, P. (1975). *Methods of Numerical Integration*. Academic Press, New York.

Dawson, D. A., and Sankoff, D. (1967). "An Inequality for Probabilities", *Proc. Amer. Math. Soc.*, **18**, pp. 504–507.

Deak, I. (1980). "Fast Procedures for Generating Stationary Normal Vectors", *J. Stat. Comput. Simul.*, **10**, pp. 225–242.

de Finetti, B. (1974). *Theory of Probability*. John Wiley.

de Neufville, R., and Stafford, J. H. (1971). *Systems Analysis for Engineers and Managers*. McGraw-Hill, New York.

der Kiureghian, A., and Liu, P.-L. (1985). "Structural Reliability Under Incomplete Probability Information", Report No. UCB/SESM-85/01. Department of Civil Engineering, University of California, Berkeley, CA.

der Kiureghian, A., and Taylor, R. L. (1983). "Numerical Methods in Structural Reliability", (in) Augusti, G., Borri, A., and Vannuchi, G. (eds.), *Proc. 4th Int. Conf. on Applications of Statistics and Probability in Soil and Structural Engineering*. Pitagora Editrice, Bologna, pp. 769–775.

Didonato, A. R., Jarnagin, M. P., and Hageman, R. K. (1980). "Computation of the Integral of the Bivariate Normal Distribution over Convex Polygons", *SIAM J. Sci. Stat. Comput.*, **1**, No. 2, pp. 179–186.

Ditlevsen, O. (1973). "Structural Reliability and the Invariance Problem", Solid Mechanics Report No. 22. University of Waterloo, Waterloo, Ontario.

Ditlevsen, O. (1979a). "Generalized Second Moment Reliability Index", *J. Struct. Mech.*, **7**, No. 4, pp. 435–451.

Ditlevsen, O. (1979b). "Narrow Reliability Bounds for Structural Systems", *J. Struct. Mech.*, **7**, No. 4, pp. 453–472.

Ditlevsen, O. (1981a). "Principle of Normal Tail Approximation", *J. Engg. Mech. Div.*, *ASCE*, **107**, No. EM6, pp. 1191–1208.

Ditlevsen, O. (1981b). Uncertainty Modelling. McGraw-Hill, New York.

Ditlevsen, O. (1982a). "The Fate of Reliability Measures as Absolutes", *Nucl. Eng. Des.*, **71**, pp. 439–440.

Ditlevsen, O. (1982b). "Systems Reliability Bounding by Conditioning", *J. Engg. Mech. Div.*, *ASCE*, **108**, No. EM5, pp. 708–718.

Ditlevsen, O. (1983a). "Fundamental Postulate in Structural Safety", *J. Engg. Mech. Div.*, *ASCE*, **109**, no. 4, pp. 1096–1102.

Ditlevsen, O. (1983b). "Gaussian Outcrossings from Safe Convex Polyhedrons", *J. Engg. Mech. Div.*, *ASCE*, **109**, No. 1, pp. 127–148.

Ditlevsen, O., and Bjerager, P. (1983). "Reliability of Highly Redundant Plastic Structures, DCAMM Report No. 263. Technical University, Denmark.
Also (1984). *J. Engg. Mech.*, *ASCE*, **110**, pp. 671–693.

Ditlevsen, O., and Madsen, H. O. (1980). "Discussion of 'Optimal Reliability Analysis by Fast Convolution' by N. C. Lind", *J. Engg. Mech. Div.*, *ASCE*, **106**, No. EM3, pp. 579–583.

Ditlevsen, O., and Madsen, H. O. (1983). "Transient Load Modeling: Clipped Normal Processes", *J. Engg. Mech. Div.*, *ASCE*, **109**, No. 2, pp. 495–515.

Dolinsky, K. (1983). "First Order Second-moment Approximation in Reliability of Structural Sytsems: Critical Review and Alternative Approach", *Struct. Safety*, **1**, No. 3, pp. 211–231.

Drezner, Z. (1978). "Computation of the Bivariate Normal Integral", *Math. Comput.*, **32**, No. 141, pp. 277–279.

Drury, C. G., and Fox, J. G. (eds) (1975). *Human Reliability in Quality Control*. Taylor and Francis, London.

Dunnett, C. W., and Sobel, M. (1955). "Approximations to the Probability Integral and Certain Percentage Points of a Multivariate Analogue of Students *t*-Distribution", *Biometrika*, **42**, pp. 258–260.

Elderton, W. P., and Johnson, M. L. (1969). *Systems of Frequency Curves*. Cambridge University Press.

Ellingwood, B. (1977). "Statistical Analysis of R. C. Beam-Column Interaction", *J. Struct. Div.*, *ASCE*, **103**, No. ST7, pp. 1377–1388.

Ellingwood, B., and Culver, C. (1977). "Analysis of Live Loads in Office Buildings", *J. Struct. Div.*, *ASCE*, **103**, no. ST8, pp. 1551–1560.

Ellingwood, B., Galambos, T. V., MacGregor, J. C., and Cornell, C. A. (1980). "Development of a Probability Based Load Criteria for American National Standard A58", NBS Special Publication No. 577. National Bureau of Standards, US Department of Commerce, Washington, DC.

Ellyin, F., and Chandrasekhar, P. (1977). "Probabilistic Dynamic Response of Beams and Frames", *J. Engg. Mech. Div.*, *ASCE*, **103**, No. EM3, pp. 411–421.

Entroy, H. C. (1960). "The Variation of Works Test Cubes", Research Report No. 10. Cement and Concrete Association, UK.

Feller, W. (1957). *An Introduction to Probability Theory and Its Applications*, Vol. 1 (2nd edn). John Wiley.

Ferry-Borges, J., and Castenheta, M. (1971). *Structural Safety*. Laboratoria Nacional de Engenhera Civil, Lisbon.

Fiessler, B. (1979). "Das Programmsystem FORM zur Berechnung der Versagenswahrscheinlichkeit von Komponenten von Tragsystemen", Berichte zur Zuverlässigkeits Theorie der Bauwerke. No. 43. Technical University, Munich.

Fiessler, B., Hawranek, R., and Rackwitz, R. (1976). "Numerische Methoden für probabilistische Bemessungsverfahren und Sicherheitsnachweise", Berichte zur Sicherheitsteorie de Bauwerke No. 14. Technical University, Munich.

Fiessler, B., Neumann, H.-J., and Rackwitz, R. (1979). "Quadratic Limit States in Structural Reliability", *J. Engg. Mech. Div.*, *ASCE*, **105**, No. EM4, pp. 661–676.

Fishburn, P. C. (1964). Decision and Value Theory. John Wiley.

Fisher, J. W., Galambos, T. V., Kulak, G. L., and Ravindra, M. K. (1978). "Load and Resistance Factor Design Criteria for Connectors, *J. Struct. Div.*, *ASCE*, **104**, No. ST9, pp. 1427–1441.

Fisher, J. W., and Struik, J. H. A. (1974). *Guide to Design Criteria for Bolted and Riveted Joints*. John Wiley, New York.

Flint, A. R., Smith, B. W., Baker, M. J., and Manners, W. (1981). "The Derivation of Safety Factors for Design of Highway Bridges", (in) *Proc. Conf. on The New Code for the Design of Steel Bridges*. Granada Publishing.

Frangopol, D. M. (1985). "Sensitivity Studies in Reliability Based Analysis of Redundant Structures", *Struct. Safety*, **3**, No. 1, pp. 13–22.

Freeman, H. (1963). *An Introduction to Statistical Inference*. Addison-Wesley, Reading, MA.

Freudenthal, A. M. (1956). "Safety and the Probability of Structural Failure", *Trans. ASCE*, **121**, pp. 1337–1397.

Freudenthal, A. M. (1961). "Safety, Reliability and Structural Design", *J. Struct. Div.*, *ACSE*, **87**, No. ST3, pp. 1–16.

Freudenthal, A. M. (1964). "Die Sicherheit der Baukonstruktionen", *Acta Tech. Hung.*, **46**, pp. 417–446.

Freudenthal, A. M. (1975). "Structural Safety, Reliability and Risk Assessment", (in) Freudenthal, A. M., *et al.*, (eds), *Reliability Approach in Structural Engineering*. Maruzen, Tokyo.

Freudenthal, A. M., Garrelts, J. M., and Shinozuka, M. (1966). "The Analysis of Structural Safety", *J. Struct. Div.*, *ASCE*, **92**, No. ST1, 267–325.

382 **References**

Galambos, T. V., and Ravindra, M. K. (1978). "Properties of Steel for Use in LRFD", *J. Struct. Div., ASCE*, **104**, No. ST9, pp. 1459–1468.

Garson, R. C. (1980). "Failure Mode Correlation in Weakest-link Systems", *J. Struct. Div., ASCE*, **106**, No. ST8. pp. 1797–1810.

Gaver, D. P., and Jacobs, P. (1981). "On Combination of Random Loads", *J. Appl. Maths., SIAM*, **40**, No. 3, pp. 454–466.

Gollwitzer, S., and Rackwitz, R. (1983). "Equivalent Components in First-order System Reliability", *Reliab. Engng.*, **5**, pp. 99–115.

Gomes, L., and Vickery, B. J. (1976). "Tropical Cyclone Gust Speeds along the Northern Australian Coast", *Civ. Engg. Trans. Inst. Engrs. Aust.*, **CE 18**, No. 2, pp. 40–48.

Gorman, M. R. (1979). "Reliability of Structural Systems", Report No. 79–2. Department of Civil Engineering, Case Western Reserve University, Cleveland, OH.

Gorman, M. R. (1981). "Automatic Generation of Collapse Mode Equations", *J. Struct. Div., ASCE*, **107**, No. ST7, pp. 1350–1354.

Gorman, M. R. (1984). "Structural Resistance Moments by Quadrature", *Struct. Safety*, **2**, pp. 73–81.

Grant, L. H., Mizra, S. A., and MacGregor, J. G. (1978). "Monte Carlo Study of Strength of Concrete Columns", *J. Amer. Conc. Inst.*, **75**, No. 8, pp. 348–358.

Grigoriu, M. (1975). "On the Maximum of the Sum of Random Process Load Models", Internal Project Working Document No. 1. Department of Civil Engineering, Massachusetts Institute of Technology, Cambridge, MA.

Grigoriu, M. (1982). "Methods for Approximate Reliability Analysis", *Struct. Safety*, **1**, No. 2, pp. 155–165.

Grigoriu, M. (1983). "Approximate Analysis of Complex Reliability Problems", *Struct. Safety*, **1**, No. 4, pp. 277–288.

Grigoriu, M. (1984). "Crossings of Non-Gaussian Translation Processes", *J. Engg. Mech. Div., ASCE*, **110**, No. 6, pp. 610–620.

Grimmelt, M. J., and Schuëller, G. I. (1982). "Benchmark Study on Methods to Determine Collapse Failure Probabilities of Redundant Structures", *Struct. Safety*, **1**, pp. 93–106.

Grimmelt, M. J., Schuëller, G. I., and Murotsu, Y. (1983). "On the Evaluation of Collapse Probabilities", (in) *Proc. 4th ASCE–EMD Speciality Conf. on Recent Advances in Engineering Mechanics*, Vol. II. pp. 859–862.

Guiffre, N., and Pinto, P. E. (1976). "Discretisation from a Level II Method", Bulletin d'Information No. 112. Comité Européen du Béton, Paris, pp. 158–189.

Gumbel, E. J. (1958). *Statistics of Extremes*. Columbia University Press, New York.

Gupta, S. S. (1963). "Probability Integrals of Multivariate Normal and Multivariate *t*", *Ann. Math. Stat.*, **34**, pp. 792–828.

Hagen, J. (ed.) (1983). *Deterrence Reconsidered*. Sage Publications.

Hallam, M. G., Heaf, N. J., and Wootton, L. R. (1978). *Dynamics of Marine* Structures (2nd edn). CIRIA Underwater Engineering Group, London.

Hammersley, J. M., and Handscomb, D. C. (1964). *Monte Carlo Methods*. John Wiley, New York.

Harbitz, A. (1983). "Efficient and Accurate Probability of Failure Calculation by Use of the Importance Sampling Technique", (in) Augusti, G., Borri, A., and Vannuchi, O. (eds), *Proc. 4th Int. Conf. on Applications of Statistics and Probability in Soil and Structural Engineering*. Pitagora Editrice, Bologna, pp. 825–836.

Harrington, M. V., and Melchers, R. E. (1985). "Time Dependent Structural Reliability", *Civ. Engg. Trans. Inst. Engrs. Aust.*, **CE27**, No. 1, pp. 130–135.

Harris, D. H., and Chaney, F. B. (1969). *Human Factors in Quality Assurance*. John Wiley.

Harris, I. R. (1971). "The Nature of Wind", (in) *"The Modern Design of Wind Sensitive Structures*. Construction Industry Research and Information Association, London.

Harris, M. E., Corotis, R. B., and Bova, C. J. (1981). "Area Dependent Processes for Structural Live Loads", *J. Struct. Div.*, ASCE, **107**, No. ST5, pp. 857–872.

Hasofer, A. M. (1974). "The Upcrossing Rate of a Class of Stochastic Processes", (in) Williams, E. J. (ed.), *Studies in Probability and Statistics*. North-Holland, Amsterdam, pp. 151–159.

Hasofer, A. M. (1984). "Objective Probabilities for Unique Objects", (in) M. Grigoriu (ed.), *Risk, Structural Engineering and Human Error*. University of Waterloo Press, Waterloo, Ontario, pp. 1–16.

Hasofer, A. M., and Lind, N. C. (1974). "Exact and Invariant Second-moment Code Format", *J. Engg. Mech. Div.*, ASCE, **100**, EM1, pp. 111–121.

Hasselmann, K., *et al.* (1973). "Measurements of the Wind Wave Growth and Swell Decay during the Joint North Sea Wave Project (JONS-WAP)", Ergänzungsheft zur Deutsche Hydrographischen Zeitschrift, Reihe A(8), No. 12.

Hastings, C., Jr. (1955). *Approximations for Digital Computers*. Princeton University Press, Princeton, NJ.

Haugen, E. B. (1968). *Probabilistic Approaches to Design*. JohnWiley.

Hawrenek, R., and Rackwitz, R. (1976). "Reliability Calculation for Steel Columns", Bulletin d'Information No. 112. Comité Européen du Béton, Paris, pp. 125–157.

Henley, E. J., and Kumamoto, H. (1981). *Reliability Engineering and Risk Assessment*. Prentice-Hall, Englewood Cliffs, NJ.

Heyman, J. (1971). *Plastic Design of Frames*, Vol. 2. Cambridge University Press.

Hohenbichler, M. (1980). *Abschätzungen für die versagenswahrscheinlichkeiten von Seriensystemen*. Technical University, Munich.

Hohenbichler, M., and Rackwitz, R. (1981). "Non-normal Dependent Vectors in Structural Safety", *J. Engg. Mech. Div.*, *ASCE*, **107**, No. EM6, pp. 1227–1237.

Hohenbichler, M., and Rackwitz, R. (1983a). "First-order Concepts in Systems Reliability", *Struct. Safety*, **1**, No. 3, pp. 177–188.

Hohenbichler, M., and Rackwitz, R. (1983b). "Reliability of Parallel Systems under Imposed Uniform Strain", *J. Engg. Mech. Div.*, *ASCE*, **109**, No. 3, pp. 896–907.

Holmes, P., Chaplin, J. R., and Tickell, R. G. (1983). "Wave Loading and Structure Response", (in) *Design of Offshore Structures*. Thomas Telford, London, pp. 3–12.

Holmes, P., and Tickell, R. G. (1979). "Full Scale Wave Loading on Cylinders", (in) *Proc. 2nd Int. Conf. on Behaviour of Offshore Structures, London, Vol. 3*. (British Hydromechanics Research Association, Cranfield, Berks., pp. 746–761.

Horne, M. R., and Price, P. H. (1977). "Commentary on the Level 2 Procedure", Rationalization of Safety and Serviceability Factors in Structural Codes, Report No. 63. Construction Industry Research and Information Association, London, pp. 209–226.

Hunter, D. (1976). "An Upper Bound for the Probability of a Union", *J. Appl. Prob.*, **13**, pp. 597–603.

Hunter, D. (1977). "Approximate Percentage Points of Statistics Expressible as Maxima", *TIMS Stud. Management Sci.*, **7**, pp. 25–36.

Ingles, O. G. (1979). "Human Factors and Error in Civil Engineering", (in) *Proc. 3rd Int. Conf. on Applications of Statistics and Probability in Soil Structural Engineering*. Sydney, pp. 402–417.

Johnson, N. L., and Kotz, S. (1972). *Distributions in Statistics: Continuous Multivariate Distributions*. John Wiley, New York.

Johnston, B. G., and Opila, F. (1941). "Compression and Tension Tests of Structural Alloys", *ASTM, Proc.*, **41**, pp. 552–570.

Joos, D. W., Sabri, Z. A., and Hussein, A. A. (1979). "Analysis of Gross Error Rates in Operation of Commercial Nuclear Power Stations", *Nucl. Engg. Des.*, **52**, 265–300.

Julian, O. G. (1957). "Synopsis of First Progress Report of Committee on Safety Factors", *J. Struct. Div.*, *ASCE*, **83**, No. ST4, pp. 1316.1–1316.22.

Kahn, H. (1956). "Use of Different Monte Carlo Sampling Techniques", (in) H. A. Meyer (ed.), *Proc. Symp. on Monte Carlo Methods*. John Wiley, New York, pp. 149–190.

Kaimal, J. C., Wyngaard, J. C., Izumi, Y., and Cote, O. R. (1972). "Spectral Characteristics of Surface Layer Turbulence", *Q.J. Roy. Meteorol. Soc.*, **98**, pp. 563–589.

Karadeniz, H., van Manen, S., and Vrouwenvelder, A. (1984). "Probabilistic Reliability Analysis for the Fatigue Limit State of Offshore Structures", *Bull. Tech. Bur. Veritas*, pp. 203–219.

Knappe, O. W., Schuëller, G. J., and Wittmanm, F. H. (1975). "On the

Probability of Failure of a Reinforced Concrete Beam in Flexure", (in) *Proc. 2nd Int. Conf. on Applications of Statistics and Probabilities in Soil Structural Engineering*. Aachen, pp. 153–170.

Knoll, F. (1985). "Quality, Whose Job?", *Introductory Report, Symp. on Safety and Quality Assurance of Civil Engineering Structures*, Tokyo, Report No. 50. IABSE, London, pp. 59–64.

Kounias, E. G. (1968). "Bounds for the Probability of a Union, with Applications", *Amer. Math. Stat.*, **39**, No. 6, pp. 2154–2158.

Kupfer, J., and Rackwitz, R. (1980). "Models for Human Error and Control in Structural Reliability", *Final Report, 11th Congr.* IABSE, London, pp. 1019–1024.

Larrabee, R. D., and Cornell, C. A. (1979). "Upcrossing Rate Solution for Load Combinations", *J. Struct. Div., ASCE*, **105**, No. ST1, pp. 125–132.

Larrabee, R. D., and Cornell, C. A. (1981). "Combination of Various Load Processes", *J. Struct. Div., ASCE*, **107**, No. ST1, pp. 223–239.

Lay, M. G. (1979). "Implications of Probabilistic Methods in Steel Structures", (in) *Proc. 3rd Int. Conf. on Applications of Statistics and Probability in Soil and Structural Engineering*, Vol. 3. Sydney, pp. 145–156.

Legerer, F. (1970). "Code Theory — A New Branch of Engineering Science", (in) Lind, N. C. (ed.), *Structural Reliability and Codified Design*, SM Study No. 3. University of Waterloo, Waterloo, Ontario, pp. 113–127.

Leicester, R. H., and Beresford, R. D. (1977). "A Probabilistic Model for Serviceability Specifications", (in) *Proc. 6th Australas. Conf. on the Mechanics and Structures of Materials*. Christchurch, pp. 407–413.

Lighthill, J. (1978). *Waves in Fluids*. Cambridge University Press.

Lin, T. S., and Corotis, R. B. (1985). "Reliability of Ductile Systems with Random Strengths", *J. Struct. Engg., ASCE*, **111**, No. 6, pp. 1306–1325.

Lin, Y. K. (1970). "First Excursion Failure of Randomly Excited Structures, II", *AIAA J.*, **8**, No. 10, pp. 1888–1890.

Lind, N. C. (1969). "Deterministic Formats for the Probabilistic Design of Structures", (in) N. Kachaturian (ed.), *An Introduction to Structural Optimization*, SM Study No. 1. University of Waterloo, Waterloo, Ontario, pp. 121–142.

Lind, N. C. (1972). *Theory of Codified Structural Design*. University of Waterloo, Waterloo, Ontario.

Lind, N. C. (1976a). "Approximate Analysis and Economics of Structures", *J. Struct. Div., ASCE*, **102**, No. ST6, pp. 1177–1196.

Lind, N. C. (1976b). "Application to Design of Level I Codes", Bulletin d'Information No. 112. Comité Européen du Béton, Paris, pp. 73–89.

Lind, N. C. (1977). "Formulation of Probabilistic Design", *J. Engg. Mech. Div., ASCE*, **103**, No. EM2, pp. 273–284.

Lind, N. C. (1979). "Optimal Reliability Analysis by Fast Convolution", *J. Engg. Mech. Div., ASCE*, **105**, No. EM3, pp. 447–452.

Lind, N. C. (1983). "Models of Human Error in Structural Reliability", *Struct. Safety*, **1**, No. 3, pp. 167–175.

Lind, N. C., and Davenport, A. G. (1972). "Towards Practical Application of Structural Reliability Theory", (in) *Probabilistic Design of Reinforced Concrete Buildings*, Special Publication No. 31. American Concrete Institute.

Lindley, D. V. (1972). *Bayesian Statistics, A Review*. Society of Industrial and Applied Mathematics.

Longuet-Higgins, M. S. (1952). "On the Statistical Distribution of the Heights of Sea Waves", *J. Marine Sci.* **11**, pp. 245–266.

Luthans, F. (1981). *Organizational Behaviour* (3rd edn). McGraw-Hill, New York.

Lyse, I., and Keyser, C. C. (1934). "Effect of Size and Shape of Test Specimens upon the Observed Physical Properties of Structural Steel", *ASTM*, Proc. **34**, Part II, pp. 202–210.

Ma, H.-F., and Ang, A. H.-S. (1981), "Reliability Analysis of Redundant Ductile Structural Systems", Structural Research Series No. 494. Department of Civil Engineering, University of Illinois, Urbana, IL.

MacGregor, J. G. (1976). "Safety and Limit States Design for Reinforced Concrete", *Can. J. Civ. Engg.*, **3**, No. 4, pp. 484–513.

Madsen, H. O. (1982). "Deterministic and Probabilistic Models for Damage Accumulation due to Time Varying Loading", DIALOG 5-82. Danish Engineering Academy, Lyngby, Denmark.

Madsen, H., Kilcup, R., and Cornell, C. A. (1979). "Mean Upcrossing Rate for Stochastic Load Processes", (in) *Probabilistic Mechanics and Structural Reliability*. ASCE, New York, 54–58.

Mann, N. R., Schafer, R. E., and Singpurwalla, N. D. (1974). *Methods for Statistical Analysis of Reliability and Life Data*. John Wiley, New York.

Matousek, M., and Schneider, J. (1976). "Untersuchungen zur Struktur des Sicherheitsproblems bei Bauwerken", Bericht No. 59. Institut für Baustatik und Konstruktion, Eidgenössiche Technische Hochschule, Zurich.

Mayer, H. (1926). *Die Sicherheit der Bauwerke*. Springer, Berlin.

McGuire, R. K., and Cornell, C. A. (1974). "Live Load Effects in Office Buildings", *J. Struct. Div.*, ASCE, **100**, No. ST7, pp. 1351–1366.

Meister, S. (1966), "Human Factors in Reliability", (in) Ireson, W. G. (ed.), *Reliability Handbook*. McGraw-Hill, New York.

Melbourne, W. H. (1977). "Probability Distributions Associated with the Windloading of Structures", *Civ. Engg. Trans. Inst. Engrs. Aust.*, **CE 19**, No. 1, pp. 58–67.

Melchers, R. E. (1977). "Influence of Organization on Project Implementation", *J. Constr. Div.*, ASCE, **103**, No. CO4, pp. 611–625.

Melchers, R. E. (1978). "The Influence of Control Processes in Structural Engineering", *Proc. Inst. Civ. Engrs.*, **65**, Part 2, pp. 791–807.

Melchers, R. E. (1979). "Selection of Control Levels for Maximum Utility of Structures" (in) *Proc. 3rd Int. Conf. on Applications of Statistics and Probability in Soil and Structural Engineering*. Sydney, pp. 839–849.

Melchers, R. E. (1980). "Societal Options for Assurance of Structural Performance", *Final Report*, 11*th Congr*. IABSE, London, pp. 983–988.

Melchers, R. E. (1981). "On Bounds and Approximations in Structural Systems Reliability", Research Report No. 1/1981. Department of Civil Engineering, Monash University.

Melchers, R. E. (1983a). "Reliability of Parallel Structural Systems", *J. Struct. Div.*, *ASCE*, **109**, No. 11, pp. 2651–2665.

Melchers, R. E. (1983b). "Static Theorem Approach to the Reliability of Parallel Plastic Structures", (in) Augusti, G., Borri, A., and Vannuchi, G. (eds.), *Proc. 4th Int. Conf. on Applications of Statistics and Probability in Soil and Structural Engineering*, Vol. 2, Pitagora Editrice, Bologna, pp. 1313–1324.

Melchers, R. E. (1984). "Efficient Monte-Carlo Probability Integration", Research Report No. 7/1984. Department of Civil Engineering, Monash University.

Melchers, R. E., Baker, M. J., and Moses, F. (1983). "Evaluation of Experience. Quality Assurance within the Building Process", Report No. 47, IABSE, London, pp. 21–38.

Melchers, R. E., and Harrington, M. V. (1984). "Human Error in Structural Reliability I — Investigation of Typical Design Tasks", Research Report No. 2/1984. Department of Civil Engineering, Monash University.

Melchers, R. E. and Tang, L. K. (1983). "Reliability of Structural Systems with Stochastically Dominant Modes", Research Report No. 2/1983. Department of Civil Engineering, Monash University.

Melchers, R. E., and Tang, L. K. (1984). "Dominant Failure Modes in Stochastic Structural Systems", *Struct. Safety*, **2**, pp. 127–143.

Melchers, R. E. and Tang, L. K. (1985a). "Failure Modes in Complex Stochastic Systems", *Proc. 4th Int. Conf. on Structural Safety and Reliability*, Vol. 1. pp. 97–106.

Melchers, R. E., and Tang, L. K. (1985b). "Reliability Analysis of Multi-Member Structures", NUMETA'85, *Proc. Int. Conf. on Numerical Methods in Engineering Theory and Applications*. A. A. Balkema, pp. 763–772.

Milton, R. C. (1972). "Computer Evaluation of the Multivariate Normal Integral", *Technometrics*, **14**, No. 4, pp. 881–889.

Mirza, S. A., and MacGregor, J. G. (1979a). "Variations in Dimensions of Reinforced Concrete Members", *J. Struct. Div.*, *ASCE*, **105**, No. ST4, p. 751–766.

Mirza, S. A., and MacGregor, J. G. (1979b). "Variability of Mechanical Properties of Reinforcing Bars", *J. Struct. Div.*, *ASCE*, **105**, No. ST5, pp. 921–937.

Mitchell, G. R., and Woodgate, R. W. (1971). "Floor Loadings in Office Buildings — The Results of a Survey", Report No. CP 3/71. Building Research Station, Garston, Herts.

Morison, J. R., O'Brien, M. P., Johnston, J. W., and Schaaf, S. A. (1950). "The Force exerted by Surface Waves on Piles", *Pet. Trans., AIME,* **189**, pp. 149–154.

Moses, F. (1974). "Reliability of Structural Systems", *J. Struct. Div., ASCE,* **100**, No. ST9, 1813–1820.

Moses, F. (1982). "System Reliability Development in Structural Engineering", *Struct. Safety,* **1**, No. 1, pp. 3–13.

Moses, F., and Kinser, D. E. (1967). "Analysis of Structural Reliability", *J. Struct. Div., ASCE,* **93**, No. ST5. pp. 147–164.

Murdock, L. J. (1953). "The Control of Concrete Quality", *Proc. Inst. Civ. Eng.,* **2**, Part 1, No. 4, pp. 426–453.

Murotsu, Y., Okada, H., and Matsuzaki, S. (1985). "Reliability Analysis of Frame Structures Under Combined Load Effects", (in) *Proc. 4th Int. Conf. on Structural Safety and Reliability,* Vol. 1. pp. 117–128.

Murotsu, Y., Okada, H., Yonesawa, M., and Kishi, M. (1983). "Identification of Stochastically Dominant Failure Models in Frame Structures", (in) Augusti, G., Borri, A., and Vannuchi, O. (eds.), *Proc. 4th Int. Conf. on Applications of Statistics and Probability in Soil and Structural Engineering.* Pitagora Editrice, Bologna, pp. 1325–1338.

Murotsu, Y., Yonesawa, M., Oba, F., and Niwa, K. (1977). "Methods for Reliability Analysis and Optimal Design of Structural Systems", *Proc. 12th Int. Symp. on Space Technology and Science.* Tokyo, pp. 1047–1054.

NBS (1953). *Probability Tables for the Analysis of Extreme Value Data,* Applied Mathematics Series No. 22. National Bureau of Standards, Washington, DC.

NBS (1959). *Tables of the Bivariate Normal Probability Distribution and Related Functions,* Applied Mathematics Series No. 50. National Bureau of Standards, Washington, DC.

Nessim, M. A., and Jordaan, I. J. (1983). "Decision-making for Error Control in Structural Engineering", (in) Augusti, G., Borri, A., and Vannuchi, O. (eds.), *Proc. 4th Int. Conf. on Applications of Statistics and Probability in Soil and Structural Engineering.* Pitagora Editrice, Bologna, pp. 713–727.

Newland, D. E. (1984). "*An Introduction to Random Vibrations and Spectral Analysis* (2nd edn.) Longman.

Nowak, S. A. (1979). "Effects of Human Error on Structural Safety", *J. Amer. Conc. Inst.,* **76**, No. 9, pp. 959–972.

Nowak, S. A. and Lind, N. C. (1979). "Practical Bridge Code Calibration", *J. Struct. Div., ASCE,* **105**, No. ST12, pp. 2497–2510.

NRCC (1977). *National Building Code of Canada.* National Research Council of Canada, Ottawa.

Osborne, A. F. (1957). *Applied Imagination: Principles and Practices of Creative Thinking.* Scribners, New York.

Oswald, G. F., and Schuëller, G. I. (1983). "On the Reliability of Deteriorating Structures", (in) Augusti, G., Borri, A., and Vannuchi, O. (eds),

Proc. 4th Int. Conf. on Applications of Statistics and Probability in Soil and Structural Engineering. Pitagora Editrice, Bologna, pp. 597–608.

Otway, H. J., Battat, M. E., Lohrding, R. K., Turner, R. D., and Cubitt, R. L. (1970). "A Risk Analysis of the Omega West Reactor", Report No. LA 4449. Los Alamos Scientific Laboratory, University of California, Los Alamos, CA.

Owen, D. B. (1956). "Tables for Computing Bivariate Normal Probabilities", *Ann. Math. Stat.*, **27**, pp. 1075–1090.

Paloheimo, E., and Hannus, M. (1974). "Structural Design based on Weighted Fractiles", *J. Struct. Div.*, ASCE, **100**, No. ST7, pp. 1367–1378.

Papoulis, A. (1965). *Probability, Random Variables and Stochastic Processes*. McGraw-Hill, New York.

Parkinson, D. B. (1980). "Computer Solution for the Reliability Index", *Eng. Struct.*, **2**, pp. 57–62.

Parzen, E. (1962). *Stochastic Processes*. Holden–Day.

Pearson, E. S., and Johnson, N. L. (1968). *Tables of the Incomplete Beta Function* (2nd edn). Cambridge University Press.

Pham, L., Holmes, J. D., and Leicester, R. H. (1983). "Safety Indices for Wind Loading in Australia", *J. Wind Engg. Indust. Aerodyn.*, **14**, pp. 3–14.

Pier, J.-C., and Cornell, C. A. (1973). "Spatial and Temporal Variability of Live Loads", *J. Struct. Div.*, ASCE, **99**, No. ST5, pp. 903–922.

Pierson, W. J., and Moskowitz, L. (1964). "A Proposed Spectral Form for Fully Developed Wind Seas Based on the Similarity Theory of S. A. Kitaigorodskii", *J. Geophys. Res.*, **69**, No. 24, pp. 5181–5190.

Popper, K. R. (1959). *The Logic of Scientific Discovery*. Basic Books.

Prest, A. R., and Turvey, R. (1965). "Cost Benefit Analysis: A Survey", *Econ. J.*, pp. 685–735.

Pugsley, A. G. (1962). *Safety of Structures*. Edward Arnold, London.

Pugsley, A. G. (1973). "The Prediction of Proneness to Structural Accidents", *Struct. Eng.*, **51**, No. 6, 195–196.

Pugsley, A., *et al.*, (1955). "Report on Structural Safety", *Struct. Eng.*, **33**, No. 5, pp. 141–149.

Rackwitz, R. (1976). "Practical Probabilistic Approach to Design", Bulletin d'Information No. 112. Comité Européen du Béton, Paris.

Rackwitz, R. (1977). "Note on the Treatment of Errors in Structural Reliability", Berichte zur Sicherheitstheorie der Bauwerke, SFB 96, Heft 21. Technical University, Munich.

Rackwitz, R. (1984). "Failure Rates for General Systems Including Structural Components", *Reliab. Eng.*, **9**, pp. 229–242.

Rackwitz, R., and Fiessler, B. (1978). "Structural Reliability under Combined Random Load Sequences", *Comput. Struct.*, **9**, pp. 489–494.

Ramachandran, K. (1984). "Systems Bounds: A Critical Study", *Civ. Engg. Syst.*, **1**, pp. 123–128.

Rand Corporation (1955). *A Million Random Digits with 1,000,000 Normal Deviates*. Free Press, New York.

Rao, N. R. N., Lohrman, M., and Tall, L. (1966). "The Effect of Strain Rate on the Yield Stress of Structural Steels", *ASTM, J. Mater.*, **1**, No. 1, pp. 241–262.

Ravindra, M. K., and Galambos, T. V. (1978). "Load and Resistance Factor Design for Steel", *J. Struct. Div.*, ASCE, **104**, No. ST9, pp. 1337–1353.

Ravindra, M. K., Heany, A. C., and Lind, N. C. (1969). "Probabilistic Evaluation of Safety Factors", *Final Report, Symp. on Concepts of Safety of Structures and Methods of Design.* IABSE, London, pp. 36–46.

Reid, S. G., and Turkstra, C. J. (1980). "Serviceability Limit States — Probabilistic Description", Report No. ST80–1, McGill University, Montreal.

Rice, S. O. (1944). "Mathematical Analysis of Random Noise", *Bell System Tech. J.*, **23**, pp. 282–332; (1945), **24**, pp. 46–156. Reprinted in Wax, N., (1954). *Selected Papers on Noise and Stochastic Processes.* Dover Publications.

Rosenblatt, M. (1952). "Remarks on a Multivariate Transformation", *Ann. Math. Stat.*, **23**, pp. 470–472.

Rosenblueth, E. (1985a). Discussion of Ditlevsen, O. "Fundamental Postulate in Structural Safety", *J. Engg. Mech.*, ASCE, **111**, No. 1, p. 109.

Rosenblueth, E. (1985b). "On Computing Normal Reliabilities", *Struct. Safety*, **2**, No. 3, pp. 165–167.

Rowe, W. D. (1977). *An Anatomy of Risk.* John Wiley.

Rubinstein, R. Y. (1981). *Simulation and the Monte Carlo Method.* John Wiley, New York.

Rüsch, H., and Rackwitz, R. (1972). "The Significance of the Concept of Probability of Failure as Applied to the Theory of Structural Safety", (in) *Development — Design — Construction.* Held and Francke Bauaktiengesellschaft, Munich.

Rüsch, H., Sell, R., and Rackwitz, R. (1969). "Statistical Analysis of Concrete Strength", Deutscher Ausschuss für Stahlbeton No. 206 (in German) Berlin.

Russell, L. R., and Schuëller, G. I. (1974). "Probabilistic Models for Texas Gulf Coast Hurricane Occurrences", *J. Pet. Tech.*, pp. 279–288.

Rzhanitzyn, R. (1957). "It is Necessary to Improve the Standards of Design of Building Structures", in Allan, D. E. (transl.), *A Statistical Method of Design of Building Structures*, Technical Translation No. 1368. National Research Council of Canada, Ottawa.

Sarpkaya, T., and Isaacson, M. (1981). *Mechanics of Wave Forces on Offshore Structures.* Van Nostrand Reinhold, New York.

Schijve, J. (1979). "Four Lectures on Fatigue Crack Growth", *Engg. Fract. Mech.*, **11**, pp. 167–221.

Schittkowskii, K. (1980). *Nonlinear Programming Codes: Information, Tests, Performance"*, Lecture Notes in Economics and Mathematical Systems No. 183. Springer, Berlin.

Schneider, J. (1981). "Organization and Management of Structural Safety during Design, Construction and Operation of Structures", (in) Moan, T. and Shinozuka, M. (eds.), *Structural Safety and Reliability*. Elsevier, Amsterdam, pp. 467–482.

Schneider, J. (ed.) (1983). "Quality Assurance within the Building Process", Report No. 47. IABSE, London.

Schuëller, G. I. (1981). *Einführung in die Sicherheit und Zuverlässigkeit von Tragwerken*. W. Ernst, Berlin.

Schuëller, G. I., and Choi, H. S. (1977). "Offshore Platform Risk Based on A Reliability Function Model, "(in) *Proc. 9th Offshore Technology Conf.* pp. 473–482.

Schuëller, G. I., Hirtz, H., and Booz, G. (1983). "The Effect of Uncertainties in Wind Load Estimation on Reliability Assessments', *J. Wind Engg. Indust. Aerodyn.*, **14**, pp. 15–26.

Schwarz, R. F. (1980). "Beitrag zur Bestimmung der Zuverlässigheit nichtlinearer Strukturen unter Berücksichtigung kombinierter Stochastischer Einwirkungen", Doctoral Thesis. Technical University, Munich.

Sexsmith, R. G., and Lind, N. C. (1977). "Policies for Selection of Target Safety Levels", (in) *Proc. 2nd Int. Conf. on Structural Safety and Reliability*, Technical University, Munich. Werner, Düsseldorf, pp. 149–162.

Shellard, H. C. (1958). "Extreme Wind Speeds over Great Britain and Northern Ireland", *Meteorol. Mag.*, **87**, pp. 257–265.

Sheppard, W. F. (1900). "On the Calculation of the double integral expressing normal correlation", *Trans. Camb. Phil. Soc.*, **19**, pp. 23–66.

Shinozuka, M. (1983). "Basic Analysis of Stuctural Safety", *J. Struct. Div.*, *ASCE*, **109**, No. 3, pp. 721–740.

Shooman, M. L. (1968). *Probabilistic Reliability: An Engineering Approach*. McGraw-Hill, New York.

Shreider, Y. A. (ed.) (1966) *The Monte Carlo Method*. Pergamon Press, Oxford.

Sibley, P. G., and Walker, A. C. (1977). "Structural Accidents and their Causes", *Proc. Inst. Civ. Engrs.*, **62**, Part 1, pp. 191–208.

Sigurdsson, G. Sørensen, J. O., and Thoft-Christensen, P. (1985). "Development of Applicable Methods for Evaluating the Safety of Offshore Structures (Part 1)", Report No. 8501. Institute of Building Technology and Structural Engineering, Aalborg University Centre, Aalborg, Denmark.

Simiu, E., Bietry, J., and Filliben, J. J. (1978). "Sampling Errors in Estimation of Extreme Winds", *J. Struct. Div.*, *ASCE*, **104**, No. ST3, pp. 491–501.

Simiu, E., and Filliben, J. J. (1980). "Weibull Distributions and Extreme Wind Speeds", *J. Struct. Div.*, *ASCE*, **106**, No. ST12, pp. 491–501.

Simiu, E., and Scanlan, R. U. (1978). *Wind Effects on Structures, An Introduction to Wind Engineering*. John Wiley, New York.

Simpson, R. H., and Riehl, H. (1981). *The Hurricane and its Impact.* Louisiana State University Press.

Skov, K. (1976). "The Calibration Procedure applied by the NKB Safety Committee", Bulletin d'Information No. 112. Comité Européen du Béton, Paris, pp. 108–124.

Slepian, D. (1962). "The One-Sided Barrier Problem in Gaussian Noise", *Bell Syst Tech. J.*, **41**, No. 2, pp. 463–501.

Sokolnikoff, I. S., and Redheffer, R. M. (1958). *Mathematics of Physics and Modern Engineering.* McGraw-Hill, New York.

Stevenson, J., and Moses, F. (1970). "Reliability Analysis of Frame Structures", *J. Struct. Div.*, *ASCE*, **96**, No. ST11, pp. 2409–2427.

Stewart, M. G., and Melchers, R. E. (1985). "Human Error in Structure Reliability, IV, Efficiency in Design Checking, Report No. 3/1985. Department of Civil Engineering, Monash University.

Stroud, A. H. (1971). *Appropriate Calculation of Multiple Integrals.* Prentice-Hall, Englewood Cliffs, NJ.

Tall, L. (ed.) (1964). *Structural Steel Design.* Ronald Press, New York.

Tall, L., and Alpsten, G. A. (1969). "On the Scatter of Yield Strength and Residual Stresses in Steel Members", (in) *Final Reports*, *Symp. on Concepts of Safety of Structures and Methods of Design.* IABSE, London, pp. 151–163.

Tang, K. L., and Melchers, R. E. (1984a). "Multinormal Distribution Function in Structural Reliability", Research Report No. 6/1984. Department of Civil Engineering, Monash University.

Tang, L. K., and Melchers, R. E. (1984b). "Reliability of Structural Systems with General Member Behaviour", Research Report, No. 1/1984. Department of Civil Engineering, Monash University.

Thoft-Christensen, P., and Sørensen, J. D. (1982). "Calculation of Failure Probabilities of Ductile Structures by the Unzipping Method", Report No. 8208. Institute for Building Technology and Structural Engineering, Aalborg University Centre, Aalborg, Denmark.

Thom, H. C. S. (1968). "New Distributions of Extreme Winds in the United States", *J. Struct. Div.*, *ASCE*, **94**, No. ST7, pp. 1787–1802.

Tickell, R. G. (1977). "Continuous Random Wave Loading on Structural Members", *Struct. Engr.*, **55**, No. 5, pp. 209–222.

Tribus, M. (1969). *Rational Descriptions, Decisions and Designs* (2nd edn). Macmillan, New York.

Turkstra, C. J. (1970). "Theory of Structural Design Decisions Study No. 2". Solid Mechanics Division, University of Waterloo, Waterloo, Ontario.

Turkstra, C. J., and Daley, M. J. (1978), "Two-moment Structural Safety Analysis", *Can. J. Civ. Engg.*, **5**, p. 414–426.

Turkstra, C. J., and Madsen, H. O. (1980). "Load Combinations in Codified Structural Design", *J. Struct. Div.*, *ASCE*, **106**, No. ST12, pp. 2527–2543.

Vanmarcke, E. H. (1973). "Matrix Formulation of Reliability Analysis and Reliability-based Design", *Comput. Struct.*, **3**, 757–770.

Vanmarcke, E. H. (1975). "On the Distribution of the First-passage Time for Normal Stationary Processes", *J. Appl. Mech.*, *ASME*, **42**, pp. 215–220.

Vanmarcke, E. H. (1983). *Random Fields*. Massachusetts Institute of Technology Press, Cambridge, MA.

Veneziano, D. (1974). "Contributions to Second Moment Reliability", Research Report No. R74–33. Department of Civil Engineering, Cambridge, MA.

Veneziano, D. (1976). "Basic Principles and Methods of Structural Safety", Bulletin d'Information No. 112. Comité Européen du Béton, Paris, pp. 212–288.

Veneziano, D., Grigoriu, M., and Cornell, C. A. (1977). "Vector-process Models for System Reliability", *J Engg. Mech. Div.*, *ASCE*, **103**, No. EM3, pp. 441–460.

Vrouwenvelder, A. (1983). "Monte Carlo Importance Sampling — Application to Structural Reliability Analysis", TNO-IBBC, Report No. B-83-529/62.6.0402. Rijswijk, Netherlands.

Wadsworth, G. P., and Bryan, J. G. (1974). *Applications of Probability and Random Variables* (2nd edn). McGraw-Hill, New York.

Walker, A. C. (1981). "Study and Analysis of the First 120 Failure Cases", (in) *Structural Failures in Buildings*. Institute of Structural Engineers, pp. 15–39.

Walker, G. R. (1984). "Discussion of Dorman C. M. L. (1984)", *Civ. Engg. Trans. Inst. Engrs. Aust.*, **CE26**, No. 2, p. 140–141.

Warner, R. F., and Kabaila, A. P. (1968). "Monte Carlo Study of Structural Safety", *J. Struct. Div.*, *ASCE*, **94**, No. ST12, pp. 2847–2859.

Warr, P. B. (1971). *Psychology at Work*. Penguin Books, Harmondsworth.

Watwood, V. B. (1979). "Mechanism Generation for Limit Analysis of Frames", *J. Struct. Div.*, *ASCE*, **109**, No. ST1, pp. 1–15.

Waugh, C. B. (1977). "Approximate Models for Stochastic Load Combination", Report No. R77-1, 562. Department of Civil Engineering, Massachusetts Institute of Technology, Cambridge, MA.

Weibull, W. (1939). "A Statistical Theory of the Strength of Materials", *Proc. Roy. Swed. Inst. Engng. Res.*, No. 151.

Weigel, R. L. (1964). Oceanographical Engineering. Prentice-Hall, Englewood Cliffs, NJ.

Wen, Y.-K. (1977a). "Statistical Combination of Extreme Loads', *J. Struct. Div.*, *ASCE*, **103**, No. ST5, pp. 1079–1093.

Wen, Y.-K. (1977b). "Probability of Extreme Load Combination", (in) *Proc. 4th Int. Conf. on Structural. Mechanics of Reactor Technnology*, San Francisco.
Also *J. Struct, Div.*, *ASCE*, **104**, No. ST10, pp. 1675–1676.

Wen, Y.-K. (1981). "A Clustering Model for Correlated Load Processes", *J. Struct. Div.*, *ASCE*, **107**, No. ST5, pp. 965–983.

Wickham, A. (1985). "Reliability Analysis Techniques for Structures with Time-dependent Strength Parameters", (in) *Proc. 4th Int. Conf. on Structural Safety and Reliability*, Vol. 3. pp. 543–552.

Yang, J.-N. (1975). "Approximation to First Passage Probability", *J. Engg. Mech. Div.*, ASCE, **101**, No. EM4, pp. 361–372.

Yura, J. A., Galambos, T. V. and Ravindra, M. K. (1978). "The Bending Resistance of Steel Beams", *J. Struct. Div.*, *ASCE*, **104**, No. ST9, pp. 1355–1370.

Zaremba, S. K. (1968). "The Mathematical Basis of Monte-Carlo and Qausi-Monte-Carlo Methods", *SIAM Rev.*, **10**, No. 3, pp. 303–314.

Index